Contemporary Immunology

Series Editor
Noel R. Rose
Johns Hopkins University, Baltimore, Maryland, USA

For further volumes:
http://www.springer.com/series/7682

Michael P. Cancro
Editor

BLyS Ligands and Receptors

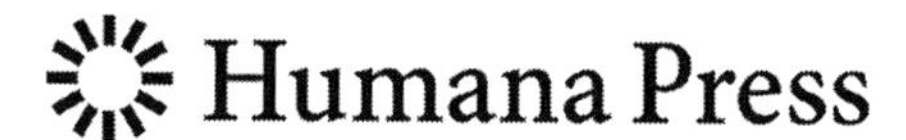

Editor
Michael P. Cancro
University of Pennsylvania
School of Medicine
36th and Hamilton Walk
Philadelphia PA 19104-6082
USA
cancro@mail.med.upenn.edu

ISBN 978-1-60327-012-0 e-ISBN 978-1-60327-013-7
DOI 10.1007/978-1-60327-013-7

Library of Congress Control Number: 2009933798

Printed on acid-free paper

springer.com

Preface

Discovery of the BLyS (also known as BAFF) family of ligands and receptors has yielded a paradigm shift in our view of B-lymphocyte selection, survival, activation, and homeostasis. Previously, the B-cell antigen receptor (BCR) was viewed as the sole mediator of these parameters, in which BCR signals were not only dominant but were also linearly related to consequent outcomes. However, appreciating that BLyS signaling is an equal partner in establishing and maintaining B-cell pools indicated that additional regulatory complexity – apparently based on population density and homeostatic demands – had to be included in models of B-cell behavior. This mounting interest was amplified by evidence of a clear relationship to autoimmunity. The resulting flurry of research activity has yielded a wealth of information and insights, impacting basic concepts in B-cell tolerance and activation as well as revealing novel translational strategies for autoimmunity, neoplasia, and transplant tolerance. This book includes 12 chapters that together yield an overview of these advances and ideas.

The initial excitement generated by associations with humoral autoimmunity, coupled with profound B lineage phenotypes in knockout mouse models, prompted immediate questions: What do these receptors and cytokines look like, how do they interact, what cells express them, and how does this inform our understating of their biology? Indeed, probing the structural features of BLyS family ligands and receptors has afforded substantial insight, as have studies directed toward understanding the basic biological actions of these molecules. These features are detailed in the first several chapters. The structural features of BLyS family ligands and receptors are detailed in Chapter 1, with emphasis on how these are related to biological interactions and activities. The next several chapters explore how these molecules function in B-cell development, selection, and activation and detail current thinking about signal transduction and downstream mediators of the various receptor ligand pairs.

How advancing knowledge of BLyS family molecules impacts our understanding of human disease states, as well as whether these molecules may serve as therapeutic targets, remains areas of intense scrutiny. Because the associations of BLyS with systemic lupus erythematosis and rheumatoid arthritis were recognized early, these questions have been extensively interrogated from the standpoint of humoral autoimmunity. Chapters 7 through 10 provide an overview of this burgeoning area, from both mechanistic and translational perspectives. Finally, because of their preferential expression in B lineage cells and their impact on survival, similar basic and translational questions exist about the roles of BLyS family members in B-cell neoplasia. The last two chapters examine these relationships.

Although this volume reveals the wealth of concepts and possibilities catalyzed by the last 10 years' research on the BLyS family, it is unlikely the plateau has yet been reached. Indeed, implications for the manipulation of pre-immune and antigen-experienced B-cell pools promise even more during the next decade, as current thoughts reach maturity and are extended to transplant tolerance and vaccine development.

Philadelphia, Pennsylvania, USA Michael P. Cancro

Acknowledgements

The conscientious and timely work of all contributing authors – even to the point of making updates to their manuscripts as this volume goes to press – is remarkable. The quality of this volume reflects these efforts, and I am most grateful. I also acknowledge the expert administrative assistance of Ms. Eileen McCann, as well as the pre-publication and editorial expertise of Ms. Samantha Lonuzzi of Humana Press.

Contents

Contributors

Stephen M. Ansell Division of Hematology and Internal Medicine, Mayo Clinic, Rochester, MN 55905, USA

Gail A. Bishop Departments of Microbiology and Internal Medicine, The University of Iowa, and the Iowa City VAMC, Iowa City, IA 52242, USA

Richard J. Bram Departments of Immunology and Pediatric and Adolescent Medicine, Mayo Clinic, Rochester, MN 55905, USA

Robert Brink Garvan Institute of Medical Research, Darlinghurst, NSW 2010, Australia

Iris Castro Department of Microbiology and Immunology, Vanderbilt University School of Medicine, Nashville, TN 37232, USA

Andrea Cerutti Department of Pathology and Laboratory Medicine, Graduate Program of Immunology and Microbial Pathogenesis, Weill Medical College of Cornell University, New York, USA

Kang Chen Department of Pathology and Laboratory Medicine, Graduate Program of Immunology and Microbial Pathogenesis, Weill Medical College of Cornell University, New York, USA

William A. Figgett The Autoimmunity Research Unit, Garvan Institute of Medical Research, Darlinghurst, NSW, Australia

Richard J. Ford Department of Hematopathology, The University of Texas M.D. Anderson Cancer Center, Houston, Texas, USA

Lingchen Fu Department of Hematopathology, The University of Texas M.D. Anderson Cancer Center, Houston, Texas, USA

David A. Fulcher Department of Immunology, Institute of Clinical Pathology and Medical Research, Westmead Hospital, Westmead 2145, Australia

Sandra Gardam Garvan Institute of Medical Research, Darlinghurst, NSW 2010, Australia

Joanne M. Hildebrand Department of Microbiology, The University of Iowa, Iowa City, IA 52242, USA

Kristen L. Hoek Department of Microbiology and Immunology, Vanderbilt University School of Medicine, Nashville, TN 37232, USA

Susan L. Kalled Biogen Idec, Inc., 12 Cambridge Center, Cambridge, MA 02142, USA

Wasif N. Khan Department of Microbiology and Immunology, University of Miami, FL 33136, USA

Yen-chiu Lin-Lee Department of Hematopathology, The University of Texas M.D. Anderson Cancer Center, Houston, Texas, USA

Fabienne Mackay The Autoimmunity Research Unit, Garvan Institute of Medical Research, Darlinghurst, NSW, Australia

Xavier Mariette Rhumatologie, Institut Pour la Santé et la Recherche Médicale (INSERM), U802, Université Paris-Sud 11, Hôpital Bicêtre, Assistance Publique-Hôpitaux de Paris (AP-HP), Le Kremlin Bicêtre, France

Flavius Martin Department of Immunology, Genentech Inc., South San Francisco, CA 94080, USA

Lachy McLean Department of Tissue Growth and Repair, Genentech Inc., South San Francisco, CA 94080, USA

Anne J. Novak Division of Hematology and Internal Medicine, Mayo Clinic, Rochester, MN 55905, USA

Lan V. Pham Department of Hematopathology, The University of Texas M.D. Anderson Cancer Center, Houston, Texas, USA

Pascal Schneider Department of Biochemistry, University of Lausanne, CH-1066 Epalinges, Switzerland

Dhaya Seshasayee Department of Immunology, Genentech Inc., South San Francisco, CA 94080, USA

Nicholas P. Shinners Department of Microbiology and Immunology, Vanderbilt University School of Medicine, Nashville, TN 37232, USA

William Stohl Division of Rheumatology, Department of Medicine, University of Southern California Keck School of Medicine, Los Angeles, CA 90033, USA

Archie Tamayo Department of Hematopathology, The University of Texas M.D. Anderson Cancer Center, Houston, Texas, USA

Stuart G. Tangye Immunology and Inflammation Program, Garvan Institute of Medical Research, Darlinghurst, NSW 2010, Australia

Pali Verma The Autoimmunity Research Unit, Garvan Institute of Medical Research, Darlinghurst, NSW, Australia

Ping Xie Department of Internal Medicine, The University of Iowa, Iowa City, IA 52242, USA

Chapter 1
The Beautiful Structures of BAFF, APRIL, and Their Receptors

Pascal Schneider

Abstract The TNF family ligands BAFF/BLyS and APRIL play important roles in the homeostasis and function of B cells. BAFF binds to the receptors BAFF-R/BR3, BCMA, and TACI, whereas APRIL interacts with BCMA, TACI, and sulfated side chains of proteoglycans. BAFF and APRIL are initially synthesized as membrane-bound proteins that can be released into soluble forms by proteolytic processing. Both cytokines display the characteristic homotrimeric structure of the TNF family, but BAFF is further able to oligomerize under certain conditions into a virus-like particle containing 20 trimers. All three receptors for BAFF are smaller than canonical TNF receptors and form compact but extensive interactions with the ligands. Similarities and differences in these interactions account for the observed binding specificities to BAFF and APRIL. In contrast to TACI and BCMA, proteoglycans interact with APRIL at distinct sites and may serve to concentrate APRIL at defined anatomical locations. Multimeric forms of BAFF and APRIL are required for signaling through TACI, whereas BAFF-R responds to all forms of BAFF, suggesting that many different types of BAFF and APRIL may be functionally relevant to control distinct aspects of their biology.

Keywords Structure · B cells · Oligomerization · Antibody · Survival · Receptor · Ligand · TNF

Abbreviations

APRIL	A proliferation-inducing ligand
BAFF	B-cell-activating factor of the TNF family
BCMA	B-cell maturation antigen
TACI	Transmembrane activator and CAML-interactor.

P. Schneider (✉)
Department of Biochemistry, University of Lausanne, Boveresses 155, CH-1066 Epalinges, Switzerland

M.P. Cancro (ed.), *BLyS Ligands and Receptors,* Contemporary Immunology, DOI 10.1007/978-1-60327-013-7_1,

1.1 Introduction

The cytokines B-cell-activating factor of the TNF Family (BAFF, also known as BLyS or TALL-1) and a proliferation-inducing ligand (APRIL) regulate several aspects of B-cell function and homeostasis. BAFF-deficient mice have little mature B cells in the periphery and impaired humoral responses, whereas BAFF transgenic mice present with B-cell hyperplasia, hyperglobulinemia, and autoimmunity [1–3]. The phenotype of APRIL-deficient mice is less severe, with deficient IgA responses to mucosal immunization, whereas APRIL transgenic mice have elevated IgA levels and can develop chronic lymphoblastic leukemia-like tumors derived from the B1 B cell population [4, 5].

BAFF binds to three receptors, BAFF-R, BCMA, and TACI, whereas APRIL binds to BCMA and TACI only (reviewed in [6, 7]) (Fig. 1.1). The phenotype of BAFF-R-deficient mice is similar to that of BAFF-null mice, with marked decrease of mature B cells [8, 9]. BCMA-null mice are essentially normal, but with impaired survival of bone marrow plasma cells [10]. Finally, TACI is essential for humoral responses to T-independent type-2 antigens, but also displays B-cell hyperplasia, indicating that TACI exerts a negative role on the peripheral B-cell population [11–13]. It is unclear whether the negative role of TACI on B cells is direct or indirect.

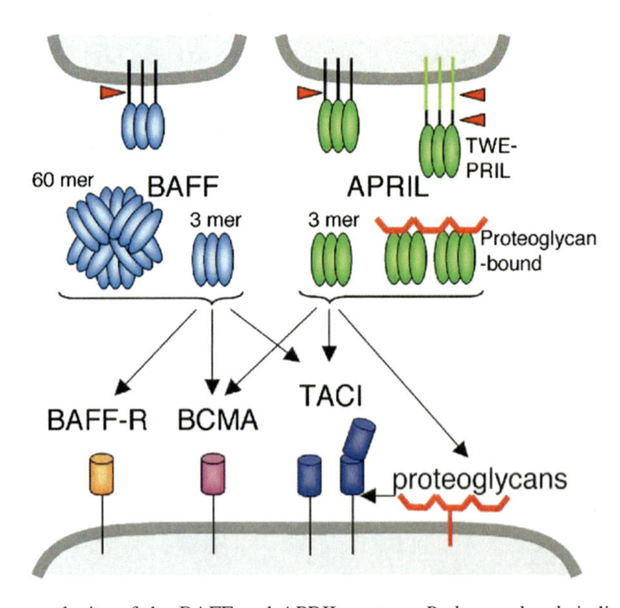

Fig. 1.1 The complexity of the BAFF and APRIL system. *Red arrowheads* indicate furin consensus cleavage sites. Glycosaminoglycan side chains of proteoglycans are shown as thick wavy *red lines*. Two splicing isoforms of human TACI differing in the presence or absence of the first cystein-rich domain are depicted

In human, mutations in TACI are found in about 8% of patients with common variable immunodeficiency, a condition characterized by low antibody levels and susceptibility to infections [14–17]. BAFF is also an essential B-cells survival factor in birds, although BAFF-producing cells are different in birds and mammals [18]. Curiously, APRIL appears to be missing in birds. In contrast, fishes do not only express BAFF and APRIL, but also express a third related ligand named BALM (for BAFF-APRIL-like molecule), which is absent in mammals [19]. In this chapter, structural aspects of BAFF, APRIL, and their receptors will be reviewed.

1.2 Multiple Forms of BAFF and APRIL

BAFF and APRIL are remarkable by the diversity of their mature forms that result from both splicing and post-translational events. The numerous splice variants have been reviewed previously [7]. Briefly, delta-BAFF lacks exon 3 and acts as a dominant negative inhibitor of the active full-length form [20]. Similar exon-skipping events occur in human APRIL, but their functional impacts have not been characterized. Moreover, an intergenic splicing between *tweak* and *april* genes generate TWE–PRIL, a hybrid ligand containing the full receptor-binding domain of APRIL [21] (Fig. 1.1). Finally, the use of an alternative splice acceptor site generates two isoforms of mouse APRIL differing by just one amino acid (Ala120). These two isoforms only display subtle differences in receptor binding and the shorter isoform does not exist in human [22].

BAFF and APRIL are both synthesized as membrane-bound proteins that can be processed to soluble forms by proteolytic processing at a furin consensus site [23, 24] (Fig. 1.2). APRIL is efficiently processed within the cell, and whether APRIL can be expressed as a membrane-bound form remains to be demonstrated [25]. However, TWE–PRIL was suggested to exist in a membrane-bound form [21].

Soluble BAFF can assemble as virus-like particles [26], and soluble APRIL can bind to proteoglycans [27, 28] (Fig. 1.1). Formation of BAFF–APRIL heteromers has also been reported in sera of patients with rheumatoid arthritis [29].

1.3 Structure of BAFF and APRIL

1.3.1 Structural Comparison to Other TNF Family Members

TNF and related ligands assemble as homotrimers [30] and BAFF and APRIL are no exception to the rule [31–33] (Fig. 1.3). Each APRIL and BAFF protomers are composed of two β-sheets containing strands AA′HCF and BB′GDE. Strands A, C, E, F, and H are the most conserved and form much of the trimer interface. This interface is rich in hydrophobic interactions that appear to be the main forces driving

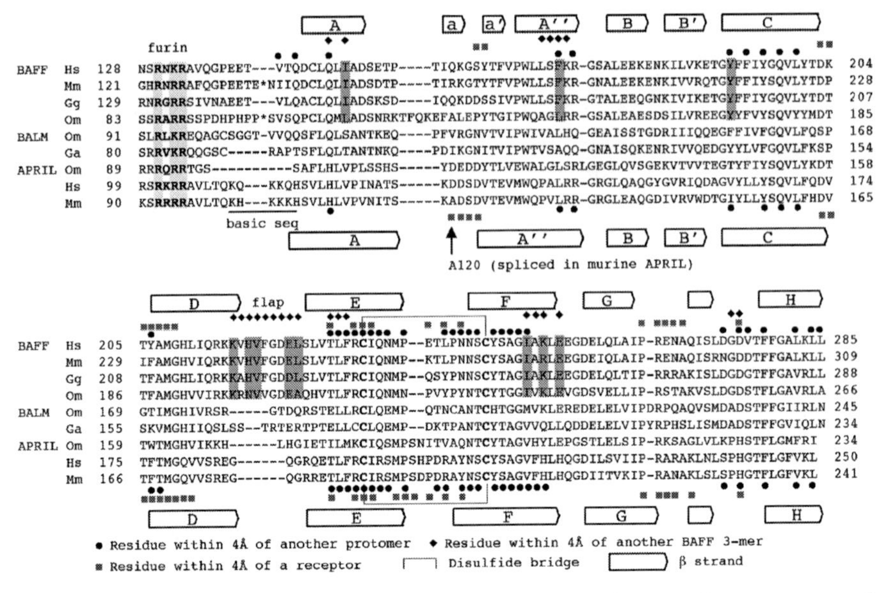

Fig. 1.2 Sequence alignment of BAFF, BALM, and APRIL. The alignment shows sequences of the C-terminal portions of BAFF, BALM, and APRIL of different species. β-strands as found in the crystal structure of human BAFF and mouse APRIL are schematized above and below the alignment, respectively. Hs: *Homo sapiens* (human). Mm: *Mus musculus* (mouse). Gg: *Gallus gallus* (chicken). Om: *Oncorhynchus mykiss* (rainbow trout). Ga: *Gasterosteus aculeatus* (three-spine stickleback, a fish). The R-X-R/K-R furin consensus cleavage site is indicated (furin). *Black dots*: residues that are located within 4 Å of another ligand protomer: These residues correspond roughly to residues involved in trimer formation. *Gray squares*: residues that are located within 4 Å of a receptor (BAFF-R or BCMA for human BAFF; BCMA or TACI for mouse APRIL). These residues correspond roughly to those involved in receptor binding. *Black diamonds*: residues that are located within 4 Å of another BAFF trimer. BAFF residues that are involved in trimer–trimer interactions for the formation of BAFF 60-mer are *shaded gray*. Cystein residues in β-sheets E and F involved in the formation of a disulfide bridge are shown in *bold* and linked by a *bracket*. The residue A120 of mouse APRIL that can be alternatively spliced is indicated by an *arrow*. The basic sequence of human and mouse APRIL involved in proteoglycan interaction is underlined (basic seq). "*" between the furin site and β-sheet A of mouse and trout BAFF indicate insertions of 29-and 20-amino acid residues, respectively

trimer formation. The BB′GDE sheet is more solvent exposed and most of its strands are less conserved in sequence among different TNF family ligands [30]. BAFF and APRIL have an internal disulfide bridge linking strands E and F (Fig. 1.2). This later feature is shared with the TNF family ligands BALM, EDA, and Tweak.

1.3.2 BAFF Forms Virus-Like Particles

In some studies, BAFF crystallized as a virus-like particle containing 20 trimers [26, 34, 35]. The BAFF 60-mer contains not only 3-fold symmetry axes

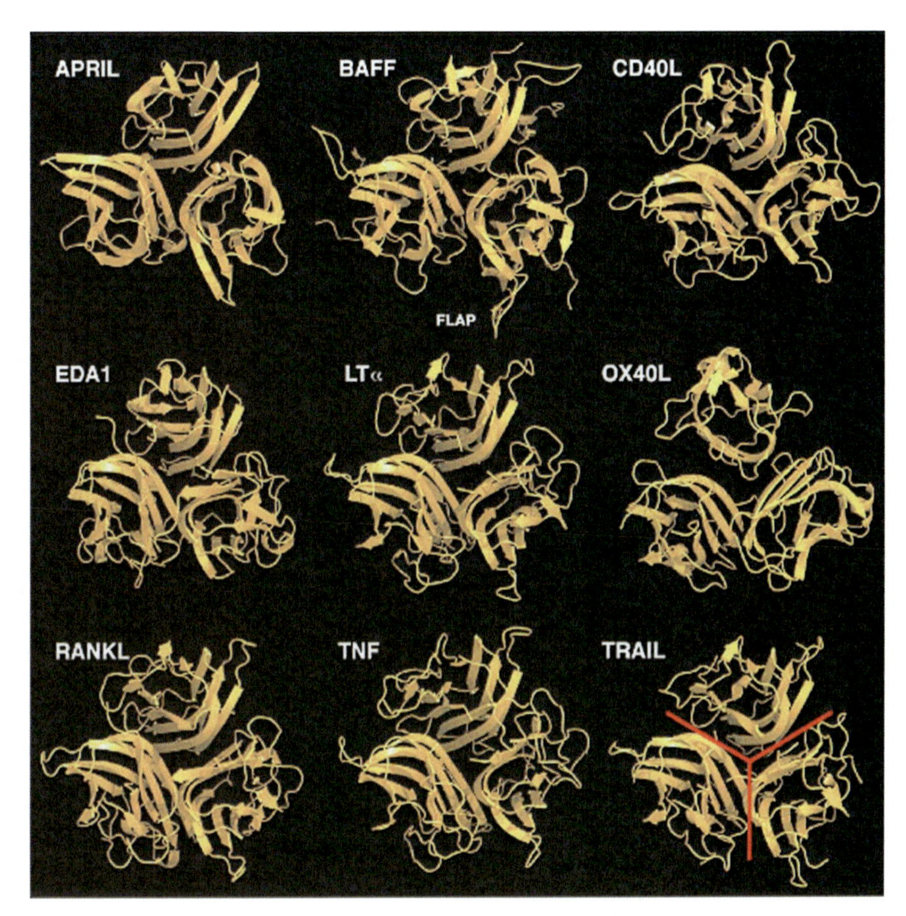

Fig. 1.3 Structural homology in the TNF family. Ribbon representation of APRIL (mouse; pdb atomic coordinate file 1XU1), BAFF (human; 1OQE), CD40L (human; 1ALY), EDA1 (human; 1RJ7), lymphotoxin-α (human; LTα; 1TNR), OX40L (mouse; 2HEW), RANKL (mouse; 1JTZ), TNF (human; 1TNF), and TRAIL (human; 1D4V). Structures are viewed along the 3-fold symmetry axis, which is highlighted in the structure of TRAIL. The unusually long DE loop of BAFF is designated "flap"

corresponding to those of each BAFF 3-mer, but also 5-fold symmetry axes (Fig. 1.4; colored ring of 5 BAFF 3-mers). In this structure, BAFF trimers interact with each other through extended sets of interactions involving the unusually long DE-loop, also known as the "flap." There are numerous interactions in this region involving salt bridges and hydrophobic interactions (Fig. 1.4, *lower panels*). In one published structure, two citrate and three magnesium ions interact with the portion of the flap facing the inside of the 60-mer structure [32]. The flap and most of the key residues involved in trimer–trimer interactions are conserved in BAFF from vertebrates (Fig. 1.2). These features are not conserved in APRIL, BALM, and other TNF family members (Fig. 1.2). The high number of charged interactions

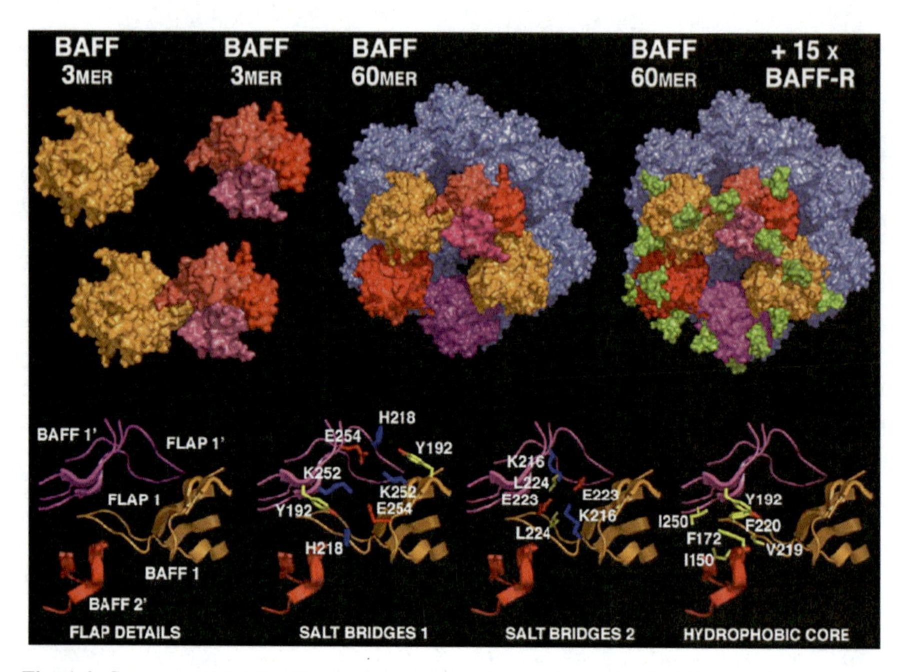

Fig. 1.4 Structure of BAFF 60-mer. *Upper left pictures*: Two individual BAFF trimers linked or not by the flap region. *Upper middle picture*: Space filling representation of BAFF 60-mer (1OTZ). Five trimers are colored in *warm tones*, and the remaining 15 are shown in *pale blue*. *Upper right picture*: Idem, but with 15 out of the 60 co-crystallized BAFF-R shown in *green* (1OTZ and 1POT). *Bottom pictures*: details of a trimer–trimer interaction, viewed from the center of the 60-mer (1OTZ). BAFF 1 and FLAP 1 belong to a given BAFF trimer, while BAFF 1′, FLAP 1′, and BAFF 2′ belong to an adjacent BAFF trimer. Residues involved in three distinct groups of interactions are shown (salt bridges 1, salt brides 2, and hydrophobic core)

required for BAFF 60-mer assembly explains the acid sensitivity of this structure [26, 36] and the different outcome of crystallization performed at acidic or neutral pH [26, 32, 33, 37]. Mutation of His218 to Ala entirely prevented 60-mer formation in human BAFF [36].

1.3.3 APRIL Binds to Proteoglycans

Proteoglycans are heavily glycosylated O-linked glycoproteins that are either soluble or membrane bound [38]. They have extended, negatively charged oligosaccharide side chains collectively designated as glycosaminoglycans. Negative charges originate from anionic monosaccharides (such as glucuronic or iduronic acids) and from more or less extensive sulfation events. Heparin is a protein-free glycosaminoglycan of the heparan sulfate type.

Human and mouse APRIL contain a short basic amino acid sequence preceding β-strand A (Fig. 1.2). This basic sequence, which is absent in BAFF, is required but not sufficient to bind heparin and proteoglycans [27, 28]. Additional basic residues scattered on the same surface of APRIL are also required for binding to heparin [27]. It is therefore likely that negatively charged glycosaminoglycans bind to a basic surface of APRIL, as illustrated in the model shown in Fig. 1.5. It is noteworthy that the proteoglycan-binding site of APRIL is clearly distinct from the binding sites of TACI and BCMA (Fig. 1.5).

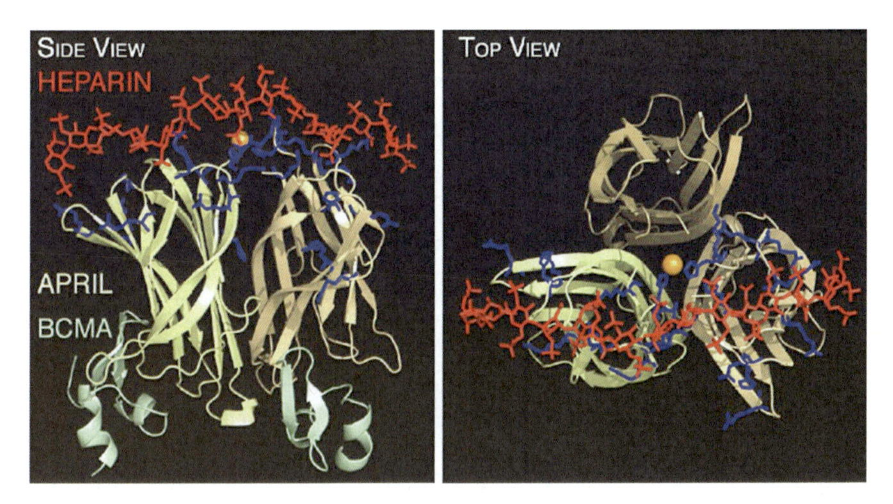

Fig. 1.5 Model of APRIL bound to heparin and BCMA. The structure of heparin (1FQ9) was manually positioned on that of the APRIL–BCMA complex (1XU2). Side chains of all basic residues in the *upper portion* of APRIL are shown in *blue*. In the side view (*left picture*), the BCMA-expressing cell would be at the *bottom* of the figure. The *orange ball* is a nickel atom probably originating from the purification process

1.4 Receptor Binding

1.4.1 Structure of BCMA, TACI, and BAFF-R

TNF receptor family members are generally type I transmembrane proteins with cystein-rich domains (CRDs) in their extracellular domains. These CRDs can be further divided into small structural units called modules [30]. For example, TNF-R1 contains four CRDs corresponding to a total of eight modules. BCMA, TACI, and BAFF-R differ significantly from canonical TNF receptors. First, they are type III transmembrane proteins lacking a signal peptide. Second, TACI contains two CRDs, BCMA has only one, and BAFF-R has just no canonical CRD. Third, they contain a module not present in other TNF receptors and thus are not easy to recognize as such based on the primary sequence only. BAFF-R, BCMA, and TACI share a structurally conserved β-hairpin (corresponding to module A1 of the third CRD of TNF-R1) immediately followed by a short, one-turn helix (Fig. 1.6). This determines much of the ligand-binding properties. After these conserved structural features, receptors diverge: TACI and BCMA have an additional helix, but that adopts different orientations [39] (Fig. 1.7). The second module of BAFF-R is truncated and the one remaining cystein residue at the end of the one-turn helix pairs with another cystein residue in the β-hairpin (Fig. 1.6).

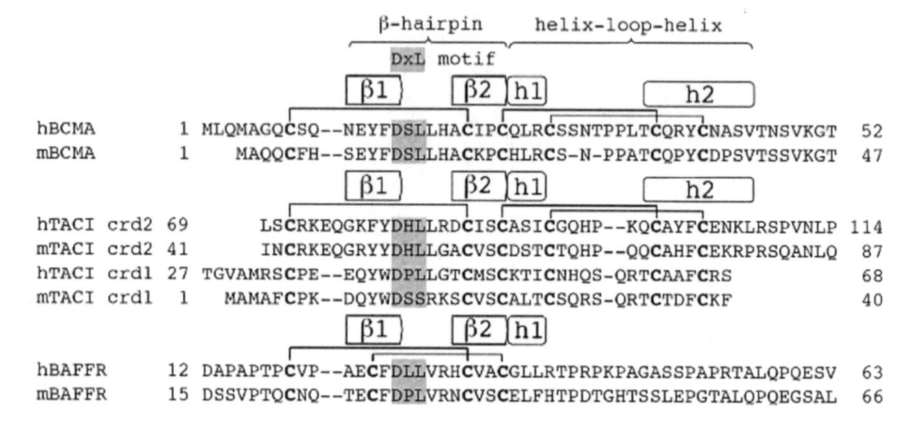

Fig. 1.6 Sequence alignment of BCMA, TACI, and BAFF-R. β-sheets and helices found in the ligand-binding domain of human BAFF, human TACI CRD2, and human BAFF-R are shown above the sequences. Connectivity of disulfide bridges is indicated by *brackets*, and cystein residues are shown in *bold*. Both CRDs of TACI are shown individually. The DxL motif that is crucial for ligand binding is shaded

TACI is remarkable in that it contains two CRDs that probably arose from exon duplication [39]. CRD2 binds BAFF and APRIL with high affinity, whereas CRD1 also binds, but with low affinity. In human, a short form of TACI lacking CRD1 is produced by alternative splicing [39] (Fig. 1.1). It is noteworthy that TACI, like APRIL, binds to proteoglycans [40] (Fig. 1.1). The site of this interaction has not been mapped.

1.4.2 Comparison of Ligand–Receptor Binding in the TNF Family

The TNF receptor family members TNF-R1, OX40, and DR5 are elongated proteins that contact their ligands at the interface between two protomers [41–44] (Fig. 1.8). The interaction extends over the entire length of the ligand and involves residues of both ligand protomers (Fig. 1.8). In contrast, the contact site of BAFF-R on BAFF is localized and involves mainly a single BAFF protomer (Fig. 1.8). The same is true for BCMA and TACI binding to BAFF and APRIL (Fig. 1.8). The surface buried by interaction with an extended receptor (like TNF-R1) or a compact receptor (like

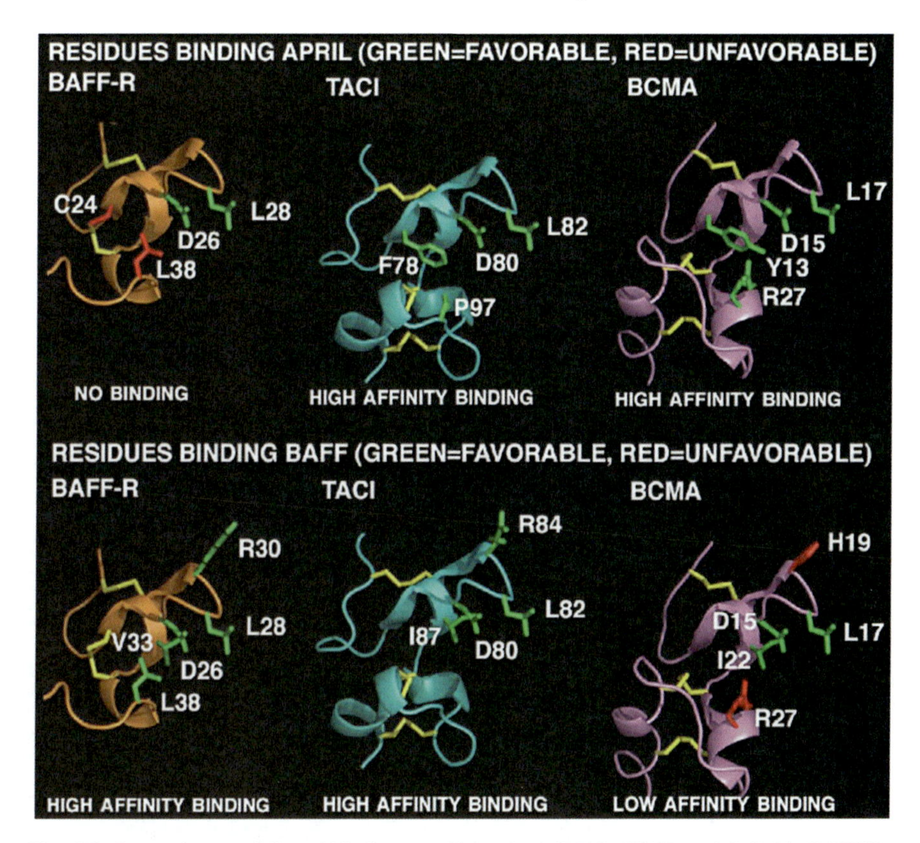

Fig. 1.7 Determinants of ligand-binding specificity in BAFF-R, TACI, and BCMA. BAFF-R (1OQE; *orange*), TACI (1XU1; *cyan*), and BCMA (1OQD; *purple*) are shown twice. The *upper part* of the figure shows residues determining binding specificity to APRIL, and the *lower part* of the figure those residues important for binding to BAFF. These residues were identified by shotgun alanine scanning [37, 39, 46]. Side chains that are required for interactions are shown in *green*, whereas those that prevent or weaken interactions are shown in *red*. Disulfide bridges are colored in *yellow*

BAFF-R) is approximately the same, so that both types of interaction can be of high affinity. As mentioned above, the binding of BCMA or TACI to APRIL does not overlap with the proteoglycan-binding site of APRIL (Fig. 1.5). In BAFF 60-mer, the receptor-binding sites remain accessible on the outer surface of the virus-like cluster (Fig. 1.4). In this structure, however, the receptor-binding pocket of BAFF is delimitated by the flap region of the neighboring BAFF trimer, raising the question of how the long form of TACI does bind this structure. If the CRD1–CRD2 junction of TACI is flexible enough, the high affinity CRD2 will probably bind BAFF 60-mer. If this junction is rigid, only the low-affinity CRD1 of TACI may have access to the binding site.

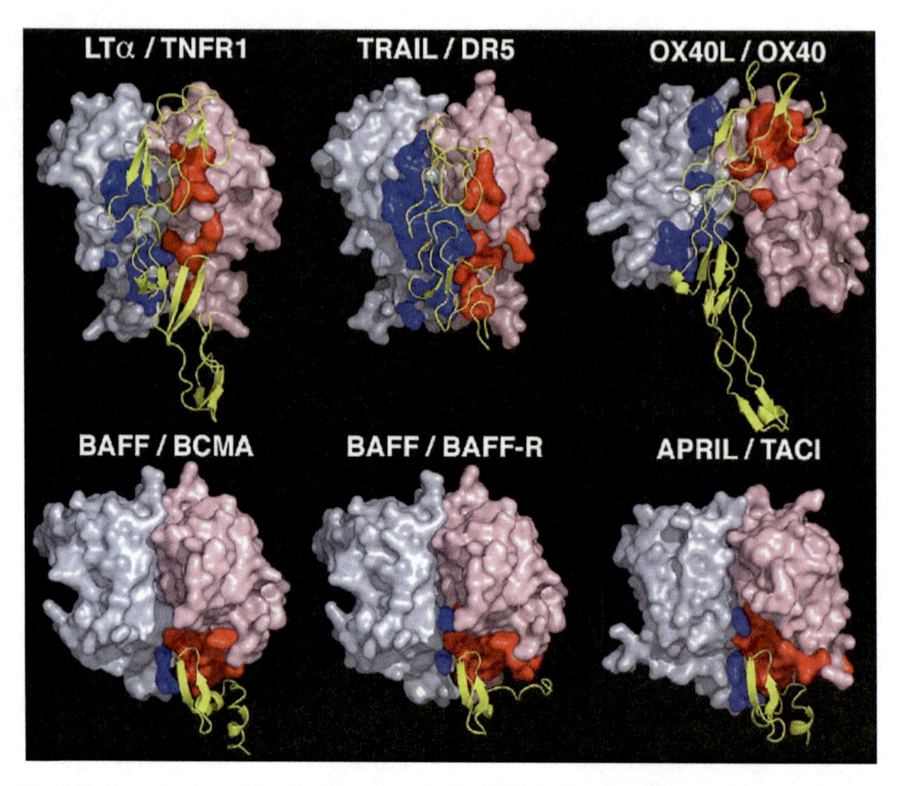

Fig. 1.8 Receptor–ligand-binding interfaces in the TNF family. Structures of six complexes of a ligand co-crystallized with its receptor are shown. The complexes are lymphotoxin-α/TNF-R1 (1TNR), TRAIL/DR5 (1DU3), OX40L/OX40 (2HEV), BAFF/BCMA (1OQD), BAFF/BAFF-R (1OQE), and APRIL/TACI (1XU1). Receptors are shown as *yellow ribbons*, whereas ligands are shown in the space filling representation, with one protomer in *pale blue* and another in *pink*. Ligand residues within 4 Å of the receptor are shown in *blue* and *red*. They provide an estimation of residues involved in the interaction. In this representation, the transmembrane domain of the receptors would be at the *bottom* of the figure

1.4.3 Structural Determinants of Binding Specificity

The β-hairpins of BCMA, BAFF-R, and TACI contain conserved aspartate and leucine residues known as the DxL motif (Fig. 1.6). A synthetic, constrained peptide containing just six residues of BAFF-R hairpin, including the DxL motif, is sufficient for BAFF binding [45]. In an experimental procedure called shotgun alanine scanning, residues of the receptors were systematically screened for their contribution to ligand binding [37, 39, 46]. This method takes into account the potentially confounding effect of decreased expression of mutant receptors and detects whether a given residue is beneficial, neutral, or detrimental for the interaction. As expected, mutations of the DxL motif abolished binding. Other important residues were ligand dependent. For example, binding to APRIL requires an aromatic residue in the hairpin (F78 of TACI; Y13 of BCMA), which is not required for BAFF binding (Fig. 1.7). This residue is absent in BAFF-R that does not bind APRIL. Conversely, an arginine residue in the hairpin favors binding to BAFF, but is not required for binding to APRIL (R30 of BAFF-R; R84 of TACI). BCMA, which binds BAFF with low affinity, has a suboptimal residue in this position (H19). The one-turn helix following the hairpin is also important in determining the specificity of the interactions. APRIL requires an arginine at the end of this helix (R27 of BCMA), whereas BAFF needs a leucine (L38 of BAFF-R). Thus, L38 of BAFF-R favors BAFF binding but prevents APRIL binding. Conversely, R27 of BCMA favors APRIL binding and weakens BAFF binding. TACI adopts a different structure in this region, with P97 being the spatially corresponding residue that shows a preference for APRIL. In summary, BCMA, TACI, and BAFF-R have an intrinsic tendency to bind BAFF and APRIL through the DxL motif, which is positively or negatively regulated in a ligand-dependent manner by various other residues. This explains the specificity of BAFF-R for BAFF and the preferential binding of BCMA to APRIL.

This structural knowledge was exploited to prepare an APRIL-specific BCMA by replacing a residue favorable for BAFF binding (I22) to one that enhances APRIL binding (I22K) [46] (Fig. 1.7). In another application, anti-BAFF-R antibodies were selected by phage display on BAFF-R and submitted to affinity maturation in order to recognize both human and mouse BAFF-R with high affinity [47]. One such antibody recognized the conserved DxL motif of BAFF-R and shared several of its BAFF-R-binding characteristics with BAFF. This antibody competed for BAFF binding and acted as a BAFF-R antagonist. Administration of this antibody to mice depleted various B-cell populations by the dual mechanism of inhibiting BAFF signals and mediating antibody-dependent cellular cytotoxicity [48]. A similar but less-marked effect was also observed in non-human primates [48].

1.5 Receptor Signaling

1.5.1 Signaling Overview

BAFF-R mediates much of the BAFF survival signals required to maintain the peripheral B-cell population. BAFF can be functionally replaced for this purpose by the deletion of TRAF2 or TRAF3 or by the constitutive activation of either canonical or non-canonical NF-κB pathways [49–52]. TRAF3 is the only known intracellular binding partner of BAFF-R [53]. TRAF3 also negatively regulates the kinase NIK, which is apical to the non-canonical NF-κB pathway [54]. This suggests that BAFF-R engagement interferes with the negative function of TRAF3 on the NF-κB pathway, ultimately leading to the activation of this pathway and its pro-survival effects.

1.5.2 TRAF Binding

A natural mutation affecting the last eight amino acids of BAFF-R in A/WySnJ mice almost completely abolishes BAFF-R activity [55, 56]. This mutation occurs close to a sequence conserved in both BAFF-R and BCMA that mediates binding to TRAF3 [53], strongly suggesting that TRAF3 regulates BAFF-R signaling. Indeed, mutation of the TRAF3-binding sequence abolishes BAFF-R's ability to engage non-canonical NF-κB [57]. TRAF-binding sequences in the intracellular portions of TNF receptors are linear peptides that usually bind either TRAF6 or TRAF1, 2, 3, and 5. The strict specificity of BAFF-R for TRAF3 is therefore unusual, but the determinants for this specificity are now well understood [57, 58]. The C-terminal portion of TRAF3 adopts a mushroom-like structure with the coiled-coil domain forming the stem and the homotrimeric TRAF-C domain forming the cap (Fig. 1.9). The TRAF-binding sequence of BAFF-R wraps around the TRAF-C domain

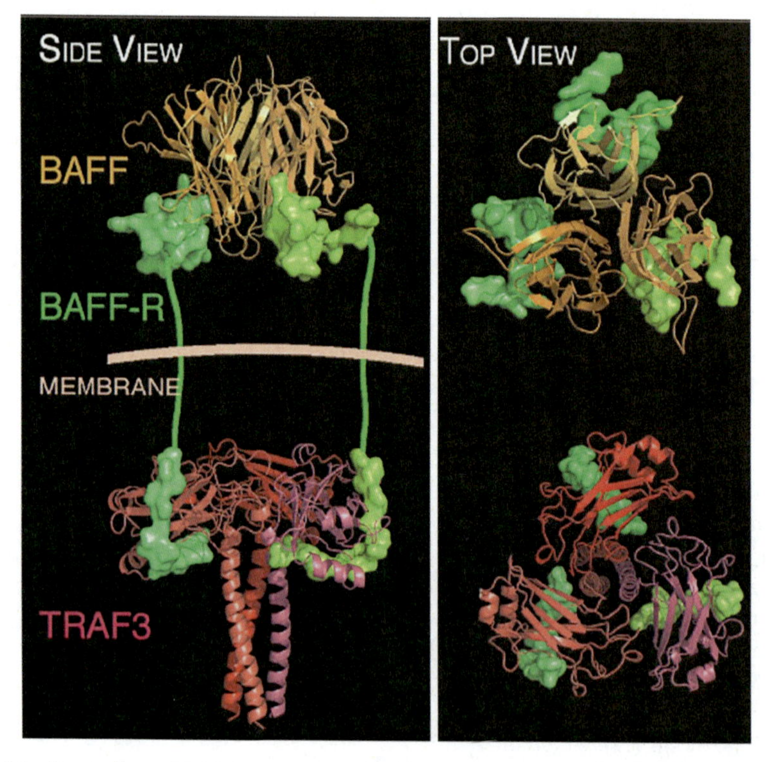

Fig. 1.9 Comparison of BAFF–BAFF-R and BAFF-R–TRAF3 complexes. Structure of a BAFF trimer (*golden ribbon*) bound to BAFF-R (*green surface*; 1OQE) and of the TRAF-binding sequence of the intracellular domain of BAFF-R (*green surface*) bound to the C-terminal portion of TRAF3 (*purple ribbons*, 2GKW). The cell membrane and the missing portion of BAFF-R are schematized as thick *pink* and *green* lines, respectively

[58] (Fig. 1.9). The overall similarity in size and symmetry of the extracellular BAFF–BAFF-R complex and of the intracellular BAFF-R–TRAF3 complex is striking. How BAFF-R exerts functional changes on TRAF-3 upon recruitment is not understood.

1.6 Oligomerization-Dependent Signaling

With the exception of delta-BAFF, whose transgenic overexpression demonstrated a dominant negative effect on B-cell homeostasis [20], little is known regarding the physiological function of the various forms of BAFF and APRIL. The analysis of chimeric mice expressing BAFF either in bone marrow-derived cells or in non-bone marrow, radiation-resistant cells indicated that radiation-resistant cells can produce sufficient amounts of BAFF for maintaining the peripheral B-cell population, whereas BAFF produced by bone marrow-derived cells may only provide a more local support to B cells [59]. In any case, this study demonstrated the existence of different sources of BAFF with apparently different physiological effects. Whether various cell types produce different forms of BAFF, such a membrane-bound or soluble BAFF, remains to be studied.

A number of in vitro studies suggest that ligand oligomerization may modulate biological activity. In the original report, BAFF 60-mer had a modest 2-fold co-stimulatory effect on human blood B cells that was not observed with a flap-deletion mutant [26]. In mouse B cells, BAFF 60-mer was only moderately more active than BAFF 3-mer at co-stimulating B-cell proliferation [36]. The differential effects of BAFF 3-mer and 60-mer were analyzed in more detail using mouse cells [60]. The co-stimulatory effect of BAFF did not rely on increased proliferation, but rather on increased B-cell survival, consistent with its relatively minor effects. Survival was the result of a mixed signal originating from both BAFF-R and TACI. Whereas BAFF-R could be stimulated by both BAFF 3-mer and BAFF 60-mer, TACI responded exclusively to BAFF 60-mer or other multimeric forms of BAFF. Thus, BAFF-mediated B-cell co-stimulation is the result of two distinct signals: one originating from BAFF-R in response to any form of BAFF and one originating from TACI in response to multimeric forms of BAFF [60]. In plasmablasts, BAFF-mediated survival depended on TACI with little contribution of BAFF-R and also required oligomeric forms of BAFF [60]. In a similar way, TACI also responded to cross-linked APRIL, but not to APRIL 3-mer [27, 60]. Importantly, TACI bound BAFF 3-mer and 60-mer equally well, but only signaled in response to the latter. It is possible that binding of BAFF 3-mer to TACI is sufficient to recruit a single trimeric TRAF (see Fig. 1.9), which may not be able to signal as such. Indeed, cross-linking of TRAF2 and TRAF6 is required for efficient activation of the NF-κB pathway [61]. Oligomeric forms of BAFF may allow signaling through TACI by enabling recruitment of several TRAFs in close proximity.

Interestingly, survival of human monocytes ex vivo was enhanced by BAFF in a TACI-dependent manner [62]. This effect was only observed with some commercial

BAFF preparations but not others, although all of them supported B-cell survival [62]. These puzzling observations could be readily explained if the active preparations contained a proportion of oligomeric BAFF able to signal through TACI. This difference would not be detected on B cells that mainly respond through BAFF-R.

The finding that BAFF-R and TACI can respond to different forms of BAFF but transmit similar survival signals can help explain the paradoxical role of TACI in B-cell biology. TACI-deficient mice indeed have impaired humoral responses despite their elevated B-cell number [12, 13]. Plasmablasts elicited in response to T-independent type-2 antigens appear to rely heavily on TACI for their survival [63], but cannot respond to BAFF 3-mer for lack of BAFF-R. Thus, deletion of TACI would penalize plasmablasts without preventing the majority of BAFF-R-expressing B cells to receive survival signals: these may even receive more, as TACI-deficient mice have elevated levels of circulating BAFF [60], leading to the observed enlargement of the B-cell pool.

It is believed that APRIL is mainly released as a soluble cytokine [25]. This form of APRIL can, however, not activate TACI and is also probably poorly active on BCMA [60]. Soluble APRIL released in vivo accumulates on proteoglycans [64]. It is tempting to speculate that proteoglycan-bound APRIL is a biologically active form particularly important to support survival of antibody-secreting cells in specialized environments such as the bone marrow or the intestinal mucosa [65, 66].

1.7 Conclusions and Open Questions

From a structural point of view, the BAFF–APRIL system is certainly one of the best characterized within the TNF family. Taken together, structural studies on BAFF, APRIL, and their receptors have provided three highly unexpected surprises. The first one was the unique mode of receptor binding on a single ligand protomer, the second one was the discovery of the oligomeric form of BAFF 60-mer, and the third one was the interaction of APRIL and TACI with glycosaminoglycans. These studies have not only provided a detailed molecular basis for the observed binding specificities between ligands and receptors, but also hinted at different mechanisms of action of trimeric and oligomeric BAFF and APRIL. The remaining open questions regarding structure are relatively few. They include the mapping of the proteoglycan-binding site(s) of TACI, the mode of binding of full-length TACI on BAFF 60-mer, the structural impact of exon 3 deletion in BAFF, and the possibility of ligand-independent pre-association of TACI.

The analysis of receptor-deficient mice has taught us a lot on the role of these different receptors. However, little is known on the relative contributions of membrane-bound BAFF, soluble BAFF 3-mer, and soluble BAFF 60-mer to the physiological functions of BAFF. With the recent discovery that these forms likely engage different sets of receptors, there is no doubt that the analysis of knock-in mice deficient for on or the other forms of BAFF and APRIL will yield interesting information on how these ligands control the various aspects of B-cell biology they are implicated in.

Acknowledgment This work was supported by the Swiss National Foundation for Scientific Research.

References

1. Gross JA, Dillon SR, Mudri S, et al. TACI-Ig neutralizes molecules critical for B cell development and autoimmune disease. impaired B cell maturation in mice lacking BLyS. Immunity 2001;15(2):289–302.
2. Mackay F, Woodcock SA, Lawton P, et al. Mice transgenic for BAFF develop lymphocytic disorders along with autoimmune manifestations. J Exp Med 1999;190(11):1697–710.
3. Schiemann B, Gommerman JL, Vora K, et al. An essential role for BAFF in the normal development of B cells through a BCMA-independent pathway. Science 2001;293(5537):2111–4.
4. Planelles L, Carvalho-Pinto CE, Hardenberg G, et al. APRIL promotes B-1 cell-associated neoplasm. Cancer Cell 2004;6(4):399–408.
5. Castigli E, Scott S, Dedeoglu F, et al. Impaired IgA class switching in APRIL-deficient mice. Proc Natl Acad Sci U S A 2004;101(11):3903–8.
6. Dillon SR, Gross JA, Ansell SM, Novak AJ. An APRIL to remember: novel TNF ligands as therapeutic targets. Nat Rev Drug Discov 2006;5(3):235–46.
7. Bossen C, Schneider P. BAFF, APRIL and their receptors: structure, function and signaling. Semin Immunol 2006;18(5):263–75.
8. Sasaki Y, Casola S, Kutok JL, Rajewsky K, Schmidt-Supprian M. TNF family member B cell-activating factor (BAFF) receptor-dependent and -independent roles for BAFF in B cell physiology. J Immunol 2004;173(4):2245–52.
9. Shulga-Morskaya S, Dobles M, Walsh ME, et al. B cell-activating factor belonging to the TNF family acts through separate receptors to support B cell survival and T cell-independent antibody formation. J Immunol 2004;173(4):2331–41.
10. O'Connor BP, Raman VS, Erickson LD, et al. BCMA is essential for the survival of long-lived bone marrow plasma cells. J Exp Med 2004;199(1):91–8.
11. Seshasayee D, Valdez P, Yan M, Dixit VM, Tumas D, Grewal IS. Loss of TACI causes fatal lymphoproliferation and autoimmunity, establishing TACI as an inhibitory BLyS receptor. Immunity 2003;18(2):279–88.
12. von Bulow GU, van Deursen JM, Bram RJ. Regulation of the T-independent humoral response by TACI. Immunity 2001;14(5):573–82.
13. Yan M, Wang H, Chan B, et al. Activation and accumulation of B cells in TACI-deficient mice. Nat Immunol 2001;2(7):638–43.
14. Pan-Hammarstrom Q, Salzer U, Du L, et al. Reexamining the role of TACI coding variants in common variable immunodeficiency and selective IgA deficiency. Nat Genet 2007;39(4):429–30.
15. Salzer U, Chapel HM, Webster AD, et al. Mutations in TNFRSF13B encoding TACI are associated with common variable immunodeficiency in humans. Nat Genet 2005;37(8):820–8.
16. Castigli E, Wilson S, Garibyan L, et al. Reexamining the role of TACI coding variants in common variable immunodeficiency and selective IgA deficiency. Nat Genet 2007;39(4):430–1.
17. Castigli E, Wilson SA, Garibyan L, et al. TACI is mutant in common variable immunodeficiency and IgA deficiency. Nat Genet 2005;37(8):829–34.
18. Kothlow S, Morgenroth I, Graef Y, et al. Unique and conserved functions of B cell-activating factor of the TNF family (BAFF) in the chicken. Int Immunol 2007;19(2):203–15.
19. Glenney GW, Wiens GD. Early diversification of the TNF superfamily in teleosts: genomic characterization and expression analysis. J Immunol 2007;178(12):7955–73.

20. Gavin AL, Duong B, Skog P, et al. deltaBAFF, a splice isoform of BAFF, opposes full-length BAFF activity in vivo in transgenic mouse models. J Immunol 2005;175(1):319–28.
21. Pradet-Balade B, Medema JP, Lopez-Fraga M, et al. An endogenous hybrid mRNA encodes TWE-PRIL, a functional cell surface TWEAK-APRIL fusion protein. Embo J 2002;21(21):5711–20.
22. Bossen C, Ingold K, Tardivel A, et al. Interactions of tumor necrosis factor (TNF) and TNF receptor family members in the mouse and human. J Biol Chem 2006;281(20):13964–71.
23. Moore PA, Belvedere O, Orr A, et al. BLyS: member of the tumor necrosis factor family and B lymphocyte stimulator. Science 1999;285(5425):260–3.
24. Schneider P, MacKay F, Steiner V, et al. BAFF, a novel ligand of the tumor necrosis factor family, stimulates B cell growth. J Exp Med 1999;189(11):1747–56.
25. Lopez-Fraga M, Fernandez R, Albar JP, Hahne M. Biologically active APRIL is secreted following intracellular processing in the Golgi apparatus by furin convertase. EMBO Rep 2001;2(10):945–51.
26. Liu Y, Xu L, Opalka N, Kappler J, Shu HB, Zhang G. Crystal structure of sTALL-1 reveals a virus-like assembly of TNF family ligands. Cell 2002;108(3):383–94.
27. Ingold K, Zumsteg A, Tardivel A, et al. Identification of proteoglycans as the APRIL-specific binding partners. J Exp Med 2005;201(9):1375–83.
28. Hendriks J, Planelles L, de Jong-Odding J, et al. Heparan sulfate proteoglycan binding promotes APRIL-induced tumor cell proliferation. Cell Death Differ 2005;12(6):637–48.
29. Roschke V, Sosnovtseva S, Ward CD, et al. BLyS and APRIL form biologically active heterotrimers that are expressed in patients with systemic immune-based rheumatic diseases. J Immunol 2002;169(8):4314–21.
30. Bodmer JL, Schneider P, Tschopp J. The molecular architecture of the TNF superfamily. Trends Biochem Sci 2002;27(1):19–26.
31. Wallweber HJ, Compaan DM, Starovasnik MA, Hymowitz SG. The crystal structure of a proliferation-inducing ligand, APRIL. J Mol Biol 2004;343(2):283–90.
32. Oren DA, Li Y, Volovik Y, et al. Structural basis of BLyS receptor recognition. Nat Struct Biol 2002;9(4):288–92.
33. Karpusas M, Cachero TG, Qian F, et al. Crystal structure of extracellular human BAFF, a TNF family member that stimulates B lymphocytes. J Mol Biol 2002;315(5):1145–54.
34. Liu Y, Hong X, Kappler J, et al. Ligand-receptor binding revealed by the TNF family member TALL-1. Nature 2003;423(6935):49–56.
35. Kim HM, Yu KS, Lee ME, et al. Crystal structure of the BAFF-BAFF-R complex and its implications for receptor activation. Nat Struct Biol 2003;10(5):342–8.
36. Cachero TG, Schwartz IM, Qian F, et al. Formation of virus-like clusters is an intrinsic property of the tumor necrosis factor family member BAFF (B cell activating factor). Biochemistry 2006;45(7):2006–13.
37. Gordon NC, Pan B, Hymowitz SG, et al. BAFF/BLyS receptor 3 comprises a minimal TNF receptor-like module that encodes a highly focused ligand-binding site. Biochemistry 2003;42(20):5977–83.
38. Couchman JR. Syndecans: proteoglycan regulators of cell-surface microdomains? Nat Rev Mol Cell Biol 2003;4(12):926–37.
39. Hymowitz SG, Patel DR, Wallweber HJ, et al. Structures of APRIL-receptor complexes: like BCMA, TACI employs only a single cysteine-rich domain for high affinity ligand binding. J Biol Chem 2005;280(8):7218–27.
40. Bischof D, Elsawa SF, Mantchev G, et al. Selective activation of TACI by syndecan-2. Blood 2006;107(8):3235–42.
41. Banner DW, D'Arcy A, Janes W, et al. Crystal structure of the soluble human 55 kd TNF receptor-human TNF beta complex: implications for TNF receptor activation. Cell 1993;73(3):431–45.

42. Mongkolsapaya J, Grimes JM, Chen N, et al. Structure of the TRAIL-DR5 complex reveals mechanisms conferring specificity in apoptotic initiation. Nat Struct Biol 1999;6(11):1048–53.
43. Hymowitz SG, Christinger HW, Fuh G, et al. Triggering cell death: the crystal structure of Apo2L/TRAIL in a complex with death receptor 5. Mol Cell 1999;4(4):563–71.
44. Compaan DM, Hymowitz SG. The crystal structure of the costimulatory OX40-OX40L complex. Structure 2006;14(8):1321–30.
45. Kayagaki N, Yan M, Seshasayee D, et al. BAFF/BLyS receptor 3 binds the B cell survival factor BAFF ligand through a discrete surface loop and promotes processing of NF-kappaB2. Immunity 2002;17(4):515–24.
46. Patel DR, Wallweber HJ, Yin J, et al. Engineering an APRIL-specific B cell maturation antigen. J Biol Chem 2004;279(16):16727–35.
47. Lee CV, Hymowitz SG, Wallweber HJ, et al. Synthetic anti-BR3 antibodies that mimic BAFF binding and target both human and murine B cells. Blood 2006;108(9):3103–11.
48. Lin WY, Gong Q, Seshasayee D, et al. Anti-BR3 antibodies: a new class of B-cell immunotherapy combining cellular depletion and survival blockade. Blood 2007;110(12):3959–67.
49. Grech AP, Amesbury M, Chan T, Gardam S, Basten A, Brink R. TRAF2 differentially regulates the canonical and noncanonical pathways of NF-kappaB activation in mature B cells. Immunity 2004;21(5):629–42.
50. Xie P, Stunz LL, Larison KD, Yang B, Bishop GA. Tumor necrosis factor receptor-associated factor 3 is a critical regulator of B cell homeostasis in secondary lymphoid organs. Immunity 2007;27(2):253–67.
51. Sasaki Y, Derudder E, Hobeika E, et al. Canonical NF-kappaB activity, dispensable for B cell development, replaces BAFF-receptor signals and promotes B cell proliferation upon activation. Immunity 2006;24(6):729–39.
52. Enzler T, Bonizzi G, Silverman GJ, et al. Alternative and classical NF-kappa B signaling retain autoreactive B cells in the splenic marginal zone and result in lupus-like disease. Immunity 2006;25(3):403–15.
53. Xu LG, Shu HB. TNFR-associated factor-3 is associated with BAFF-R and negatively regulates BAFF-R-mediated NF-kappa B activation and IL-10 production. J Immunol 2002;169(12):6883–9.
54. Liao G, Zhang M, Harhaj EW, Sun SC. Regulation of the NF-kappaB-inducing kinase by tumor necrosis factor receptor-associated factor 3-induced degradation. J Biol Chem 2004;279(25):26243–50.
55. Thompson JS, Bixler SA, Qian F, et al. BAFF-R, a newly identified TNF receptor that specifically interacts with BAFF. Science 2001;293(5537):2108–11.
56. Yan M, Brady JR, Chan B, et al. Identification of a novel receptor for B lymphocyte stimulator that is mutated in a mouse strain with severe B cell deficiency. Curr Biol 2001;11(19):1547–52.
57. Morrison MD, Reiley W, Zhang M, Sun SC. An atypical tumor necrosis factor (TNF) receptor-associated factor-binding motif of B cell-activating factor belonging to the TNF family (BAFF) receptor mediates induction of the noncanonical NF-kappaB signaling pathway. J Biol Chem 2005;280(11):10018–24.
58. Ni CZ, Oganesyan G, Welsh K, et al. Key molecular contacts promote recognition of the BAFF receptor by TNF receptor-associated factor 3: implications for intracellular signaling regulation. J Immunol 2004;173(12):7394–400.
59. Gorelik L, Gilbride K, Dobles M, Kalled SL, Zandman D, Scott ML. Normal B cell homeostasis requires B cell activation factor production by radiation-resistant cells. J Exp Med 2003;198(6):937–45.
60. Bossen C, Cachero TG, Tardivel A, et al. TACI, unlike BAFF-R, is solely activated by oligomeric BAFF and APRIL to support survival of activated B cells and plasmablasts. Blood 2008;111(3):1004–12.

61. Baud V, Liu ZG, Bennett B, Suzuki N, Xia Y, Karin M. Signaling by proinflammatory cytokines: oligomerization of TRAF2 and TRAF6 is sufficient for JNK and IKK activation and target gene induction via an amino-terminal effector domain. Genes Dev 1999;13(10):1297–308.
62. Chang SK, Arendt BK, Darce JR, Wu X, Jelinek DF. A role for BLyS in the activation of innate immune cells. Blood 2006;108(8):2687–94.
63. Mantchev GT, Cortesao CS, Rebrovich M, Cascalho M, Bram RJ. TACI is required for efficient plasma cell differentiation in response to T-independent type 2 antigens. J Immunol 2007;179(4):2282–8.
64. Schwaller J, Schneider P, Mhawech-Fauceglia P, et al. Neutrophil-derived APRIL concentrated in tumor lesions by proteoglycans correlates with human B-cell lymphoma aggressiveness. Blood 2007;109(1):331–8.
65. Belnoue E, Pihlgren M, McGaha TL, et al. APRIL is critical for plasmablast survival in the bone marrow and poorly expressed by early life bone marrow stromal cells. Blood 2008;111:2755–64.
66. Xu W, He B, Chiu A, et al. Epithelial cells trigger frontline immunoglobulin class switching through a pathway regulated by the inhibitor SLPI. Nat Immunol 2007;8(3):294–303.

Chapter 2
BAFF Receptor Regulation of Peripheral B-Lymphocyte Survival and Development

Wasif N. Khan, Nicholas P. Shinners, Iris Castro, and Kristen L. Hoek

Abstract B-lymphocyte homeostasis depends on exogenous signals for survival during development and in immune responses to invading pathogens. These signals are continually provided by either tonic or antigen-mediated BCR signals and other trophic factors. B-cell-activating factor (BAFF) has emerged as a key growth factor for B lymphocytes. Through its interaction with a TNF-R family member, BAFF-R or BR3, BAFF promotes survival of both immature and mature B cells. BAFF/BR3 interaction also facilitates BCR-induced B-cell proliferation. Thus, dysregulation of the signals emanating from these receptors leads to autoimmune disease, whereas interference with these signals leads to B-cell immunodeficiencies. Multiple signal transduction pathways, including those involving transcription factor NF-κB, appear to play critical roles in BAFF-mediated B-cell biological responses. Recent studies have revealed that BR3 and BCR are functionally linked and that Bruton's cytoplasmic tyrosine kinase (Btk)/NF-κB signaling plays an essential role in this process. Therefore, the primary objective of this article is to discuss BR3-signaling pathways, and the cooperation with BCR signals, that regulate B-cell survival during development and activation.

Keywords Transitional B cells · B-cell antigen receptor · B-cell-activating factor receptor · NF-κB · Bruton's tyrosine kinase · Apoptosis · Tolerance · Activation

2.1 Introduction

B lymphocytes are integral to the adaptive immune system and are the main source of humoral immunity. In mammals, B lymphocytes are generated throughout life in the bone marrow (BM). Early B-cell development in the bone marrow occurs in a sequential, stepwise fashion, culminating in the expression of IgM molecules or

---

W.N. Khan (✉)
Department of Microbiology and Immunology, University of Miami, FL 33136, USA

M.P. Cancro (ed.), *BLyS Ligands and Receptors*, Contemporary Immunology,
DOI 10.1007/978-1-60327-013-7_2,

B-cell antigen receptors (BCRs) on the cell surface [1, 2]. Upon expression of BCR, immature B cells emigrate to the spleen as early transitional (transitional stage 1 or T1) B cells. Of the 10–20 million immature (IgM^{+}) B cells generated daily in the bone marrow, only a small proportion ultimately enter the mature peripheral B-cell pool [3–5].

In contrast to T lymphocytes, whose negative and positive selection is tightly regulated in the context of MHC restriction before release from the thymus, removal of self-reactive B cells is accomplished through numerous selection processes that occur in both the bone marrow and the periphery [6–8]. In this regard, immature B cells in the BM that express self-reactive BCR are deleted, anergized, or given the opportunity to undergo receptor editing [9–15]. Self-reactive transitional B-cell clones that escape the BM and migrate to the spleen are subject to further checkpoints for deletion, while maturation of select non-self-reactive clones occurs presumably through a positive selection process. The keys to these selection processes are the heterogeneity of transitional B cells and their distinct responses to environmental cues [16–22]. Early transitional B cells are proposed to be targeted for negative selection [23, 24]. However, it remains controversial as to how or whether cell subpopulations (or clones) within the later stages of the transitional B-lymphocyte compartment undergo positive selection to become mature B cells [7, 19, 23–25]. Further, precisely how the B-cell developmental program, which includes negative and positive selection steps, is executed at the molecular level remains unclear. Accumulating evidence points to a critical role for BAFF/BR3 in facilitating BCR-directed control of transitional B cells into both mature follicular (Fo) and marginal zone (MZ) B cells. Thus, signaling through both BCR and BR3 orchestrates intracellular events that, under normal physiological conditions, selectively allow maturation of only non-self-reactive transitional cells into mature B lymphocytes.

2.2 Peripheral B-Cell Development

Upon arrival in the spleen, transitional B cells undergo a maturation process characterized by a sequential series of discrete stages that can be identified based upon the cell surface expression of developmental markers. In recent years, several models that describe peripheral B-cell development have been proposed. The model originally proposed by Loder and Carsetti included three subsets termed T1, T2, and mature, in which the T1 subset serves as the precursor for T2 B cells which in turn serve as precursors for mature B cells [26]. Subsequently, the Allman group suggested that in addition to T1, T2, and mature B cells, an additional transitional subset, termed T3, exists [17]. The Loder group defined their three subsets by the following expression pattern of cell surface markers: T1 (IgM^{high} IgD^{dull} CD23^{-} and CD21^{low}); T2 (IgM^{high} IgD^{high} CD23^{+} and CD21^{high}); and mature (IgM^{int} IgD^{high} CD23^{+} CD21^{int}). The model proposed by the Allman group defined splenic B cells by the following markers: T1 (AA4^{+} IgM^{high} IgD^{dull} CD23^{-}); T2 (AA4^{+} IgM^{high} IgD^{high} CD23^{+}); T3 (AA4^{+} IgM^{low} IgD^{high} CD23^{+}); and mature (AA4^{-} IgM^{int}

IgD^{high} $CD23^{+}$). The identification of mature follicular and marginal zone B cells as phenotypically and functionally discrete subsets has led to further distinctions in the model for development of peripheral B cells. Recent reports also indicate that T3 B cells do not constitute a developmentally distinct subpopulation, but rather anergic mature B cells [27–29]. Additionally, the Allman group has recently characterized what they term marginal zone precursors (T2-preMZ) whose phenotypic characteristics resemble some of the $CD21^{hi}$ T2 cells within the T2 subset originally described by the Loder group. Together, these data argue for the elimination of T3 B cells from contention as a distinct developmental stage within transitional B cells, and for a model that is more complex than originally believed. Thus, splenic B cells can be divided into at least five functionally distinct developmental subpopulations: immature transitional type 1 (T1) B cells, transitional type 2 (T2), and pre-marginal zone (T2-preMZ) B cells, as well as mature follicular (Fo) and marginal zone (MZ) B cells [16, 17, 20, 25, 26, 30, 31].

In the current model for peripheral B-cell development, the T1 subset serves as a precursor for T2 B cells, which in turn serve as precursors for either mature Fo B cells or for T2-preMZ B cells, the immediate precursors of mature MZ B cells [25] (Fig. 2.1). Using a combination of the most current schemes to identify B-cell subpopulations, detailed phenotypic characteristic of each B-cell subset are as follows: T1 ($AA4^{+}$ IgM^{hi} HSA^{hi} IgD^{-} $CD21^{-}$ $CD23^{-}$); T2 ($AA4^{+}$ IgM^{hi} HSA^{hi} IgD^{hi}

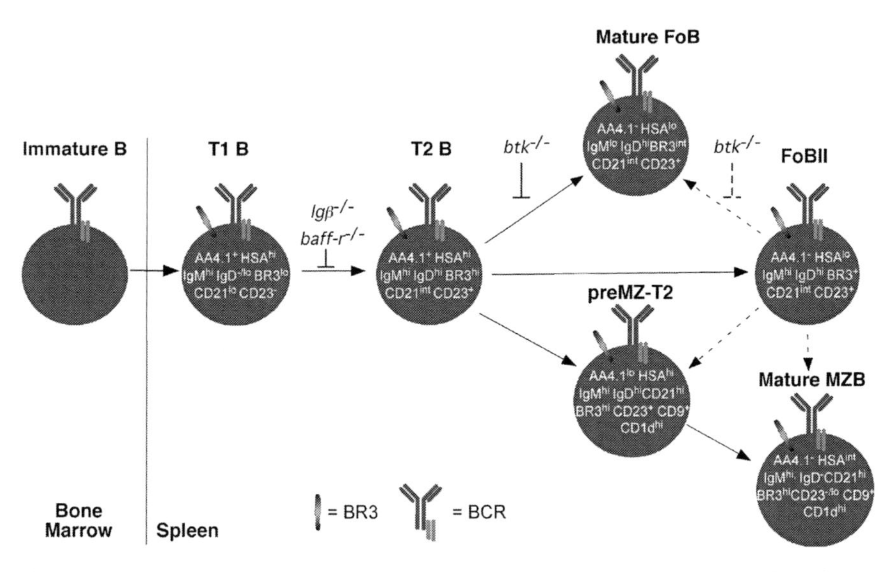

Fig. 2.1 A model of peripheral B-lymphocyte development. Upon surface expression of IgM (BCR), immature B cells leave the bone marrow and emigrate to the spleen as T1 B cells. In the spleen, T1 B cells differentiate into T2 B cells, which in turn serve as precursors for either marginal zone precursors (preMZ) or mature follicular (FoB) B cells. An additional follicular B-cell population (FoB II) has recently been identified (Cariappa et al. and Fig. 2.3) [16]; however, the differentiation potential of this population has yet to be definitively determined (designated by *dashed lines*)

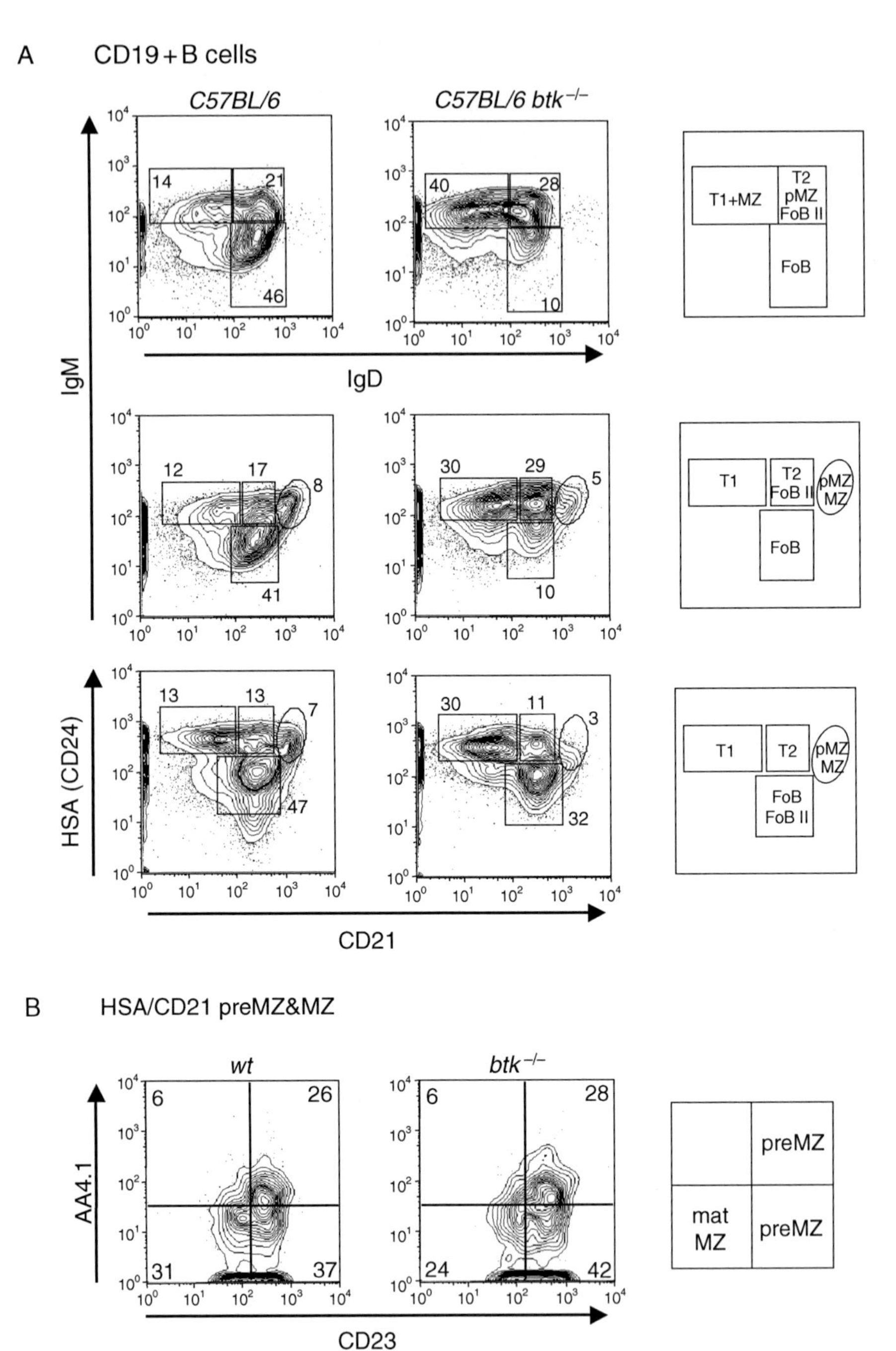

Fig. 2.2 Phenotypic characterization of B-cell subsets in the spleen. Freshly isolated C57BL/6 wild-type and C57BL/6 *btk*$^{-/-}$ splenocytes from 10-week-old mice were labeled with antibodies directed against CD19, CD93 (AA4.1), CD21, CD23, CD24 (HSA), IgM, and IgD to identify

CD21^{int} CD23^{+}); T2-preMZ (AA4$^{+/-}$ IgM^{hi} IgD^{hi} CD21^{hi} CD23^{+} CD9^{+} CD1d^{hi}); Fo (AA4^{-} IgM^{lo} HSA^{lo} IgD^{hi} CD21^{int} CD23^{+}); and MZ B cells (AA4^{-} IgM^{hi} HSA^{int} IgD^{lo} CD21^{hi} CD23^{lo} CD9^{+} CD1d^{hi}) (Fig. 2.2) [16, 17, 20, 25, 26, 30–33]. Additionally, a recent report has suggested that another mature follicular stage, termed FoB II (AA4^{-} HSA^{low} IgM^{hi} IgD^{hi} CD23^{+} CD21^{int}), may also exist [16]. Consistent with these studies our results also show that the FoB II cells differ from "traditional" FoB cells as they are generated in the absence of Btk-mediated signals and continue to express higher levels of surface IgM as compared to FoB cells (Fig. 2.3). Thus, this population of B cells has been proposed to potentially serve as immediate precursors for both mature Fo and T2-preMZ B cells, although the differentiation potential of these cells has not been determined. The T2 population we refer to in this discussion is based on exclusion of all the other B cell populations (T1, mature FoB, T2-preMZ, FoB II, and MZB) that have been currently identified [20, 34].

In the current model of peripheral B-cell development, T1 B cells can develop into T2 B cells, which in turn are thought to serve as the precursor to mature B cells. In support of this model, initial studies of auto-reconstitution of peripheral B cells by sublethal irradiation indicated that HSA^{high} immature splenic B cells occupied the spleen prior to HSA^{low} mature B cells [3, 4]. Further, bromodeoxyuridine (Brdu) labeling of HSA^{high} immature splenic B cells resulted in rapid labeling kinetics similar to that of BM-derived immature B cells, while HSA^{low} mature splenic B cells were quite different in their distinctly slow labeling kinetics [4]. These initial studies provided evidence that splenic B cells are comprised of phenotypically and biologically distinct subsets. Indeed, immature and mature B-cell subpopulations display distinct responses to similar stimuli. For example, immature B cells undergo apoptosis in response to BCR engagement, whereas mature B cells proliferate under similar stimulatory conditions [1, 3–5, 35–37]. Recent studies also corroborate these findings and further define that BCR cross-linking results in differential responses from purified transitional B-cell subpopulations and mature B cells. In response to BCR cross-linking in vitro, T1 B cells die, while some of the T2 B cells survive and display the phenotypic characteristics of mature B cells [21, 22]. BCR cross-linking also results in increased phosphorylation of the anti-apoptotic protein Akt and MAPK, ERK1/2 as well as increased expression of Bcl-xL in T2 B cells compared to T1 B cells [21]. Additionally, we have found that BCR cross-

Fig. 2.2 (continued) B-cell subsets. (**a**) In order to distinguish T1, T2, FoB, and MZB (mature + preMZ) cells, CD19^{+} B cells within the live lymphocyte gate are displayed by IgM and IgD expression (*top panels*), IgM and CD21 expression (*middle panels*), or HSA and CD21 expression (*bottom panels*). IgM^{lo} mature FoB are absent in *btk*$^{-/-}$ B cells. (**b**) CD21^{hi} MZ and preMZ from the HSA/CD21 MZ gate were further distinguished by CD23 and AA4.1 expression. Mature MZ B cells are CD23^{lo} and AA4.1^{-} (*lower left quadrant*), while preMZ are CD23^{+} and AA4.1^{+} or AA4.1^{-} (*upper* and *lower right quadrants*). We have previously demonstrated that mature MZ express intermediate levels of HSA [20]

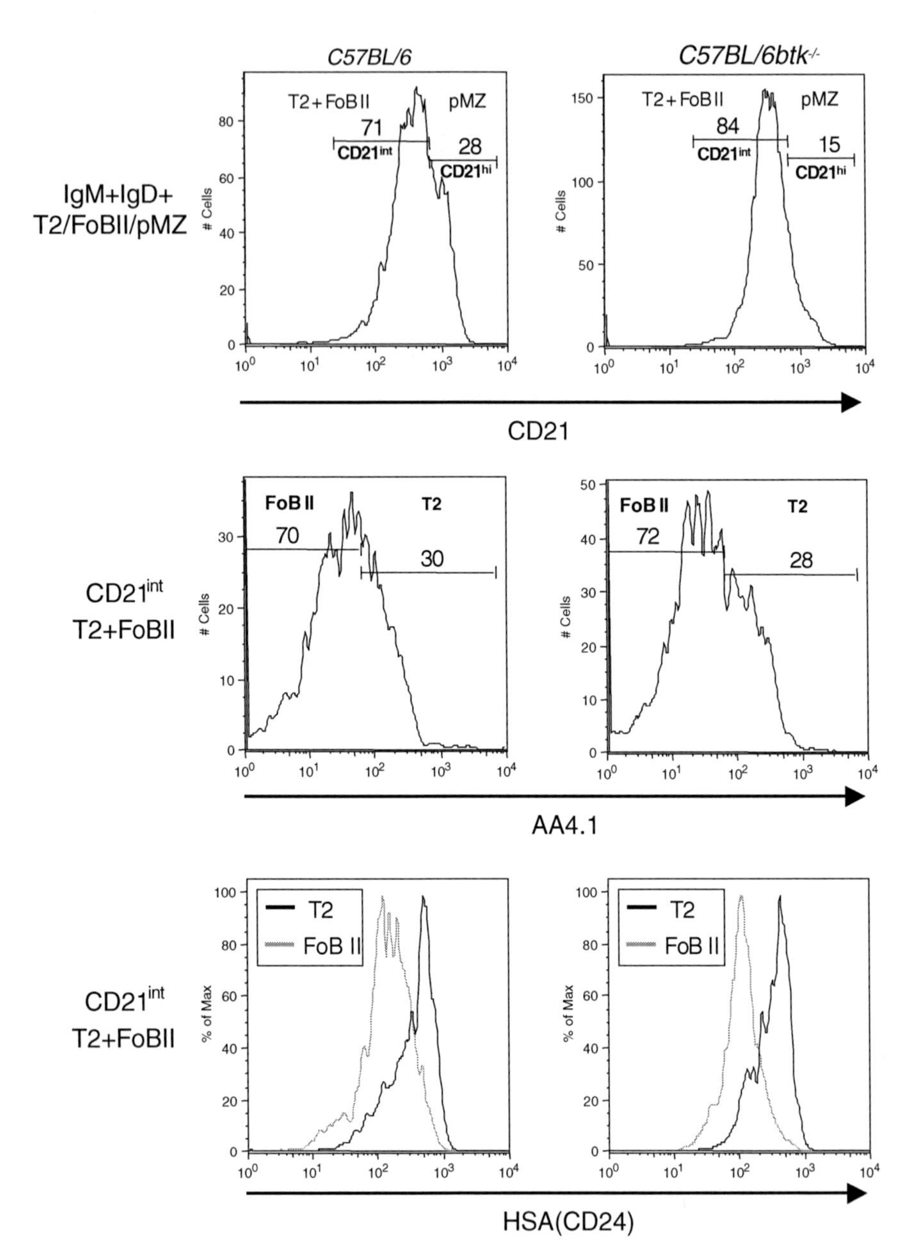

Fig. 2.3 Confirmation of FoB II subset identification. Freshly isolated C57BL/6 wild-type and C57BL/6 *btk*$^{-/-}$ splenocytes from 10-week-old mice were labeled as in Fig. 2.2 for B-cell subset identification. CD19^{+} B cells from the IgM/IgD T2/preMZ/FoB II gate in Fig. 2.2 were displayed as a histogram for CD21 expression (*top panels*). CD21^{hi} cells represent the T2-preMZ population, and are not analyzed further. CD21^{int} cells were displayed as a histogram for AA4.1 expression (*middle panels*). AA4.1^{+} cells represent the immature T2 population, while AA4.1^{-} cells represent the mature FoB II population. We found a significantly larger proportion/number

linking induced the production DAG and IP_3 in T2 and mature Fo B cells, but not T1 B cells [20]. Further, transcription factor NF-κB is only weakly activated in T1 when compared with T2 and mature B cells [34]. These observations suggest that the BCR may deliver distinct signals at discrete stages or checkpoints of peripheral B-cell development which may play a critical role in implementing peripheral B-cell developmental and activation program and preserve self-tolerance.

2.3 BAFF/BR3 in B-Cell Survival and Development

BAFF (also known as BLyS, TALL-1, THANK, zTNF4, or TNFSF13c) is a member of the tumor necrosis factor (TNF) family that functions as a major pro-survival factor for peripheral B cells [37–44]. A type II transmembrane protein, BAFF is secreted by monocytes, macrophages, dendritic cells, growth factor-stimulated neutrophils, and activated T cells after proteolytic processing [45–54]. Additionally, stromal cells have been found to be an in vivo source of BAFF that contributes to peripheral B-cell survival [55]. There is also a splice variant of BAFF (ΔBAFF), which may serve as a negative regulator of BAFF function by forming multimers with BAFF [56, 57]. Although our knowledge of BAFF expression among various cell types provides biological evidence for its role in peripheral B-cell development and survival, the biochemical events triggered in BAFF-stimulated B cells remain poorly understood. The effects of BR3 expression profile in distinct B-cell populations (discussed below) on the biochemical signaling also remain unclear.

BAFF binds three distinct receptors of the TNF-R family: B-cell maturation antigen (BCMA); transmembrane activator and cyclophilin ligand interactor (TACI); and BAFF receptor 3 (BR3, or BR3) [58]. Although BCMA RNA is readily detectable in spleen and lymph nodes, as well as in B-cell lines, cell surface detection of BCMA protein on B cells remains elusive. Currently, it is believed that BCMA function is likely restricted to germinal center B cells or terminally differentiated plasma cells [59–63]. TACI is found on the cell surface of splenic MZ and T2 B cells, while there is little to none detected on T1 or Fo B cells [61]. BR3 is detectable in all splenic B-cell subsets, being most highly expressed on T2 and MZ B cells, followed by Fo, and least on T1 B cells (Fig. 2.1) [30, 64].

Fig. 2.3 (continued) (approximately 30%) of $IgM^+IgD^+CD21^{int}AA4.1^+$ T2 B cells than previously described [16]. Expression of HSA in $CD21^{int}AA4.1^-$ FoB II B cells is significantly lower than in $CD21^{int}AA4.1^+$ T2 B cells (*bottom panels*), thus FoB II cells represent a "mature" phenotype. This population of cells likely represents the increased percentage of follicular B cells observed in *btk*$^{-/-}$ B cells stained with HSA/CD21 compared to *btk*$^{-/-}$ B cells stained with IgM/CD21 in Fig. 2.2

Gene-targeted deletion studies indicate that loss of TACI, BCMA, or both does not disrupt peripheral B-cell development, while loss of BAFF or BR3 or mice that contain a dysfunctional BR3 (A/WySnJ) results in a severe deficiency of T2 and mature B cells [55, 65, 66]. Additionally, injection of normal mice with a receptor fusion protein that sequesters soluble BAFF results in the rapid loss of B cells [67, 68]. The defects in B-cell survival and development of Fo B cells caused by loss of BR3 function in A/WySnJ mice can be restored by transgenic expression of the anti-apoptotic gene Bcl-xL [69]. Consistently, enforced expression of Bcl-xL or Bcl-2 in BR3-deficient or TACI-Ig transgenic (BAFF function-inhibited) animals also restored splenic B-cell survival and the Fo B-cell compartment [66, 69, 70]. However, mature B-cell reconstitution was not complete, as neither of these transgenes was able to restore the development of mature MZ B cells in BR3 mutant mice [66, 69, 70]. Conversely, transgenic mice that overexpress BAFF exhibit an increase in B-cell number and a breakdown in B-cell tolerance that leads to autoimmunity [71]. Further, elevated levels of BAFF in the serum of patients with autoimmune diseases such as systemic lupus erythematosis (SLE), rheumatoid arthritis (RH), and Sjogren's implicate BAFF as a survival factor that allows peripheral B cells to escape negative selection [72].

One of the earliest reports to indicate the normal function of BAFF suggested that BAFF acts as a survival factor during peripheral B-cell development. In this study, culture with BAFF promoted immature T2 B-cell survival, while inclusion of anti-IgM concurrent with BAFF promoted both T2 B-cell survival and the formation of phenotypically mature B cells [73]. The survival effect provided by BAFF occurs largely in the absence of cellular proliferation, although BAFF has been shown to induce entry into early G1 [44, 73]. These findings suggest that signals from the BCR direct proliferation and maturation of T2 to mature B cells, while BAFF/BR3 interactions function to promote survival of transitional B cells and facilitate BCR-induced proliferation and maturation. Consistent with this hypothesis, peripheral B-cell development is impaired between the T1 and T2 stages in mice deficient for either BAFF or BR3. Further, the highest levels of BR3 are expressed at the T2 B-cell stage. In this regard, a recent report has suggested that a fraction of T2 B cells may actually undergo homeostatic proliferation in response to BAFF in vivo [30].

2.4 Mechanisms of BR3-Mediated B-Cell Survival

Recent studies have begun to reveal some of the downstream effectors utilized by BAFF/BR3 to orchestrate B-cell survival. The current understanding is that BAFF-mediated B-cell survival is regulated by at least five pathways that result in attenuation of apoptosis: (1) activation of classical and alternative NF-κB; (2) activation of the anti-apoptotic proteins Mcl-1 and Pim2; (3) downregulation of the pro-apoptotic protein Bim; (4) nuclear exclusion of PKCδ; and (5) PKCβ-dependent activation of Akt [38, 74–77].

Early studies revealed that NF-κB, a family of transcription factors associated with survival signaling and immune function, was activated following culture of cells with BAFF [78, 79]. Initial studies investigating BAFF treatment of B cells indicated that BAFF/BR3 signaling activates the alternative NF-κB pathway via the processing of p100 in an IKKα-dependent manner [38, 80]. These early reports indicated that activation of the alternative NF-κB pathway is critical for BAFF-induced B-cell survival. However, subsequent studies, including our own, have demonstrated that the classical NF-κB pathway is also activated in response to BAFF [38, 39, 44, 75, 81, 82]. Initial studies demonstrated that rapid and transient p50 activity is required to initiate early BAFF-mediated survival signals including activation of Bcl-xL [39]. Further, it was found that loss of IKKγ led to a block in B-cell development beyond the T1 stage, which is reminiscent of loss of BR3, and that constitutively active IKK2 (IKKβ) completely restored the peripheral B-cell compartment in BR3-deficient animals, supporting a role for classical NF-κB signaling in BAFF-mediated B-cell survival and differentiation [81]. Our recent results reveal that BAFF-mediated activation of the classical NF-κB pathway is impaired in cells deficient for the protein tyrosine kinase Btk, suggesting that Btk may serve to couple BR3 to classical NF-κB activity [82]. Because Btk is the target for mutations in B-cell deficiency diseases, X-linked agammaglobulinemia (XLA) [83–85] in man and X-linked immunodeficiency (*Xid*) [86–88] in mouse, these findings suggest that defective BR3 signaling may contribute XLA and *Xid* phenotypes.

Members of the Bcl-2 family of proteins have also been analyzed as potential pro-survival targets in BAFF-mediated signaling. While Bcl-2 is upregulated by BAFF in B cells [38], it appears that most anti-apoptotic genes of the Bcl-2 family and their encoded proteins (e.g., Bcl-2, Bcl-XL, and Bfl1/A1) are either not regulated or are only modestly regulated by BAFF [39, 44, 74, 89]. However, BAFF has been shown recently to utilize another member of the Bcl-2 family, Mcl-1, to facilitate B-cell survival. Mcl-1 gene transcription, translation, and post-translational modifications all occur downstream of BAFF via non-redundant mechanisms involving Pim2 and the Akt/mTOR pathways [90]. In this regard, increased Pim2 expression has been found following BAFF treatment and Pim2-mediated survival signals have been found to be dependent upon IKKα and p52 [75, 91]. Accumulating evidence also suggests that BAFF regulates Bim, a pro-apoptotic protein of the Bcl-2 family. Initial analyses of BAFF transgenic and Bim-deficient mice revealed similar phenotypes [71, 92], suggesting that survival signals from BAFF may result in downregulation of the pro-apoptotic protein Bim. Indeed, stimulation with BAFF has been found to result in ERK-dependent phosphorylation and subsequent downregulation of Bim concomitant with B-cell survival [74]. Additionally, B cells from double transgenic HEL-Ig/sHEL mice that lack Bim do not require BAFF-mediated signals for survival, yet depend on BAFF for differentiation [93].

Stimulation of B cells by mitogenic or inflammatory stimuli results in proteolytic cleavage of PKCδ. This processing event results in the formation of a catalytically active fragment that translocates to the nucleus and phosphorylates histone H2B, thereby initiating cell death [76, 94–96]. Gene-targeted deletion of PKCδ results in an increased B-cell compartment and a B-cell survival advantage when compared to wild-type controls similar to the phenotype observed in BAFF transgenic animals [71, 76, 97, 98]. That peripheral tolerance is broken in both of these models suggests a functional relationship between BAFF/BR3 signaling and PKCδ. A recent report has demonstrated that the development of mature B cells occurs independently of BR3-mediated signals in PKCδ-deficient animals and that in vitro culture of B cells with BAFF results in cell survival coincidental with nuclear exclusion of catalytically active PKCδ [76]. Nuclear exclusion of PKCδ occurs via activation of classical NF-κB activity at the level of IKKβ, as cells expressing constitutively active IKKβ do not localize PKCδ to the nucleus, even in the absence of BR3 [81]. It is unclear whether this effect occurs by activation of the classical pathway through IκBα degradation or whether it is independent of NF-κB DNA binding and transactivation function. Further, analysis of mice with compound deficiency of PKCδ and Btk (*pkcδ*$^{-/-}$ x *btk*$^{-/-}$) revealed a B-cell survival defect similar to that observed in Btk-deficient B cells, suggesting that the survival defects in these compound mutant mice most likely represent two independent pathways that regulate B-cell survival [76]. Additionally, the same group has also implicated PKCβ in BAFF-mediated activation of the anti-apoptotic protein, Akt. However, it appears that BAFF-activated PKCβ regulates total B-cell metabolic fitness rather than B-cell survival [77].

2.5 Mechanisms of BAFF-Mediated Activation of Classical and Alternative NF-κB Pathways

The transcription factor NF-κB is one of the key regulators of innate and adaptive immunity [99, 100]. In the adaptive immune system, NF-κB regulates survival, development, and activation of lymphocytes [101, 102]. The NF-κB/Rel family contains five members: RelA (p65), RelB, c-Rel, NF-κB1 (p50/p105), and NF-κB2 (p52/p100) [100]. All members can form homodimers and heterodimers except RelB, which preferentially partners with p52. NF-κB/Rel dimers are kept inactive via their association with inhibitory proteins belonging to the inhibitor of NF-κB (IκB) family or the p100 and p105 precursor proteins [103]. Upon cellular activation via cell surface receptors including BCR, TCR, TLR, IL-1, TNF-R, and the BR3, dimeric NF-κB/Rel transcription factors are released from their inhibitors and translocate to the nucleus where they regulate the expression of multiple cellular genes [104, 105].

There are two pathways to NF-κB activation: the classical (canonical) pathway and the alternative (non-canonical) pathway [106, 107]. The classical pathway proceeds via the activation of a large IκB kinase (IKK) complex, which contains two catalytic subunits (IKKα and IKKβ) and a regulatory subunit (IKKγ, also called NEMO, IKKAP1, or Fip-3) [103, 108–112]. Upon activation, the IKK catalytic subunits, primarily IKKβ, target IKBα for site-specific phosphorylation at two conserved serine residues (S^{32} and S^{36}), thus earmarking the inhibitor for degradation by the ubiquitin–proteasome pathway [113–115]. Functional release of IKBα unmasks the nuclear localization signal on Rel dimers and permits their translocation to the nuclear site of action and transcriptional activation of target genes [116]. Recent studies suggest that IKBα can also inactivate nuclear NF-κB by binding Rel proteins and exporting them to the cytoplasm [117, 118]. The alternative pathway involves NF-κB-inducing kinase (NIK) and activation of IKKα, which phosphorylates p100. Phosphorylated p100 is then proteolytically processed into p52 [106]. Like IκBs, p100 also serves as an inhibitor; it heterodimerizes with RelB, and p100 processing following phosphorylation by IKKα releases p52/RelB dimers for nuclear translocation to activate gene transcription of a distinct set of κB-responsive genes [106].

One of the earliest studies investigating the effects of BAFF-mediated signaling in B cells revealed that BAFF/BR3 signaling activates the alternative NF-κB pathway via IKKα-dependent p100 processing [38]. A subsequent study revealed that processing of p100 was abrogated in A/WySnJ (BR3 mutant) mice or mice injected with BAFF-neutralizing antibodies, further supporting the role for activation of the alternative NF-κB pathway in response to BAFF-mediated signaling [80]. However, primary B cells lacking p50 or p52 or both, which may affect both NF-κB pathways, display defective in vitro survival in the presence of soluble BAFF [38, 39, 75], suggesting that both classical and alternative pathways are involved in BAFF-mediated B-cell survival. The significance of the alternative pathway in BAFF-dependent B-cell survival is further supported by our studies showing that B cells isolated from mice deficient in RelB, which is primarily targeted by the alternative pathway, display survival defects in response to BAFF in vitro [82]. In this regard, constitutive activity of IKKβ has been shown to restore B-cell development in BR3-deficient animals [81], while blockade of the classical NF-κB pathway has been shown to inhibit the survival of chronic lymphocytic leukemia B cells cultured with BAFF [119]. Further, BAFF-mediated immunoglobulin class-switching and autoantibody production in autoreactive B cells has been linked to activation of the classical NF-κB pathway [75]. Together, these observations have prompted intensive investigations of biological role of BR3 signaling in NF-κB activation.

Functional significance of the classical NF-κB pathway in B-cell survival is evident from the observed B-cell deficiency in mice with gene-targeted deletions of certain NF-κB subunits, IKKβ and IKKγ, as well as in animals that express a constitutive repressor of IκBα (IκBΔN) [39, 44, 81, 120–122]. Consistently, B-cell expansion, particularly that of MZ B cells, is observed in mice with constitutively active IKKβ [81]. That classical NF-κB activity also plays a role in BAFF-mediated signaling during B-cell development and survival is made apparent by the complete restoration of the B-cell compartment in BR3-deficient mice that express constitutively active IKKβ [81]. However, the mechanisms that control activation of the classical NF-κB pathway downstream of BR3 remain elusive.

Recently, we and others have demonstrated that BR3 activates the classical NF-κB pathway as evidenced by phosphorylation and degradation of IκBα in response to BAFF [39, 82, 119]. Phosphorylation of IκBα in response to BAFF has been shown to be dependent on IKKβ activity, as treatment with UTC (an IKKβ-specific inhibitor) prevented BAFF-mediated IκBα phosphorylation [119]. Additionally, we have recently demonstrated that degradation of IκBα in response to BAFF requires Btk, a cytoplasmic tyrosine kinase that has previously been shown to be required for BCR-mediated degradation of IκBα [82, 123, 124]. The dependence of BR3 signaling, which leads to IκBα phosphorylation and NF-κB activation, on Btk was unexpected, given that antigen receptors (BCR and TCR) typically employ distinct membrane-proximal molecular events from TNF-R family receptors, although the signals from both antigen and TNF receptors ultimately converge on the IKK signalosome to activate classical NF-κB [106]. Unlike TNF-R signaling, stimulation of the BCR initiates a cascade of tyrosine phosphorylation resulting in Btk-dependent activation of PLC-γ2 [125, 126]. In BCR-stimulated cells, the IKK complex is recruited to lipid rafts and activated through PKCβ, CARMA1, Bcl-10, and MALT1 [127]. However, it is unclear how Btk couples the TNF-R family receptor, BR3 to NF-κB. Recent studies indicating that PLC-γ2 is required for BR3-mediated activation of NF-κB suggest an involvement of the Btk substrate, PLC-γ2, in BAFF-induced NF-κB activation [128]. In this regard, a recent report demonstrated that PKCβ is activated in response to BAFF and targets the Aκt pathway. PKCβ is activated by the Btk/PLC-γ2 pathway in response to BCR stimulation. Thus, a potential mechanism by which the Btk/PLC-γ2 pathway activates classical NF-κB downstream of BR3 may involve PKCβ. More investigations are required to resolve this issue.

Members of the TNF-R family have been shown to activate NF-κB by utilizing tumor necrosis factor receptor-associated factors (TRAFs) [129–131]. Currently, TRAF3 is the only known TRAF to bind sequences within the cytoplasmic tail of the BR3 as well as NF-κB-inducing kinase (NIK) that is known to regulate alternative NF-κB activation [40, 132, 133]. Ubiquitination-dependent degradation of TRAF3 is necessary for BR3-induced activation of alternative NF-κB. TRAF2, an E3 ligase, may facilitate signal-dependent ubiquitination and degradation of TRAF3. Alternatively, TRAF2 may promote TRAF3 degradation via cIAP1 and cIAP2 E3 ligases, which are constitutively associated with TRAF2 [134]. Emerging data from our laboratory now indicate that BR3 activation results in TRAF2 tyrosine phosphorylation, which is dependent upon Btk[34]. These data support the possibility that activated Btk may phosphorylate TRAF2 and contribute to activation of NF-κB pathway by BR3. In this regard, mutational analyses have implicated a role for the tyrosine phosphorylation of TRAF2 in CD40-mediated signaling; conversion of Y484 to phenylalanine results in diminished TRAF2 and TRAF3 degradation [135]. Degradation of TRAF3 is relevant because this molecule has been shown to be a negative regulator of the alternative NF-κB pathway [136–138]. Loss of TRAF3 has been shown to result in increased NF-κB-inducing kinase (NIK) protein levels and constitutive p100 processing [137, 139]. Prior studies have also shown that activation through a number of TNF-R family members leads to down-regulation of TRAF2 by degradation mediated by lysine 48 ubiquitination by E3 ubiquitin ligase cIAP [140, 141]. Since TRAF2 is known to indirectly regulate TRAF3 degradation and BAFF stimulation results in TRAF3 degradation [142, 143], it is enticing to speculate that BAFF may activate Btk to modify TRAF2 by tyrosine phosphorylation in order to degrade TRAF3, ultimately leading to p100 processing.

B lineage-specific deletion of TRAF3 results in constitutive activation of the alternative NF-κB pathway leading to an expansion of the B-lymphocyte compartment and an increase in their viability in vitro [144]. Likewise, loss of TRAF2 results in the constitutive activation of the alternative pathway and B-cell expansion [145]. This raises the possibility that TRAF3 and TRAF2 together regulate NIK and activate alternative NF-κB pathway. However, TRAF2 and TRAF3 may support distinct functions in this process [139]. TRAF2-deficient B cells undergo constitutive p100 processing, yet unlike TRAF3-deficient B cells, TRAF2-deficient B cells cannot activate the classical NF-κB pathway [145]. Given that TRAF2 is critical for activation of the classical NF-κB pathway and inhibition of the alternative NF-κB pathway in B cells, modification of TRAF2 by Btk may play a role in the regulation of the alternative NF-κB pathway, in addition to the classical NF-κB pathway, which we have recently demonstrated [82]. Together, these findings suggest that BR3 stimulation may result in TRAF3 degradation via TRAF2/cIAP1/2 complexes, which would allow a significant accumulation of NIK protein and in turn activate IKKα-dependent p100 processing [138].

2.6 A Unified View of How BCR and BR3 Regulate B-Cell Development and Activation: BCR-Positive Regulation of BR3-Induced NF-κB Activation

The BCR and BR3 are two receptors on the surface of B cells that are absolutely required for peripheral B-cell development and survival. While BCR signaling guides the generation of functional mature B lymphocytes from transitional B cells in the spleen and maintains self-tolerance, BR3 ensures their viability and metabolic fitness. It is clear that the transitional B-cell populations display differential survival potential to BCR and/or BR3 engagement, which increases with maturation. While several signal transduction pathways may regulate apoptosis in T1 and T2 B cells in response to BCR and BAFF, the transcription factor NF-κB appears to be a major regulator of this process. This view is consistent with the B-cell phenotypes observed in mice with gene deletions of various signaling components that regulate nuclear translocation of NF-κB DNA-binding subunits, as well as the NF-κB subunits themselves [38, 82, 103, 121, 146, 147]. Therefore, one challenge in understanding differential sensitivity of T1 and T2 B cells to apoptosis during development and activation is to elucidate the molecular mechanisms that regulate the activity of NF-κB within each transitional B-cell population.

The kinetics of the two NF-κB pathways are different. Activation of the classical pathway is rapid and robust, but transient [148]. It involves rapid mobilization of membrane-proximal effectors that initiate IKKγ-dependent IKKβ activation, resulting in IκBα degradation. Activation of the alternative pathway is slow but sustained. It involves de novo synthesis of NIK, which phosphorylates and activates IKKα. In turn, IKKα targets p100 for proteolytic processing to generate the mature transcription factor, p52 [148]. Because proteolysis is an irreversible process, depletion of p100 by proteolytic processing serves as a bottleneck in the sustained activation of the alternative pathway. The transcription and translation of new p100 is at least partially under the control of the classical NF-κB pathway, as evidenced by RelA targeting of the p100 gene promoter [149]. Additionally, p100 transcription and protein synthesis require Btk and PLC-γ2, two regulators of the classical NF-κB pathway [82].

The current understanding in the field is that those receptors that activate the alternative pathway also activate the classical pathway to a varying extent. A potential reason for activation of the classical pathway downstream of the receptors that activate the alternative pathway is the dependence of the alternative pathway on production of signaling components that are induced by classical NF-κB pathway. Some receptors, including CD40 and LTβR, activate both NF-κB pathways robustly and may be considered self-sufficient in activating both NF-κB pathways. However, BR3 activates the alternative pathway more robustly than the classical NF-κB pathway, suggesting that maintenance of the alternative pathway downstream of BR3 is facilitated by another receptor that potently activates the classical pathway [82, 123, 150]. We propose that BCR signaling enables BR3 to sustain activity of the alternative pathway and this may serve to maintain B-cell survival. Indeed, we have recently discovered that disruption of BCR signaling in Btk-deficient B cells significantly reduces the ability of these B cells to activate the alternative pathway following BAFF stimulation [82]. Our recent data also indicate that the NF-κB subunit RelB, which can dimerize with p52, is also induced by BCR signaling [34]. In addition, the Cancro group and we have demonstrated that BR3 expression is induced following BCR engagement [64, 82, 151]. These findings, taken together with the initial discovery that co-culture with BAFF promoted the survival and BCR-mediated differentiation of T2 cells into phenotypically mature B cells [73], suggest that peripheral B-cell survival is regulated by a cooperative signaling between BCR and BR3. They also suggest a molecular mechanism for a functional link between BCR and BR3. According to this model, through activation of the classical pathway, BCR likely reinforces robust and sustained activation of alternative NF-κB pathway downstream of BR3 by means of inducing the expression of BR3, p100 and RelB. Thus, BCR and BR3 signaling together promote robust activation of the classical NF-κB pathway, followed by sustained BR3-mediated activation of the alternative pathway. Genetic evidence demonstrating that function of both NF-κB pathways is required for transitional B-cell maturation beyond the T1 stage also supports this model [38]. As discussed below, the two pathways have both independent and overlapping functions in the regulation of B-cell survival [38, 61, 82].

Regulation of survival by cooperative BCR- and BR3-dependent mechanisms may play a role in both preventing activation-induced cell death by BCR and protecting from BCR-induced apoptosis during B-cell development and selection. The transition of T1 into T2 cells is accompanied by a change in BCR signaling, which allows the induction of anti-apoptotic genes as well as the ability to productively respond to BR3-mediated survival signaling. We have recently discovered that cooperative BCR and BR3 signaling increases as T1 B cells mature into T2 B cells and that this increase coincides with the enhanced survival potential of T2 B cells in response to BAFF [34, 82]. Thus, these findings further suggest that BR3 function is critical for survival and proliferation beyond the T1 stage of B-cell development, a developmental checkpoint at which activity of both NF-κB pathways becomes important for further maturation [38]. The mechanisms driving this developmental switch are less clear. A potential mechanism is increased BR3 expression upon T1 cell differentiation into T2 B cells in vivo [64]. This increase in BR3 expression coincides with a change in BCR signaling in T2 cells that allows for sustained nuclear NF-κB activity and enhanced BR3 expression [34]. Additionally, we have found that BCR-mediated signaling in T2, but not in T1, B cells can induce p100 and RelB, which feeds into the alternative NF-κB pathway downstream of BR3 [34].

Based on the findings that BCR signaling regulates the expression of BR3, which in turn reinforces the alternative NF-κB pathway downstream of BR3 signaling, we envision a scenario in which tonic BCR signaling may similarly regulate BR3 function in vivo in the context of B-cell selection and maturation. In this scenario, BCR signaling modulates BR3, p100, and RelB expression as T1 B cells mature into T2 B cells. Functional expression of BR3 would also induce expression of the BCR co-receptor CD21 [152], a hallmark of maturation into T2 B cells. In contrast to T1, robust BCR signaling in T2 cells would display enhanced NF-κB signaling, which would lead to further increases in BR3 expression on T2 B cells, thus affording them

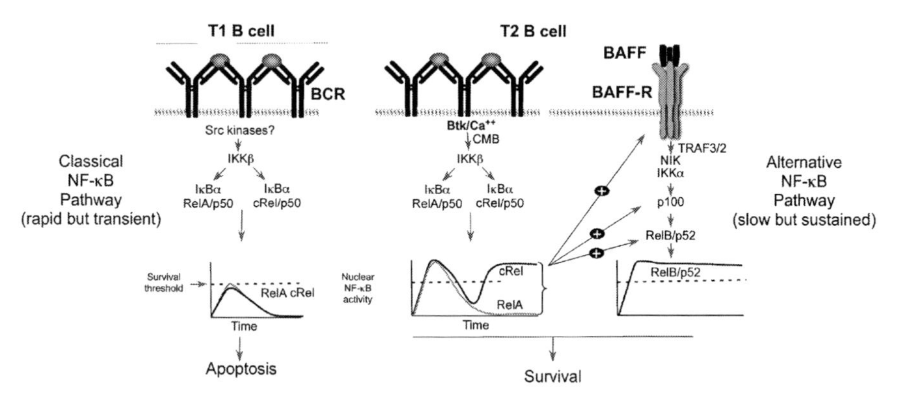

Fig. 2.4 Proposed model for the distinct survival potential of T2 compared to T1 B cells. BCR signals activate IKKβ and lead to IκBα degradation in both T1 and T2 B cells. This activation results in a transient and weak induction of NF-κB (Rel A and c-Rel) (phase I)) in T1 B cells. In contrast to T1, T2 B cells induce robust phase I NF-κB activity followed by a slower phase (phase II), which requires de novo c-Rel gene transcription and translation. T1 cells, because of reduced Btk/Ca^{2+} signaling, do not induce the second phase. However, T2 B cells induce c-Rel nuclear activity via Btk/Ca^{2+} pathway following BCR engagement. Sustained c-Rel leads to heightened expression of anti-apoptotic genes and induces the expression of BR3 and downstream components of the alternative pathway, p100 and RelB, selectively in T2 B cells. The ability to maintain c-Rel and BR3 signaling to a threshold sufficient to activate anti-apoptotic program allows T2 B cells to survive an undergo positive selection, while in T1 B cells, BCR-mediated NF-κB activity is not sustained, resulting in apoptosis and negative selection. (y axis = NF-κB, x axis = time, *dotted line* represents the survival threshold)

a distinct survival advantage that is necessary for further maturation. In this context, mature B lymphocytes would maintain survival through tonic BCR signaling as well as BR3-mediated signaling.

Taken together, these findings suggest that expression of the BR3 and induction of BR3-mediated signaling at the T2 stage may constitute a mechanism to confine the survival advantage among transitional B cells to the T2 subset, thus facilitating their positive selection and maturation into fully functional mature B cells. Since B cells in vivo are constitutively stimulated with BAFF and are likely to be engaged via the BCR, if specificity-based positive selection occurs, restriction of BR3 expression and function to T2 B cells would prevent survival of BCR-engaged T1 B cells, providing a mechanism for negative selection of T1 B cells – an important step in avoiding autoimmunity. Thus, tightly orchestrated interplay between BCR and BR3 signaling to the various combinations of NF-κB dimers facilitates peripheral B-cell development and survival, while maintaining self-tolerance (Fig. 2.4).

References

1. Meffre E, Casellas R, Nussenzweig MC. Antibody regulation of B cell development. Nat Immunol 2000;1(5):379–85.
2. Rajewsky K. Clonal selection and learning in the antibody system. Nature 1996;381:751–8.
3. Allman DM, Ferguson SE, Cancro MP. Peripheral B cell maturation. I. Immature peripheral B cells in adults are heat-stable antigenhi and exhibit unique signaling characteristics. J Immunol 1992;149(8):2533–40.
4. Allman DM, Ferguson SE, Lentz VM, Cancro MP. Peripheral B cell maturation. II. Heat-stable antigen(hi) splenic B cells are an immature developmental intermediate in the production of long-lived marrow-derived B cells. J Immunol 1993;151(9):4431–44.
5. Carsetti R, Kohler G, Lamers MC. Transitional B cells are the target of negative selection in the B cell compartment. J Exp Med 1995;181(6):2129–40.
6. Goodnow CC. Balancing immunity and tolerance: deleting and tuning lymphocyte repertoires. Proc Natl Acad Sci USA 1996;93(6):2264–71.
7. Pillai S. Two lymphoid roads diverge–but does antigen bade B cells to take the road less traveled? Immunity 2005;23(3):242–4.
8. Rolink AG, Schaniel C, Andersson J, Melchers F. Selection events operating at various stages in B cell development. Curr Opin Immunol 2001;13(2):202–7.
9. Gay D, Saunders T, Camper S, Weigert M. Receptor editing: an approach by autoreactive B cells to escape tolerance. J Exp Med 1993;177(4):999–1008.
10. Goodnow CC, Crosbie J, Adelstein S, et al. Altered immunoglobulin expression and functional silencing of self- reactive B lymphocytes in transgenic mice. Nature 1988;334(6184):676–82.
11. Hartley SB, Crosbie J, Brink R, Kantor AB, Basten A, Goodnow CC. Elimination from peripheral lymphoid tissues of self-reactive B lymphocytes recognizing membrane-bound antigens. Nature 1991;353(6346):765–9.
12. Nemazee DA, Burki K. Clonal deletion of B lymphocytes in a transgenic mouse bearing anti-MHC class I antibody genes. Nature 1989;337(6207):562–6.
13. Radic MZ, Erikson J, Litwin S, Weigert M. B lymphocytes may escape tolerance by revising their antigen receptors. J Exp Med 1993;177(4):1165–73.
14. Tiegs SL, Russell DM, Nemazee D. Receptor editing in self-reactive bone marrow B cells. J Exp Med 1993;177(4):1009–20.

15. Pelanda R, Schwers S, Sonoda E, Torres RM, Nemazee D, Rajewsky K. Receptor editing in a transgenic mouse model: site, efficiency, and role in B cell tolerance and antibody diversification. Immunity 1997;7(6):765–75.
16. Cariappa A, Boboila C, Moran ST, Liu H, Shi HN, Pillai S. The recirculating B cell pool contains two functionally distinct, long-lived, posttransitional, follicular B cell populations. J Immunol 2007;179(4):2270–81.
17. Allman D, Lindsley RC, DeMuth W, Rudd K, Shinton SA, Hardy RR. Resolution of three nonproliferative immature splenic B cell subsets reveals multiple selection points during peripheral B cell maturation. J Immunol 2001;167(12):6834–40.
18. Allman D, Srivastava B, Lindsley RC. Alternative routes to maturity: branch points and pathways for generating follicular and marginal zone B cells. Immunol Rev 2004;197: 147–60.
19. Cancro MP. Peripheral B-cell maturation: the intersection of selection and homeostasis. Immunol Rev 2004;197:89–101.
20. Hoek KL, Antony P, Lowe J, et al. Transitional B cell fate is associated with developmental stage-specific regulation of diacylglycerol and calcium signaling upon B cell receptor engagement. J Immunol 2006;177(8):5405–13.
21. Petro JB, Gerstein RM, Lowe J, Carter RS, Shinners N, Khan WN. Transitional type 1 and 2 B lymphocyte subsets are differentially responsive to antigen receptor signaling. J Biol Chem 2002;277(50):48009–19.
22. Su TT, Rawlings DJ. Transitional B lymphocyte subsets operate as distinct checkpoints in murine splenic B cell development. J Immunol 2002;168(5):2101–10.
23. Cancro MP, Kearney JF. B cell positive selection: road map to the primary repertoire? J Immunol 2004;173(1):15–19.
24. Su TT, Guo B, Wei B, Braun J, Rawlings DJ. Signaling in transitional type 2 B cells is critical for peripheral B-cell development. Immunol Rev 2004;197:161–78.
25. Pillai S, Cariappa A, Moran ST. Positive selection and lineage commitment during peripheral B-lymphocyte development. Immunol Rev 2004;197:206–18.
26. Loder F, Mutschler B, Ray RJ, et al. B cell development in the spleen takes place in discrete steps and is determined by the quality of B cell receptor-derived signals. J Exp Med 1999;190(1):75–89.
27. Melchers F. Anergic B cells caught in the act. Immunity 2006;25(6):864–7.
28. Melchers F, Rolink AR. B cell tolerance–how to make it and how to break it. Curr Top Microbiol Immunol 2006;305:1–23.
29. Teague BN, Pan Y, Mudd PA, et al. Cutting edge: transitional T3 B cells do not give rise to mature B cells, have undergone selection, and are reduced in murine lupus. J Immunol 2007;178(12):7511–5.
30. Meyer-Bahlburg A, Andrews SF, Yu KO, Porcelli SA, Rawlings DJ. Characterization of a late transitional B cell population highly sensitive to BAFF-mediated homeostatic proliferation. J Exp Med 2008;205(1):155–68.
31. Srivastava B, Quinn WJ, III, Hazard K, Erikson J, Allman D. Characterization of marginal zone B cell precursors. J Exp Med 2005;202(9):1225–34.
32. Cariappa A, Tang M, Parng C, et al. The follicular versus marginal zone B lymphocyte cell fate decision is regulated by Aiolos, Btk, and CD21. Immunity 2001;14(5):603–15.
33. Pillai S, Cariappa A, Moran ST. Marginal zone B cells. Annu Rev Immunol 2005;23:161–96.
34. Castro I, Wright JA, Damdinsuren B, Hoek KL, Carlesso G, Shinners NP, Gerstein RM, Woodland RT, Sen R, Khan WN. B cell receptor-mediated sustained c-Rel activation facilitates late transitional B cell survival through control of B cell activating factor receptor and NF-KappaB2. J Immunol. 2009 Jun 15;182(12):7729–37. PMID:19494297
35. King LB, Monroe JG. Immunobiology of the immature B cell: plasticity in the B-cell antigen receptor-induced response fine tunes negative selection. Immunol Rev 2000;176:86–104.
36. Monroe JG. B-cell antigen receptor signaling in immature-stage B cells: integrating intrinsic and extrinsic signals. Curr Top Microbiol Immunol 2000;245(2):1–29.

37. Sandel PC, Monroe JG. Negative selection of immature B cells by receptor editing or deletion is determined by site of antigen encounter. Immunity 1999;10(3):289–99.
38. Claudio E, Brown K, Park S, Wang H, Siebenlist U. BAFF-induced NEMO-independent processing of NF-kappaB2 in maturing B cells. Nat Immunol 2002;3(10):958–65.
39. Hatada EN, Do RK, Orlofsky A, et al. NF-kappaB1 p50 is required for BLyS attenuation of apoptosis but dispensable for processing of NF-kappaB2 p100 to p52 in quiescent mature B cells. J Immunol 2003;171(2):761–8.
40. Morrison MD, Reiley W, Zhang M, Sun SC. An atypical tumor necrosis factor (TNF) receptor-associated factor-binding motif of B cell-activating factor belonging to the TNF family (BAFF) receptor mediates induction of the noncanonical NF-kappaB signaling pathway. J Biol Chem 2005;280(11):10018–24.
41. Ramakrishnan P, Wang W, Wallach D. Receptor-specific signaling for both the alternative and the canonical NF-kappaB activation pathways by NF-kappaB-inducing kinase. Immunity 2004;21(4):477–89.
42. Rolink AG, Tschopp J, Schneider P, Melchers F. BAFF is a survival and maturation factor for mouse B cells. Eur J Immunol 2002;32(7):2004–10.
43. Yan M, Brady JR, Chan B, et al. Identification of a novel receptor for B lymphocyte stimulator that is mutated in a mouse strain with severe B cell deficiency. Curr Biol 2001;11(19):1547–52.
44. Zarnegar B, He JQ, Oganesyan G, Hoffmann A, Baltimore D, Cheng G. Unique CD40-mediated biological program in B cell activation requires both type 1 and type 2 NF-kappaB activation pathways. Proc Natl Acad Sci USA 2004;101(21):8108–13.
45. Hahne M, Kataoka T, Schroter M, et al. APRIL, a new ligand of the tumor necrosis factor family, stimulates tumor cell growth. J Exp Med 1998;188(6):1185–90.
46. Huard B, Arlettaz L, Ambrose C, et al. BAFF production by antigen-presenting cells provides T cell co-stimulation. Int Immunol 2004;16(3):467–75.
47. Litinskiy MB, Nardelli B, Hilbert DM, et al. DCs induce CD40-independent immunoglobulin class switching through BLyS and APRIL. Nat Immunol 2002;3(9):822–9.
48. Mackay F, Groom JR, Tangye SG. An important role for B-cell activation factor and B cells in the pathogenesis of Sjogren's syndrome. Curr Opin Rheumatol 2007;19(5):406–13.
49. Mackay F, Silveira PA, Brink R. B cells and the BAFF/APRIL axis: fast-forward on autoimmunity and signaling. Curr Opin Immunol 2007;19(3):327–36.
50. Mackay FS, Woods JA, Heringova P, et al. A potent cytotoxic photoactivated platinum complex. Proc Natl Acad Sci USA 2007;104(52):20743–8.
51. Moore JH, Parker JS, Olsen NJ, Aune TM. Symbolic discriminant analysis of microarray data in autoimmune disease. Genet Epidemiol 2002;23(1):57–69.
52. Nardelli B, Belvedere O, Roschke V, et al. Synthesis and release of B-lymphocyte stimulator from myeloid cells. Blood 2001;97(1):198–204.
53. Scapini P, Nardelli B, Nadali G, et al. G-CSF-stimulated neutrophils are a prominent source of functional BLyS. J Exp Med 2003;197(3):297–302.
54. Schneider P, MacKay F, Steiner V, et al. BAFF, a novel ligand of the tumor necrosis factor family, stimulates B cell growth. J Exp Med 1999;189(11):1747–56.
55. Gorelik L, Gilbride K, Dobles M, Kalled SL, Zandman D, Scott ML. Normal B cell homeostasis requires B cell activation factor production by radiation-resistant cells. J Exp Med 2003;198(6):937–45.
56. Gavin AL, Ait-Azzouzene D, Ware CF, Nemazee D. DeltaBAFF, an alternate splice isoform that regulates receptor binding and biopresentation of the B cell survival cytokine, BAFF. J Biol Chem 2003;278(40):38220–8.
57. Gavin AL, Duong B, Skog P, et al. DeltaBAFF, a splice isoform of BAFF, opposes full-length BAFF activity in vivo in transgenic mouse models. J Immunol 2005;175(1): 319–28.
58. Kalled SL, Ambrose C, Hsu YM. The biochemistry and biology of BAFF, APRIL and their receptors. Curr Dir Autoimmun 2005;8:206–42.

59. Avery DT, Kalled SL, Ellyard JI, et al. BAFF selectively enhances the survival of plasmablasts generated from human memory B cells. J Clin Invest 2003;112(2): 286–97.
60. Gras MP, Laabi Y, Linares-Cruz G, et al. BCMAp: an integral membrane protein in the Golgi apparatus of human mature B lymphocytes. Int Immunol 1995;7(7):1093–106.
61. Ng LG, Sutherland AP, Newton R, et al. B cell-activating factor belonging to the TNF family (BAFF)-R is the principal BAFF receptor facilitating BAFF costimulation of circulating T and B cells. J Immunol 2004;173(2):807–17.
62. Chang SK, Arendt BK, Darce JR, Wu X, Jelinek DF. A role for BLyS in the activation of innate immune cells. Blood 2006;108(8):2687–94.
63. Klein B, Tarte K, Jourdan M, et al. Survival and proliferation factors of normal and malignant plasma cells. Int J Hematol 2003;78(2):106–13.
64. Smith SH, Cancro MP. Cutting edge: B cell receptor signals regulate BLyS receptor levels in mature B cells and their immediate progenitors. J Immunol 2003;170(12): 5820–3.
65. Miller DJ, Hayes CE. Phenotypic and genetic characterization of a unique B lymphocyte deficiency in strain A/WySnJ mice. Eur J Immunol 1991;21(5):1123–30.
66. Sasaki Y, Casola S, Kutok JL, Rajewsky K, Schmidt-Supprian M. TNF family member B cell-activating factor (BAFF) receptor-dependent and -independent roles for BAFF in B cell physiology. J Immunol 2004;173(4):2245–52.
67. Gross JA, Dillon SR, Mudri S, et al. TACI-Ig neutralizes molecules critical for B cell development and autoimmune disease. impaired B cell maturation in mice lacking BLyS. Immunity 2001;15(2):289–302.
68. Schneider P, Takatsuka H, Wilson A, et al. Maturation of marginal zone and follicular B cells requires B cell activating factor of the tumor necrosis factor family and is independent of B cell maturation antigen. J Exp Med 2001;194(11):1691–7.
69. Amanna IJ, Dingwall JP, Hayes CE. Enforced bcl-xL gene expression restored splenic B lymphocyte development in BAFF-R mutant mice. J Immunol 2003;170(9):4593–600.
70. Tardivel A, Tinel A, Lens S, et al. The anti-apoptotic factor Bcl-2 can functionally substitute for the B cell survival but not for the marginal zone B cell differentiation activity of BAFF. Eur J Immunol 2004;34(2):509–18.
71. Mackay F, Woodcock SA, Lawton P, et al. Mice transgenic for BAFF develop lymphocytic disorders along with autoimmune manifestations. J Exp Med 1999;190(11): 1697–710.
72. Mackay F, Browning JL. BAFF: a fundamental survival factor for B cells. Nat Rev Immunol 2002;2(7):465–75.
73. Batten M, Groom J, Cachero TG, et al. BAFF mediates survival of peripheral immature B lymphocytes. J Exp Med 2000;192(10):1453–66.
74. Craxton A, Draves KE, Gruppi A, Clark EA. BAFF regulates B cell survival by downregulating the BH3-only family member Bim via the ERK pathway. J Exp Med 2005;202(10): 1363–74.
75. Enzler T, Bonizzi G, Silverman GJ, et al. Alternative and classical NF-kappa B signaling retain autoreactive B cells in the splenic marginal zone and result in lupus-like disease. Immunity 2006;25(3):403–15.
76. Mecklenbrauker I, Kalled SL, Leitges M, Mackay F, Tarakhovsky A. Regulation of B-cell survival by BAFF-dependent PKCdelta-mediated nuclear signalling. Nature 2004;431(7007):456–61.
77. Patke A, Mecklenbrauker I, Erdjument-Bromage H, Tempst P, Tarakhovsky A. BAFF controls B cell metabolic fitness through a PKC{beta}- and Akt-dependent mechanism. J Exp Med 2006;203(11):2551–62.
78. Kanakaraj P, Migone TS, Nardelli B, et al. Blys binds to b cells with high affinity and induces activation of the transcription factors nf-kappab and elf-1. Cytokine 2001;13(1): 25–31.

79. Mukhopadhyay A, Ni J, Zhai Y, Yu GL, Aggarwal BB. Identification and characterization of a novel cytokine, THANK, a TNF homologue that activates apoptosis, nuclear factor-kappaB, and c-Jun NH2-terminal kinase. J Biol Chem 1999;274(23):15978–81.
80. Kayagaki N, Yan M, Seshasayee D, et al. BAFF/BLyS receptor 3 binds the B cell survival factor BAFF ligand through a discrete surface loop and promotes processing of NF-kappaB2. Immunity 2002;17(4):515–24.
81. Sasaki Y, Derudder E, Hobeika E, et al. Canonical NF-kappaB activity, dispensable for B cell development, replaces BAFF-receptor signals and promotes B cell proliferation upon activation. Immunity 2006;24(6):729–39.
82. Shinners NP, Carlesso G, Castro I, et al. Bruton's tyrosine kinase mediates NF-kappaB activation and B cell survival by B cell-activating factor receptor of the TNF-R family. J Immunol 2007;179(6):3872–80.
83. Conley ME, Cooper MD. Genetic basis of abnormal B cell development. Curr Opin Immunol 1998;10(4):399–406.
84. Tsukada S, Saffran DC, Rawlings DJ, et al. Deficient expression of a B cell cytoplasmic tyrosine kinase in human X-linked agammaglobulinemia. Cell 1993;72:279–90.
85. Vetrie D, Vorchovsky I, Sideras P, et al. The gene involved in X-linked agammaglobulinemia is a member of the *src* family of protein-tyrosine kinases. Nature 1993;361:226–33.
86. Rawlings DJ. Bruton's tyrosine kinase controls a sustained calcium signal essential for B lineage development and function. Clin Immunol 1999;91(3):243–53.
87. Rawlings DJ, Saffran DC, Tsukada S, et al. Mutation of unique region of Bruton's tyrosine kinase in immunodeficient XID mice. Science 1993;261(5119):358–61.
88. Thomas JD, Sideras P, Smith CI, Vorechovsky I, Chapman V, Paul WE. Colocalization of X-linked agammaglobulinemia and X-linked immunodeficiency genes. Science 1993;261(5119):355–8.
89. Trescol-Biemont MC, Verschelde C, Cottalorda A, Bonnefoy-Berard N. Regulation of A1/Bfl-1 expression in peripheral splenic B cells. Biochimie 2004;86(4–5):287–94.
90. Woodland RT, Fox CJ, Schmidt MR, et al. Multiple signaling pathways promote B lymphocyte stimulator dependent B-cell growth and survival. Blood 2008;111(2):750–60.
91. Lesley R, Xu Y, Kalled SL, et al. Reduced competitiveness of autoantigen-engaged B cells due to increased dependence on BAFF. Immunity 2004;20(4):441–53.
92. Bouillet P, Metcalf D, Huang DC, et al. Proapoptotic Bcl-2 relative Bim required for certain apoptotic responses, leukocyte homeostasis, and to preclude autoimmunity. Science 1999;286(5445):1735–8.
93. Oliver PM, Vass T, Kappler J, Marrack P. Loss of the proapoptotic protein, Bim, breaks B cell anergy. J Exp Med 2006;203(3):731–41.
94. Ajiro K. Histone H2B phosphorylation in mammalian apoptotic cells. An association with DNA fragmentation. J Biol Chem 2000;275(1):439–43.
95. Brodie C, Blumberg PM. Regulation of cell apoptosis by protein kinase c delta. Apoptosis 2003;8(1):19–27.
96. Cheung WL, Ajiro K, Samejima K, et al. Apoptotic phosphorylation of histone H2B is mediated by mammalian sterile twenty kinase. Cell 2003;113(4):507–17.
97. Khare SD, Sarosi I, Xia XZ, et al. Severe B cell hyperplasia and autoimmune disease in TALL-1 transgenic mice. Proc Natl Acad Sci USA 2000;97(7):3370–5.
98. Miyamoto A, Nakayama K, Imaki H, et al. Increased proliferation of B cells and autoimmunity in mice lacking protein kinase Cdelta. Nature 2002;416(6883):865–9.
99. Bonizzi G, Karin M. The two NF-kappaB activation pathways and their role in innate and adaptive immunity. Trends Immunol 2004;25(6):280–8.
100. Ghosh S, May MJ, Kopp EB. NF-kappa B and Rel proteins: evolutionarily conserved mediators of immune responses. Annu Rev Immunol 1998;16:225–60.
101. Gerondakis S, Grumont R, Rourke I, Grossmann M. The regulation and roles of Rel/NF-kappa B transcription factors during lymphocyte activation. Curr Opin Immunol 1998;10(3):353–9.

102. Gerondakis S, Strasser A. The role of Rel/NF-kappaB transcription factors in B lymphocyte survival. Semin Immunol 2003;15(3):159–66.
103. Pasparakis M, Luedde T, Schmidt-Supprian M. Dissection of the NF-kappaB signalling cascade in transgenic and knockout mice. Cell Death Differ 2006;13(5):861–72.
104. Pahl HL. Activators and target genes of Rel/NF-κB transcription factors. Oncogene 1999;18:6853–66.
105. Sha WC. Regulation of immune responses by NF-kappa B/Rel transcription factor [published erratum appears in J Exp Med 1998 Feb 16;187(4):661]. J Exp Med 1998;187(2): 143–6.
106. Ghosh S, Karin M. Missing pieces in the NF-kappaB puzzle. Cell 2002;109(Suppl):S81–96.
107. Pomerantz JL, Baltimore D. Two pathways to NF-kappaB. Mol Cell 2002;10(4):693–5.
108. Delhase M, Hayakawa M, Chen Y, Karin M. Positive and negative regulation of IkappaB kinase activity through IKKbeta subunit phosphorylation [see comments]. Science 1999;284(5412):309–13.
109. DiDonato JA, Hayakawa M, Rothwarf DM, Zandi E, Karin M. A cytokine-responsive IkappaB kinase that activates the transcription factor NF-kappaB [see comments]. Nature 1997;388(6642):548–54.
110. Mercurio F, et al. IKK-1 and IKK-2: cytokine-activated I kappaB kinses essential for NF-kappaB activation. Science 1997;278:860–6.
111. Woronicz JD, Gao X, Cao Z, Rothe M, Goeddel DV. IkappaB kinase-beta: NF-kappaB activation and complex formation with IkappaB kinase-alpha and NIK. Science 1997;278(5339):866–9.
112. Zandi E, Karin M. Bridging the gap: composition, regulation, and physiological function of the IkappaB kinase complex. Mol Cell Biol 1999;19(7):4547–51.
113. Scherer DC, Brockman JA, Chen Z, Maniatis T, Ballard DW. Signal-induced degradation of I kappa B alpha requires site-specific ubiquitination. Proc Natl Acad Sci USA 1995;92(24):11259–63.
114. Brockman JA, Scherer DC, McKinsey TA, et al. Coupling of a signal response domain in I kappa B alpha to multiple pathways for NF-kappa B activation. Mol Cell Biol 1995;15(5):2809–18.
115. Brown K, Gerstberger S, Carlson L, Franzoso G, Siebenlist U. Control of I kappa B-alpha proteolysis by site-specific, signal-induced phosphorylation. Science 1995;267(5203):1485–8.
116. Baldwin AS, Jr. The NF-kappa B and I kappa B proteins: new discoveries and insights. Annu Rev Immunol 1996;14:649–83.
117. Tam WF, Sen R. IkappaB family members function by different mechanisms. J Biol Chem 2001;276(11):7701–4.
118. Tam WF, Wang W, Sen R. Cell-specific association and shuttling of IkappaBalpha provides a mechanism for nuclear NF-kappaB in B lymphocytes. Mol Cell Biol 2001;21(14):4837–46.
119. Endo T, Nishio M, Enzler T, et al. BAFF and APRIL support chronic lymphocytic leukemia B-cell survival through activation of the canonical NF-kappaB pathway. Blood 2007;109(2):703–10.
120. Li ZW, Omori SA, Labuda T, Karin M, Rickert RC. IKK beta is required for peripheral B cell survival and proliferation. J Immunol 2003;170(9):4630–7.
121. Pasparakis M, Schmidt-Supprian M, Rajewsky K. IkappaB kinase signaling is essential for maintenance of mature B cells. J Exp Med 2002;196(6):743–52.
122. Schmidt-Supprian M, Bloch W, Courtois G, et al. NEMO/IKK gamma-deficient mice model incontinentia pigmenti. Mol Cell 2000;5(6):981–92.
123. Petro JB, Khan WN. Phospholipase C-gamma 2 Couples Bruton's Tyrosine Kinase to the NF-kappa B signaling pathway in B lymphocytes. J Biol Chem 2001;276(3):1715–9.
124. Petro JB, Rahman SM, Ballard DW, Khan WN. Bruton's tyrosine kinase is required for activation of IkappaB kinase and nuclear factor kappaB in response to B cell receptor engagement. J Exp Med 2000;191(10):1745–54.

125. Humphries LA, Dangelmaier C, Sommer K, et al. Tec kinases mediate sustained calcium influx via site-specific tyrosine phosphorylation of the phospholipase Cgamma Src homology 2-Src homology 3 linker. J Biol Chem 2004;279(36):37651–61.
126. Kim YJ, Sekiya F, Poulin B, Bae YS, Rhee SG. Mechanism of B-cell receptor-induced phosphorylation and activation of phospholipase C-gamma2. Mol Cell Biol 2004;24(22): 9986–99.
127. Thome M. CARMA1, BCL-10 and MALT1 in lymphocyte development and activation. Nat Rev Immunol 2004;4(5):348–59.
128. Hikida M, Johmura S, Hashimoto A, Takezaki M, Kurosaki T. Coupling between B cell receptor and phospholipase C-gamma2 is essential for mature B cell development. J Exp Med 2003;198(4):581–9.
129. Bishop GA. The multifaceted roles of TRAFs in the regulation of B-cell function. Nat Rev Immunol 2004;4(10):775–86.
130. Bradley JR, Pober JS. Tumor necrosis factor receptor-associated factors (TRAFs). Oncogene 2001;20(44):6482–91.
131. Dempsey PW, Doyle SE, He JQ, Cheng G. The signaling adaptors and pathways activated by TNF superfamily. Cytokine Growth Factor Rev 2003;14(3–4):193–209.
132. Ni CZ, Oganesyan G, Welsh K, et al. Key molecular contacts promote recognition of the BAFF receptor by TNF receptor-associated factor 3: implications for intracellular signaling regulation. J Immunol 2004;173(12):7394–400.
133. Xu LG, Shu HB. TNFR-associated factor-3 is associated with BAFF-R and negatively regulates BAFF-R-mediated NF-kappa B activation and IL-10 production. J Immunol 2002;169(12):6883–9.
134. Rothe M, Pan MG, Henzel WJ, Ayres TM, Goeddel DV. The TNFR2-TRAF signaling complex contains two novel proteins related to baculoviral inhibitor of apoptosis proteins. Cell 1995;83(7):1243–52.
135. Haxhinasto SA, Bishop GA. Synergistic B cell activation by CD40 and the B cell antigen receptor: role of B lymphocyte antigen receptor-mediated kinase activation and tumor necrosis factor receptor-associated factor regulation. J Biol Chem 2004;279(4): 2575–82.
136. Hauer J, Puschner S, Ramakrishnan P, et al. TNF receptor (TNFR)-associated factor (TRAF) 3 serves as an inhibitor of TRAF2/5-mediated activation of the noncanonical NF-kappaB pathway by TRAF-binding TNFRs. Proc Natl Acad Sci USA 2005;102(8):2874–9.
137. He JQ, Zarnegar B, Oganesyan G, et al. Rescue of TRAF3-null mice by p100 NF-kappa B deficiency. J Exp Med 2006;203(11):2413–8.
138. Liao G, Zhang M, Harhaj EW, Sun SC. Regulation of the NF-kappaB-inducing kinase by tumor necrosis factor receptor-associated factor 3-induced degradation. J Biol Chem 2004;279(25):26243–50.
139. He JQ, Saha SK, Kang JR, Zarnegar B, Cheng G. Specificity of TRAF3 in its negative regulation of the noncanonical NF-kappa B pathway. J Biol Chem 2007;282(6):3688–94.
140. Brown KD, Hostager BS, Bishop GA. Regulation of TRAF2 signaling by self-induced degradation. J Biol Chem 2002;277(22):19433–8.
141. Li X, Yang Y, Ashwell JD. TNF-RII and c-IAP1 mediate ubiquitination and degradation of TRAF2. Nature 2002;416(6878):345–7.
142. Hostager BS, Haxhinasto SA, Rowland SL, Bishop GA. Tumor necrosis factor receptor-associated factor 2 (TRAF2)-deficient B lymphocytes reveal novel roles for TRAF2 in CD40 signaling. J Biol Chem 2003;278(46):45382–90.
143. Qing G, Qu Z, Xiao G. Stabilization of basally translated NF-kappaB-inducing kinase (NIK) protein functions as a molecular switch of processing of NF-kappaB2 p100. J Biol Chem 2005;280(49):40578–82.
144. Xie P, Stunz LL, Larison KD, Yang B, Bishop GA. Tumor necrosis factor receptor-associated factor 3 is a critical regulator of B cell homeostasis in secondary lymphoid organs. Immunity 2007;27(2):253–67.

145. Grech AP, Amesbury M, Chan T, Gardam S, Basten A, Brink R. TRAF2 differentially regulates the canonical and noncanonical pathways of NF-kappaB activation in mature B cells. Immunity 2004;21(5):629–42.
146. Grossmann M, Metcalf D, Merryfull J, Beg A, Baltimore D, Gerondakis S. The combined absence of the transcription factors rel and RelA leads to multiple hemopoietic cell defects [In Process Citation]. Proc Natl Acad Sci USA 1999;96(21):11848–53.
147. Grossmann M, O'Reilly LA, Gugasyan R, Strasser A, Adams JM, Gerondakis S. The anti-apoptotic activities of Rel and RelA required during B-cell maturation involve the regulation of Bcl-2 expression. Embo J 2000;19(23):6351–60.
148. Dejardin E. The alternative NF-kappaB pathway from biochemistry to biology: pitfalls and promises for future drug development. Biochem Pharmacol 2006;72(9):1161–79.
149. Liptay S, Schmid RM, Nabel EG, Nabel GJ. Transcriptional regulation of NF-kappa B2: evidence for kappa B-mediated positive and negative autoregulation. Mol Cell Biol 1994;14(12):7695–703.
150. Petrie RJ, Schnetkamp PP, Patel KD, Awasthi-Kalia M, Deans JP. Transient translocation of the B cell receptor and Src homology 2 domain-containing inositol phosphatase to lipid rafts: evidence toward a role in calcium regulation. J Immunol 2000;165(3):1220–7.
151. Treml LS, Carlesso G, Hoek KL, et al. TLR stimulation modifies BLyS receptor expression in follicular and marginal zone B cells. J Immunol 2007;178(12):7531–9.
152. Gorelik L, Cutler AH, Thill G, et al. Cutting edge: BAFF regulates CD21/35 and CD23 expression independent of its B cell survival function. J Immunol 2004;172(2):762–6.

Chapter 3
Regulation of B-Cell Self-Tolerance By BAFF and the Molecular Basis of Its Action

Sandra Gardam and Robert Brink

Abstract Signals delivered following the binding of BAFF to its receptor BAFF-R are essential for the survival of mature conventional B cells. In order to maintain self-tolerance, B cells that express antigen receptors with significant reactivity against self-antigens are prevented from receiving or responding to these survival signals. The great majority of B cells produced from bone marrow precursors fail to join the mature peripheral B-cell pool either due to their self-reactivity or due to a stochastic failure to receive adequate survival signals from the limiting levels of BAFF available in vivo. The tight control over BAFF expression plays an important role in enforcing self-tolerance, as is illustrated by the escape of some self-reactive B-cell clones and the production of autoantibodies that accompanies the elevation of BAFF levels in vivo. Recent experiments have identified the molecular basis for the unique dependence of B cells on survival signals delivered by BAFF. In the absence of BAFF, B-cell survival is constitutively suppressed through the cooperative actions of the TRAF2 and TRAF3 signal adapters. BAFF circumvents this activity by triggering the recruitment of TRAF3 to BAFF-R and causing the depletion of TRAF3 from the cell via a TRAF2-dependent mechanism. In this way, critical B-cell survival signals such as the alternative NF-κB pathway are activated. Sustained exposure to BAFF is normally required to maintain the activity of these pathways and B-cell survival. However, this requirement is completely removed in B cells that lack expression of TRAF2 or TRAF3.

Keywords BAFF · B cells · Self-tolerance · Signalling · NF-κB · NIK · TRAF2 · TRAF3

R. Brink (✉)
Garvan Institute of Medical Research, Darlinghurst NSW 2010, Australia

M.P. Cancro (ed.), *BLyS Ligands and Receptors,* Contemporary Immunology,
DOI 10.1007/978-1-60327-013-7_3,

3.1 Regulation of Self-Reactive B Cells by BAFF

3.1.1 The Requirement for B-Cell Self-Tolerance

A fundamental characteristic of B and T lymphocytes is the rearrangement during early lymphopoiesis of the genetic elements that ultimately encode the variable regions of their cell surface antigen receptors. The essentially random nature of this process allows for the development of lymphocyte "repertoires" with the collective ability to recognise and respond to epitopes on virtually any foreign organism or antigen. The price of this diversity is the inevitable production of lymphocyte clones with the ability to recognise components of the host itself. Activation of these cells has the potential to lead to autoimmune destruction of host tissues and must be avoided if the immune system is to protect but not harm the host. In the case of B cells, the imperative is to prevent the differentiation of self-reactive B cells into plasma cells, as these cells secrete soluble copies of the antigen receptor (antibodies). Thus it is the differentiation of self-reactive B cells into plasma cells that leads to the production of potentially pathogenic autoantibodies.

Self-tolerance is ultimately a very efficient process since the production of pathogenic autoantibodies is a relatively rare phenomenon despite the fact that up to 75% of the B cells produced in humans have significant self-reactivity [1]. Because T-cell help is usually required to drive the differentiation of B cells into plasma cells, the production of autoantibodies is avoided to a large extent by maintenance of self-tolerance within the T-cell compartment. However, cross-reactive foreign antigens and T-independent stimuli both have the potential to trigger autoantibody production independently of autoreactive T helper cells [2]. An important component of self-tolerance, therefore, is the purging from the B-cell repertoire of clones that possess significant self-reactivity. This can occur at several different stages during B-cell development depending on the expression of the self-antigen recognised by the self-reactive B cell and the nature of the interaction of self-antigen with the antigen receptor (BCR).

3.1.2 Self-Tolerance Checkpoints During B-Cell Development

3.1.2.1 Immature Bone Marrow B Cells

Both the BCR expressed on the surface of B cells and the antibodies secreted upon their differentiation into plasma cells are encoded by the immunoglobulin (Ig) heavy and light chain genes. Rearrangement of Ig variable region genes occurs in the pro- and pre-B cells located within adult bone marrow. Clones that successfully rearrange their Ig genes and commence BCR expression in the bone marrow are referred to as immature B cells (Fig. 3.1). Immature bone marrow B cells represent the first stage of B-cell development where cellular fate is shaped by the interaction of the BCR with the external antigenic environment. Foreign antigens are not usually encountered within bone marrow. However, B cells that bind strongly to widely expressed

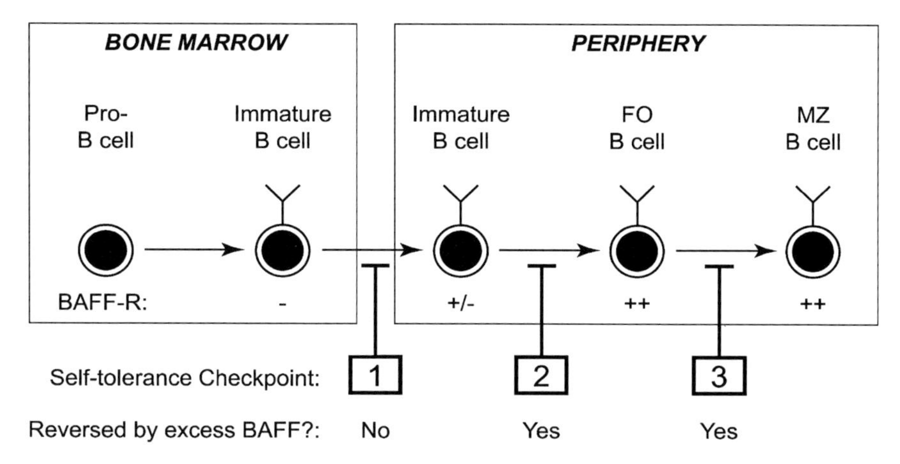

Fig. 3.1 Three major self-tolerance checkpoints during B-cell development and their potential for reversal by BAFF overexpression. A simplified model of B-cell development is shown that also indicates the three major Self-tolerance checkpoints (i.e. points at which self-reactive B cells are eliminated). Antigen receptor (BCR) expression is shown for all cells after the pro-B-cell stage and the relative levels of BAFF-R expression on these B cells is also indicated. The existence of experimental evidence for reversal of self-tolerance at each of these checkpoints is shown ("Yes" in this case indicates that some rather than all self-reactive B cells can be rescued from in vivo deletion by expression of excess BAFF – see Section 3.1.4). Immature B cells in the periphery are also referred to as "transitional" B cells and can be further subdivided into T1, T2, and T3 subsets (see Section 3.1.2.2). To avoid complexity, these subsets have not been included in this Figure. Mature B-cell subsets indicated are the follicular (FO) and marginal zone (MZ) populations

or bone marrow-specific self-antigens typically enter a process that either eliminates their self-reactivity or ultimately results in cell death (Checkpoint 1, Fig. 3.1).

Recognition of self-antigen by immature bone marrow B cells often triggers receptor editing. In this process, Ig gene rearrangements recommence in the immature B cell in an attempt to change the specificity of the self-reactive BCR [3, 4]. If successful, the newly non-self-reactive B cell is able to continue its development and join the peripheral B-cell pool. However, strongly self-reactive B cells that do not manage to change their specificity are blocked from developing further and subsequently die [5, 6] (Fig. 3.1). The elimination of self-reactive clones at this point in B-cell development appears to be the major mechanism by which B-cell self-tolerance is enforced. Thus, whilst 75% of newly generated immature bone marrow B cells are self-reactive, only 35% of the B cells that subsequently emerge from the bone marrow remain so [1].

3.1.2.2 Immature to Mature B-Cell Transition in the Periphery

B cells that survive receptor editing and deletion within the bone marrow eventually emerge from this tissue and migrate into the periphery, predominantly to the spleen (Fig. 3.1). These B cells are still not fully mature but pass through a number of immature "transitional" stages (T1, T2, T3) for 1–2 days before either entering the mature peripheral B-cell pool or dying [7–9]. The great majority of immature peripheral B cells do in fact die before they fully mature, due at least in part to the further elimination of self-reactive clones at this juncture. Cells that enter the mature B-cell pool are able to survive for weeks to months within the periphery and occupy physiological niches within the peripheral lymphoid tissues such as the primary B-cell follicles of the spleen and lymph nodes as well as the splenic marginal zone (Fig. 3.1).

As was discussed in Section 3.1.2.1, B cells that strongly interact with self-antigen in the bone marrow are eliminated either by receptor editing or by clonal deletion. However, B cells that interact more weakly with self-antigen can evade both of these fates and migrate into the periphery. Nevertheless, weaker interactions with self-antigen during early B-cell development can still have significant consequences, such as rendering the B cell unresponsive (or "anergic") to stimuli that would normally result in cellular activation [10, 11]. The induction of anergy on its own serves as an effective mechanism of B-cell self-tolerance as it greatly reduces the chances of such cells undergoing plasma cell differentiation. In practice, however, anergic self-reactive B cells also fail to mature, survive for only a few days in vivo, and are excluded from the follicular regions of peripheral lymphoid tissues in which mature B cells normally reside [11–13]. Thus, although anergic self-reactive B cells can emerge from the bone marrow, they are unable to join the long-lived mature B-cell pool. Indeed, immature T3 B cells appear to represent the final stage of differentiation for anergic self-reactive B cells [14]. Although it has not been modelled experimentally, it is likely that self-reactive B cells that are "ignorant" of their self-reactivity in the bone marrow but that bind strongly to self-antigen upon their migration into the periphery also fail to make the transition from immature to mature peripheral B cell (Checkpoint 2, Fig. 3.1). The importance of this self-tolerance checkpoint is emphasised by the observation that frequency of self-reactive B cells drops from ~35% in immature peripheral B cells to ~15% in the mature peripheral B-cell pool [1].

3.1.2.3 Marginal Zone B-Cell Development

The majority of mature B cells are said to have a follicular phenotype, characterised by high levels of surface CD23 and IgD. These cells circulate around the body through blood and lymph and, together with follicular dendritic cells, occupy the primary follicles within secondary lymphoid tissues. Within the spleen, a separate subset of mature B cells called marginal zone (MZ) B cells is also present. As their name suggests, they occupy a separate MZ niche within the spleen that lies distal to the follicle [15]. MZ B cells do not recirculate, are characterised by low expression of CD23 and IgD and high expression of CD21 and CD1d, and show a greater propensity for activation and rapid antibody production compared to follicular B cells. Although the precise lineage relationship between MZ and follicular B cells remains unclear, it appears that follicular B cells can act as precursors of MZ B cells at least under some circumstances [16, 17] (Fig. 3.1).

A number of reports have suggested that the MZ B-cell compartment is enriched for self-reactive B cells [18]. However, there are at least two instances where a particular self-reactive BCR specificity is selectively eliminated from the MZ B-cell compartment but not the mature follicular B-cell population [19, 20] (Checkpoint 3, Fig. 3.1). Evidently, the development of MZ B cells is also a B-cell self-tolerance checkpoint and possibly the one which operates with the highest stringency of the three checkpoints considered here. The ease with which MZ B cells can be activated and undergo plasma cell differentiation provides a clear rationale for this to be the case.

The overall picture to emerge, therefore, is that self-reactive B cells can be removed at three separate checkpoints points during their development depending on the strength of the interaction of their BCR with self-antigen (Fig. 3.1). In this way the self-reactive B cells that pose the greatest threat to the host are eliminated at the earliest checkpoint in the bone marrow, whereas those that are potentially less dangerous are deleted in the periphery at the immature to mature transition or specifically removed from the MZ compartment only [15].

3.1.3 Points of Action of BAFF and BAFF-R During B-Cell Development

The constitutive expression of BAFF in secondary lymphoid tissues is essential for sustaining the long-term survival of mature B cells in vivo. Thus mature B cells are rapidly deleted when their access to BAFF is blocked in adult mice [21, 22] and mature B cells are virtually absent in mice that do not express BAFF [23]. On the other hand, both the survival and numbers of mature B cells are greatly increased in transgenic mice that overexpress soluble BAFF [24, 25]. Thus, although BAFF is a potent B-cell survival factor, its normal expression in vivo is limiting and not designed to result in maximal B-cell survival.

BAFF is capable of binding to three receptors: BCMA (B-cell maturation antigen), TACI (transmembrane activator and calcium modulator and cyclophilin ligand interactor), and BAFF-R (BAFF receptor, also known as BR3), all of which are expressed on B-lineage cells at various points during development [26]. Analysis of mice specifically deficient in the expression or function of each of the three receptors for BAFF demonstrate that BAFF-R is completely responsible for delivering pro-survival signals to mature B cells by BAFF. Thus mature B cells are largely absent in mice that do not express BAFF-R or express it in a functionally inactive form [27–29], whereas mature B cells numbers are either unaffected or increased in mice lacking BCMA or TACI, respectively [23, 30–32].

In contrast to the situation for mature B cells, immature bone marrow B cells and their immediate peripheral descendents remain unaffected by both the absence of BAFF and its overexpression in vivo [23–25]. Thus B-cell development up until the T2 transitional stage does not require BAFF-dependent survival signals. Consistent with this is that fact that BAFF-R, the pro-survival BAFF receptor for mature B cells, is virtually absent from newly generated and T1 B cells and is only expressed at high levels on B cells as they near the mature B-cell transition [33] (Fig. 3.1). Not surprisingly therefore, the absence of BAFF-R expression does not effect the development of immature B cell populations in bone marrow or the periphery [28, 29].

As discussed above, over half of the immature B cells that enter the periphery do not make the transition into mature long-lived B-cell pool but instead die around the transitional T2 stage [8, 9, 34]. The fact that BAFF is both required for the transition of immature to mature B cells and is present in limiting amounts in vivo provides an explanation for this phenomenon. Thus if the levels of BAFF available in vivo cannot sustain the survival and maturation of all the B cells that enter the periphery, then competition for BAFF should indeed result in the attrition of B cells at the T2 transitional stage. Consistent with this proposition is the observation that raising the availability of BAFF in vivo results in a substantial increase in the proportion of T2 cells that enter the mature B-cell pool [33].

In addition to its role in regulating the immature to mature B-cell transition, BAFF also appears to be particularly important for the generation and maintenance of the MZ B-cell compartment. This is evident both from the particular susceptibility of MZ B cells to depletion of BAFF in adult mice [22] and the preferential expansion of this mature B-cell subset in mice that overexpress BAFF [24]. Indeed overexpression of BAFF is not only associated with the accumulation of MZ B cells in their natural location in the spleen, but also with the appearance of MZ phenotype cells in other tissues including lymph nodes, blood, and salivary glands [35].

3.1.4 BAFF and the Regulation of B-Cell Self-Tolerance Checkpoints

As well causing B-cell hyperplasia, transgenic overexpression of BAFF in mice is associated with a number of autoimmune phenomena including autoantibody production [24, 25, 35]. In addition, high levels of BAFF have been associated with a number of human autoimmune diseases [26, 35, 36]. The possibility that the overexpression of BAFF may precipitate autoimmunity by circumventing the normal processes of B-cell self-tolerance has therefore received significant attention and is summarised below in relation to the three key B-cell self-tolerance checkpoints described in Section 3.1.2.

3.1.4.1 Immature Bone Marrow B Cells

As described in Section 3.1.3, the immature bone marrow B-cell compartment develops independently of BAFF or BAFF-R and is unaffected by the overexpression of BAFF in vivo. It seems unlikely therefore, that elevation of BAFF levels would interfere with the normal elimination of strongly self-reactive B cells that occurs during this early stage of development. This has been confirmed experimentally by the demonstration that the deletion of B cells recognising membrane-bound self-antigen in the bone marrow (Checkpoint 1, Fig. 3.1) proceeds normally in transgenic mice overexpressing BAFF [20].

3.1.4.2 Immature to Mature B-Cell Transition in the Periphery

Several earlier studies have shown that self-reactive B cells that would normally be deleted at the immature to mature B-cell transition in the periphery can mature and survive if they do not have to "compete" with non-self-reactive B cells [11–13]. Since limiting BAFF expression normally regulates the transition of B cells through this developmental bottleneck, this observation suggests that self-reactive B cells may normally be deleted around the T2 stage due to reduced responsiveness to, and thus inability to compete for, the limiting BAFF-survival signals available in vivo.

This possibility has been examined by reducing the availability of BAFF in vivo via the administration to mice of a soluble version of the BCMA extracellular domain [37]. This treatment reduced the survival of all B cells but particularly affected the survival of self-reactive B cells that recognised soluble self-antigen with high affinity. In other words, these cells were indeed more dependent on BAFF for their survival than the majority of B cells. This study also showed that these self-reactive cells bound less BAFF per cell when competing non-self-reactive B cells were present [37]. This data therefore supported the idea that contact with self-antigen can reduce B-cell responsiveness to BAFF, and thus make the self-reactive B cells incapable of obtaining sufficient survival signals within a normal repertoire due to the limiting levels of BAFF present in vivo. Because these self-reactive B cells do survive and mature in the absence of competition, their responsiveness to BAFF is reduced rather than eliminated. This model predicts, therefore, that elevation of BAFF levels in vivo may indeed rescue self-reactive B cells that are normally deleted at the immature to mature B-cell transition (Checkpoint 2, Fig. 3.1).

This prediction has been directly tested using BAFF transgenic mice. In this case, self-reactive B cells recognising soluble self-antigen in the presence of competing non-self-reactive B cells were rescued from deletion at the T2 transitional B-cell stage and matured into follicular B cells in the presence of excess BAFF [20]. However, if the self-reactive B cells were deleted slightly earlier during their maturation, they were resistant to rescue by the increased levels of BAFF expressed in these mice [20]. It appears, therefore, that self-reactive B cells that are normally deleted prior to entering the mature compartment can be rescued by increased expression of BAFF, but only if their normal point of deletion is close to this transition. This is likely to be due to the fact that the expression of the pro-survival BAFF-R increases during early maturation and peaks just prior to the mature transition (Fig. 3.1), giving the cells that reach this point the best chance of responding to BAFF [20].

An interesting aspect of the deletion of self-reactive B cells at the immature to mature transition is that these cells are prevented from entering the follicle and are primarily found in the T-cell area of the spleen [11–13]. Because FDCs are located within the heart of B-cell follicles and are known to express BAFF [38], it was thought that these or some other cells localised within the follicle may provide a critical source of B-cell survival signals that cannot be accessed efficiently by self-reactive B cells excluded from the follicle. The possibility that such a mechanism may underlie the reduced ability of self-reactive B cells to compete for survival signals has been investigated [39]. It was found that self-reactive B cells that lacked expression of the chemokine receptor CCR7 were not excluded from the follicle but were still deleted prior to entering the long-lived mature B-cell pool. Thus the inability of such self-reactive B cells to compete for BAFF-mediated survival signals does not result from reduced access to BAFF brought about by follicular exclusion. It is more likely that contact with self-antigen renders these B cells intrinsically hyporesponsive to BAFF survival signals.

3.1.5 Marginal Zone B-Cell Development

Unlike the deletion of self-reactive B cells at the immature to mature B-cell transition, the prevention of self-reactive B cells from entering the MZ compartment (Checkpoint 3, Fig. 3.1) does not require competition from a non-self-reactive B-cell population [19]. This on the one hand indicates that deletion of self-reactive B cells prior to MZ differentiation is relatively stringent, an assertion supported by the relatively low avidity of self-antigen required for this form of deletion [20]. What this also means, however, is that competition for limiting BAFF is unlikely to be the mechanism for deletion of self-reactive B cells prior to their entry into the MZ compartment. Nevertheless, the potent activity of BAFF in expanding the MZ B cell compartment when it is overexpressed in vivo suggests that deletion of self-reactive B cells at this point may indeed be compromised by the presence of excess BAFF.

This question has been investigated by observing the effects of transgenic overexpression of BAFF on the fate of self-reactive B cells that recognise soluble self-antigen with relatively low affinity. Whilst these cells are normally excluded from the MZ B-cell compartment, overexpression of BAFF restored them to this compartment in similar numbers to when their self-antigen was absent [20]. As well as being more easily activated by antigen, the physiological positioning of MZ B cells next to the marginal sinus means that they are more readily exposed than follicular B cells to polyclonal stimuli such as LPS and CpG that are typically associated with blood-borne pathogens [40]. Thus the promotion into the MZ compartment by excess BAFF of self-reactive B-cell specificities that are normally restricted to the follicle may well contribute to the autoimmunity associated with BAFF overexpression.

In summary, the ability of BAFF overexpression to rescue self-reactive B cells from deletion is limited to those cells normally deleted relatively late in their maturation. The ability of self-reactive B cells to be rescued by BAFF is most likely determined by their expression of BAFF-R, which peaks around the point during B-cell maturation where BAFF-mediated rescue begins to operate [15, 20] (Fig. 3.1). Equally important, however, is the strength of the interaction between the BCR and the self-antigen. Thus, as is exemplified by the ability of BAFF overexpression to prevent the deletion of intermediate but not high-affinity self-reactive B cells from the MZ compartment [20], excess BAFF is likely to rescue only peripheral self-reactive B cells that sit relatively close to the normal thresholds (e.g. antigen avidity) that determine whether or not a self-reactive B cell will be deleted.

3.2 Signalling BAFF-Dependent B-Cell Survival

The critical role of BAFF in regulating B-cell homeostasis and self-tolerance has resulted in an extensive investigation of the signalling events triggered by BAFF in B cells. Because primary B cells require BAFF for their survival, analysis of their responses to BAFF has proved difficult. Some of the major insights have come from in vivo mouse models in which B cells either lack expression of or express constitutively active versions of key signalling molecules. These types of analyses have indicated that activation of members of the NF-κB family of transcription factors is critical. Of particular importance appears to be the activation of the alternative (NF-κB2/p52) NF-κB pathway, which is mediated entirely by BAFF signalling through BAFF-R in resting primary B cells [41, 42] (Fig. 3.2). This pathway is essential for normal B-cell survival and recent experiments have revealed that the proximal regulators of the alternative NF-κB pathway in fact determine the BAFF-dependent nature of primary B-cell survival [43]. There is some evidence that BAFF-R can activate the canonical (NF-κB1/p50) NF-κB pathway and that this pathway may also play a role in BAFF-mediated B-cell survival (Fig. 3.2). In fact, like BAFF or BAFF-R-deficient mice, mice doubly deficient for both NF-κB1 and NF-κB2 lack mature B cells [44], whereas the single knockout mice have less severe phenotypes [44–47]. The exact nature of the events downstream of the NF-κB pathways and the relative importance of the BCR and BAFF-R in activating them are yet to be fully resolved.

3.2.1 The NF-κB Signalling Pathways

The operation of the NF-κB transcription pathways in immune cells has been recently reviewed elsewhere [48]. Briefly, the NF-κB transcription factor family consists of five proteins. NF-κB1 and NF-κB2 are synthesised in long precursor forms, p105 and p100, respectively, and are subsequently processed via partial degradation by the proteasome to form the active subunits, p50 and p52, respectively. This occurs constitutively for p105, whilst p100 processing requires the delivery of specific activation signals. The remaining members of the family are the Rel proteins: RelA (also known as p65), RelB, and c-Rel. The Rel proteins contain *trans*-activation domains capable of initiating transcription when they form dimers with p50 or p52 and bind to DNA in the nucleus. In the absence of signalling, pre-formed NF-κB dimers are held inactive in the cytoplasm. Upon signal initiation, nuclear localisation signals in the dimers are revealed, facilitating their migration to the nucleus to activate gene transcription (Fig. 3.2).

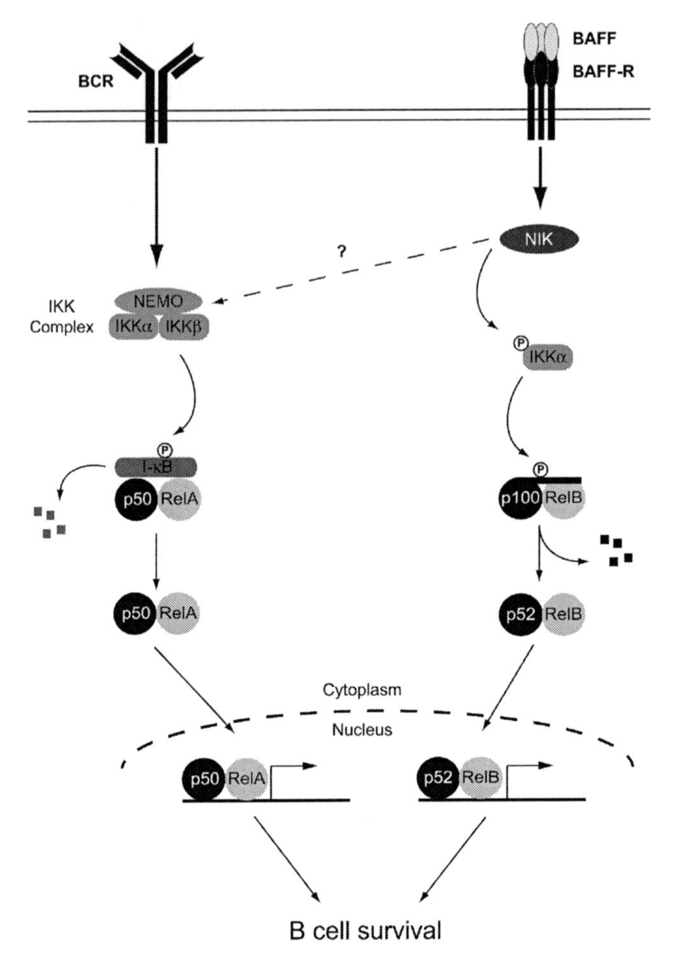

Fig. 3.2 The NF-κB signalling pathways. Canonical (*left*) and alternative (*right*) NF-κB activation in B cells contributes to B-cell survival (see Section 3.2.1 for details). Whilst BAFF-R is capable of initiating both pathways, it is probable in a physiological setting that the BCR is responsible for the majority of activation of the canonical pathway in naive primary B cells

3.2.1.1 The Canonical NF-κB Pathway

Following the constitutive processing of p105, the active subunit p50 forms dimers predominantly with RelA or c-Rel. These dimers are held inactive in the cytoplasm by inhibitors of NF-κB (Iκ-B) proteins, which mask their nuclear localisation signals. Receptors that strongly initiate canonical NF-κB signalling, such as the BCR and CD40, utilise a variety of signal adapters to cause the activation of the Iκ-B kinase (IKK) complex. This complex consists of a regulatory unit (IKKγ, also know as NEMO) and two kinases (IKKα and IKKβ) that are capable of phosphorylating Iκ-B proteins. Once phosphorylated, Iκ-B proteins are degraded by the proteasome, releasing p50/RelA or p50/c-Rel dimers that can migrate to the nucleus and initiate gene transcription (Fig. 3.2).

3.2.1.2 The Alternative NF-κB Pathway

The precursor form of NF-κB2, p100, contains a carboxy-terminal region that is rich in ankyrin repeats and bestows it with inhibitory properties similar to those of the Iκ-B proteins. Thus p100 forms dimers in the cytoplasm, usually with RelB, and prevents its own nuclear localisation prior to processing. Activation of the alternative pathway requires the serine-threonine kinase NF-κB-inducing kinase (NIK). Accumulation of NIK protein appears to be the critical event in this pathway, with increased levels of NIK subsequently resulting in phosphorylation and activation of IKKα. Neither IKKβ nor NEMO are required for activation of the alternative NF-κB pathway. Rather IKKα alone is thought to be responsible for the phosphorylation of p100 at serines 866 and 870, which then initiates its proteasomal processing to p52 [49]. Liberated p52 remains in a heterodimer with RelB and together they can translocate to the nucleus (Fig. 3.2). The alternative NF-κB pathway is activated with slower kinetics compared to the canonical pathway and is primarily triggered through members of the TNF receptor (TNFR) superfamily such as LT-βR, CD40, and BAFF-R.

3.2.2 The Contribution of the Canonical NF-κB Pathway to B-Cell Survival

BAFF is able to activate canonical NF-κB signalling, but does so weakly and primarily via TACI [50]. However, evidence does exists for a low level of activation via BAFF-R [50, 51] raising the possibility that this pathway may contribute to the ability of BAFF to promote B-cell survival. The kinetics of the NF-κB1 pathway initiated by BAFF-R are considerably slower than is observed for strong activators of this pathway such as TNF and CD40L, and it has been suggested that BAFF-R may utilise NIK to activate the canonical as well as the alternative NF-κB pathway [52] (Fig. 3.2).

The importance of canonical NF-κB pathway in B-cell survival has been demonstrated by the fact that B-cell-specific ablation of NEMO results in a paucity of mature B cells [51]. However, this phenotype is not as severe as BAFF-R deficiency, confirming that the canonical NF-κB pathway is not the only pathway downstream of BAFF-R contributing to survival. Interestingly, NEMO-deficient B cells contained decreased amounts of alternative NF-κB pathway components, p100 and RelB. Processing of this reduced amount of p100 to p52 did occur in these cells, suggesting the alternative pathway was active [51]. There is evidence that the expression of both p100 and RelB is under the influence of canonical NF-κB dimers [53–55]. As such the impaired survival of mature B cells in the absence of NF-κB1 activity could be ascribed at least in part to a role for this pathway in controlling the transcription of alternative pathway components.

To determine if canonical NF-κB signalling alone is sufficient to allow B cells to survive past the immature to mature checkpoint in the periphery, a mouse line was produced with constitutive activation of the canonical pathway in B cells via a constitutively active IKKβ protein [51]. When crossed on to the BAFF-R-deficient background, this facilitated rescue of the mature B-cell compartment in the absence of alternative NF-κB pathway activation.

Whilst activation of the canonical NF-κB pathway appears to be an important factor in the survival of primary B cells, it is difficult to determine the extent to which this is due to the actions of BAFF. B-cell survival also depends on signals that constitutively emanate from the BCR, a known activator of the canonical NF-κB pathway [56]. Given the relatively poor activation of this pathway by BAFF-R, it is probable that it is the BCR that is primarily responsible for the canonical NF-κB activity observed in resting B cells [57]. On the other hand, the primary function of BAFF/BAFF-R signalling appears to be activation of the alternative NF-κB pathway (Fig. 3.2).

3.2.3 *The Alternative NF-κB Pathway Is the Major Contributor to B-Cell Survival Downstream of BAFF-R*

BAFF-R is responsible for virtually all of the alternative NF-κB pathway activation in mature B cells [41, 42] and it is this pathway which has been most strongly been associated with B-cell survival. Thus *Nfκb2*$^{-/-}$ mice are viable but display a deficiency of peripheral B cells [47, 58] and B cells lacking NF-κB2 fail to survive when provided with BAFF *ex vivo* [42]. In addition, mature NF-κB2-deficient B cells fail to survive in mixed bone marrow chimeras when they are forced to "compete" with wild-type B cells [59]. Taken together, these results confirm that, while the canonical pathway may contribute to survival, the alternative pathway is of primary importance in promotion of B-cell survival by BAFF. At this point, however, it remains unknown what gene targets are activated by the alternative NF-κB pathway in order to promote B-cell survival.

3.2.3.1 TRAF Proteins Are Fundamental Regulators of NF-κB2 Signalling

A phenotype similar to mice, which transgenically overexpress BAFF has recently been described in mice that lack B-cell expression of the signal adapters TNFR-associated factors 2 or 3 (TRAF2 or TRAF3) [43, 60, 61]. These mice display an expanded B-cell compartment, a surfeit of marginal zone B cells and hyperactivity of NF-κB2 in B cells, leading to the proposition that TRAF2 and TRAF3 negatively regulate BAFF signalling and NF-κB2 activation. In the case of TRAF3, this conclusion is consistent with previous in vitro observations. First, TRAF3 is the only TRAF to be recruited to BAFF-R [62]. Second, overexpression studies have shown that TRAF3 inhibits NF-κB2 activation via a number of TNFR members [63]. Third, studies in transformed B-cell lines have indicated that TRAF3 constitutively interacts with NIK and catalyses its proteasomal degradation, thereby inhibiting processing of p100 [64]. Finally, TRAF3-deficient mice display early post-natal lethality [65], a phenotype that is counteracted by crossing to the *Nfκ*b2$^{-/-}$ background [66].

More unexpected was the phenotype of the mice containing TRAF2-deficient B cells. Overexpression of TRAF2 or its recruitment to members of the TNFR superfamily has been shown to activate the canonical NF-κB and JNK pathways [67–69] and a negative regulatory role for TRAF2 had not previously been described. Additionally, TRAF2 is not recruited to BAFF-R, so its involvement in signalling by this receptor has not been recognised before. However, it is clear from the conditional deletion approach [43, 61] that TRAF2 negatively regulates NF-κB2 signalling initiated by BAFF-R in primary B cells. In fact, removal of TRAF2 from B cells completely rescued mature B-cell development in BAFF-deficient mice [43]. Thus, the removal of this potent suppressor of B-cell survival pathways on its own was sufficient to activate these pathways and allow peripheral B cells to pass the immature to mature checkpoint. These B cells contained hyperactive NF-κB2 activation, but no significant changes in NF-κB1 activity [61] consistent with the proposition that the primary role for BAFF in B-cell survival is to activate the NF-κB2 pathway.

3.2.3.2 Unravelling the Proximal Signalling Events That Allow TRAFs to Suppress NF-κB2 Signalling

Production of mice with B cells doubly deficient in TRAF2 and TRAF3 revealed a phenotype not more severe than single deletion of either of these genes [43]. This indicated that TRAF2 and TRAF3 play cooperative and non-redundant roles in suppressing B-cell survival pathways. However, there is no precedent for these molecules acting in such a manner, raising the question of what molecular mechanisms underpin this activity.

TRAF2 and TRAF3 are both capable of interacting with NIK but do so at separate sites on NIK [64, 69, 70] and only TRAF3 can bind to BAFF-R. In vitro studies using transformed B-cell lines have indicated that TRAF3 binds NIK and facilitates its degradation in proteasomes [64]. How this occurs and what the role of TRAF2 remains open to question. TRAF proteins contain RING finger domains, zinc-chelated structures that can mark proteins for proteasomal degradation by catalysing the attachment of ubiquitin moieties via their lysine-48 side chains [71]. However, ubiquitin ligase activity associated with the TRAF3 RING finger domain has not been reported. The TRAF2 RING finger does have ubiquitin ligase activity but only appears to catalyse attachment of ubiquitin molecules via lysine-63, a process associated with the building of signalling scaffolds rather than proteasomal degradation [72–74]. However, TRAF2 can interact with cellular inhibitor of apoptosis protein 1 (c-IAP1), a ubiquitin ligase that is capable of catalysing lysine-48 ubiquitination [75]. Recent data showed that both c-IAP1-deficient and TRAF2-deficient MEFs contained higher amounts of NIK and increased levels of p100 processing [76]. Furthermore, c-IAP1 was shown to be capable of mediating degradation of NIK and the TRAF2-binding site present in c-IAP1 was essential for this function [77]. It is likely, therefore, that TRAF2 constitutively suppresses B-cell survival by recruiting c-IAP1 to NIK to facilitate its degradation. TRAF3 binding must also be required for this process although its precise role remains to be elucidated (Fig. 3.3a).

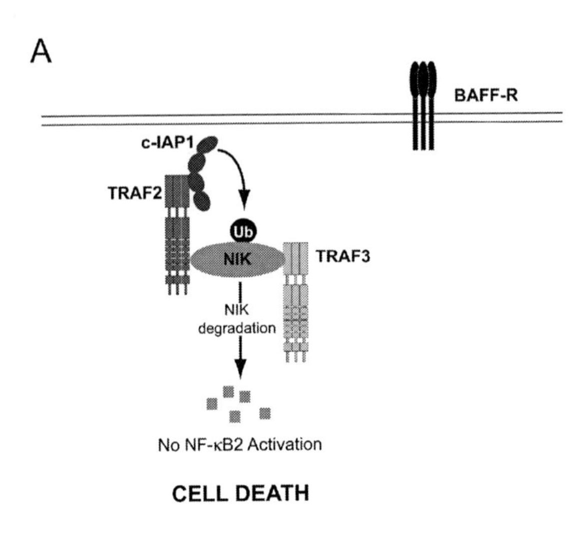

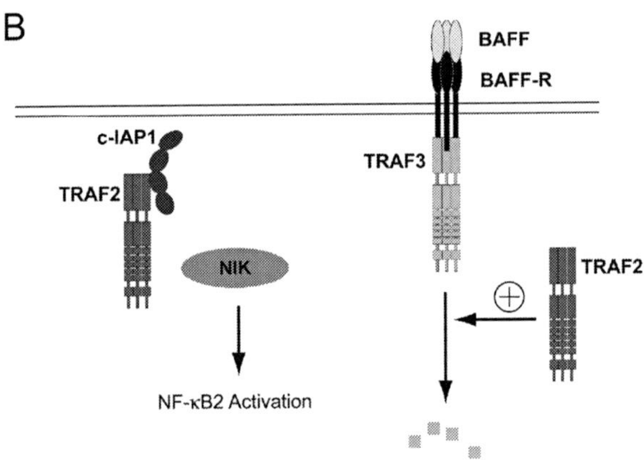

Fig. 3.3 A model of the proximal signalling events that regulate NF-κB2 activation downstream of BAFF-R. (**a**) When BAFF is absent, TRAF2, TRAF3, and c-IAP1 cooperate to ubiquitinate NIK leading to its proteasomal degradation. Under these circumstances NIK levels are too low to initiate NF-κB2 signalling and the B cell dies. (**b**) BAFF binding to BAFF-R recruits TRAF3 to the receptor and it is subsequently depleted from the cell in a TRAF2-dependent manner. Without the contribution of TRAF3, constitutive NIK degradation is reversed. NIK accumulation ultimately initiates NF-κB2 signalling and facilitates B-cell survival

3.2.3.3 BAFF Is an Obligate Survival Factor for B Cells Because It Reverses the Suppression of NF-κB2

Despite the constitutive suppression of NF-κB2 signalling by the actions of TRAF2, TRAF3, and c-IAP1, B cells can survive due to the ability of BAFF to reverse this suppression via the depletion of cellular TRAF3 (Fig. 3.3b). This was first demonstrated in transformed B-cell lines, which showed extensive proteolysis of TRAF3 upon BAFF, consequent accumulation of NIK, and thus promotion of NF-κB2 processing [64]. It was subsequently demonstrated that TRAF3 is depleted during the development of normal primary B cells concurrent with the acquisition of BAFF-R expression and dependence on BAFF for survival. Furthermore, this depletion of TRAF3 failed to occur in BAFF or BAFF-R-deficient mice [43]. An additional role for TRAF2 in this process was indicated by the fact that depletion of TRAF3 was inactivated during the development of TRAF2-deficient B cells [43].

These data therefore reveal dual roles for TRAF2 in regulating BAFF signalling in B cells. TRAF2 on the one hand appears to constitutively recruit c-IAP1 to NIK to facilitate NIK degradation and thus suppress NF-κB2 activation, a role it cooperates with TRAF3 to perform (Fig. 3.3a). BAFF-R signalling lifts this suppression by recruiting TRAF3 to its cytoplasmic domain and initiating its depletion from the cell, a process that also requires TRAF2 (Fig. 3.3b). It is possible that TRAF2 recruits c-IAP1 or another ubiquitin ligase to TRAF3 to facilitate its degradation. However, the exact mechanism by which TRAF3 is depleted from the cell and TRAF2's role in this process remain unclear.

In line with the above description of the proximal signalling events linking BAFF-R with NF-κB2 activation and thus survival in B cells, studies in human patients suffering multiple myeloma have identified mutations in a number of the key regulatory molecules discussed here [78, 79]. Thus, mutations were found which inactivated or deleted TRAF2, TRAF3, or c-IAP1, and also mutations that elevated the expression of or caused increased activation of NIK or NF-κB2. These findings lend support to the theory that these signalling events occur similarly in human and mouse B cells and do indeed impact on B-cell survival to the point of potentially contributing to B-lineage tumours when they are mutated.

3.2.4 Other Intracellular Mediators of B-Cell Survival Initiated by BAFF

Whilst NF-κB2 signalling is the primary survival pathway activated by BAFF and BAFF-R signalling, exactly how it facilitates B-cell survival is not clear. Studies have aimed to identify genes that are up-regulated by BAFF or proteins that are modified and may promote survival. Apart from TRAF3, the only signalling molecule known to be recruited to BAFF-R is the adapter protein Act1 [80]. This protein is thought to act as a negative regulator of BAFF-mediated B-cell survival but its mechanism of action also remains unclear. Either the recruitment of Act1 or the activation of NF-κB2 signalling or both may modulate the following downstream events that promote B-cell survival.

3.2.4.1 Increasing Glycolysis

BAFF signalling can initiate the phosphorylation of Akt [81, 82], which promotes cell survival by increasing glucose uptake and glycolysis [83]. Microarray analysis has also revealed that BAFF stimulation of mature B cells causes the upregulation of a panel of glycolytic enzymes [81], thus promoting metabolism of glucose and other nutrients [82]. This may represent one direct method by which BAFF signalling can facilitate the survival of B cells. Both phosphoinositide 3-kinase (PI3K) and protein kinase C β (PKCβ) were shown to be important in activating Akt following BAFF stimulation [81].

3.2.4.2 Modulation of Pro- and Anti-apoptotic Proteins

A number of studies have indicated that BAFF upregulates the expression of anti-apoptotic members of the Bcl-2 family of proteins in B cells, including Bcl-2, Bcl-xL, and A1/Bfl-1 [33, 84, 85]. At the same time, BAFF signalling also downregulates the expression of pro-apoptotic family member Bim and so counteracts the upregulation of this molecule induced by BCR signalling [86]. These combined actions of BAFF almost certainly play an important role in sustaining B-cell survival. The importance of Bim downregulation in particular is suggested by the fact that BAFF-transgenic and *Bim*$^{-/-}$ mice both exhibit B-cell hyperplasia and autoimmunity [24, 25, 86] and that *Bim*$^{-/-}$ B cells are relatively resistant to antigen-induced cell death [87]. Many of the anti-apoptotic Bcl-2 family proteins are regulated by NF-κB transcription factors [33, 85, 88, 89] suggesting that these proteins may represent important targets of the NF-κB pathways stimulated by BAFF.

PKCδ is a pro-apoptotic enzyme that exerts its effects in the nucleus. Mice lacking PKCδ expression exhibit dramatic B-cell hyperplasia and systemic autoimmunity [90], again similar to BAFF-transgenic mice. Further analysis showed that B cells that did not express PKCδ failed to undergo peripheral deletion in response to soluble self-antigen [91]. A connection with BAFF-mediated survival signals was subsequently established when it was found that the pro-apoptotic translocation of PKCδ to the B-cell nucleus could be inhibited by BAFF [92]. Thus, BAFF signalling cannot only regulate the levels of pro- and anti-apoptotic proteins but can potentially modulate their functions within B cells.

3.3 Conclusions

It is clear that the intracellular signalling pathways triggered by BAFF play a critical role in regulating B-cell homeostasis and can have significant effects on the enforcement of B-cell self-tolerance. Whilst not discussed in detail here, it is likely that improper activation of these signalling pathways also makes a significant contribution to cancer in the B-cell lineage [26]. There still remains much to learn about the critical events involved in the regulation of BAFF-dependent survival signals. A challenge for the future will be to determine whether our increasing knowledge of these pathways can be used to develop more effective treatments for autoimmune and neoplastic diseases involving B cells.

References

1. Wardemann H, Yurasov S, Schaefer A, Young JW, Meffre E, Nussenzweig MC. Predominant autoantibody production by early human B cell precursors. Science 2003;301:1374–7.
2. Goodnow CC, Adelstein S, Basten A. The need for central and peripheral tolerance in the B cell repertoire. Science 1990;248:1373–9.
3. Radic MZ, Erikson J, Litwin S, Weigert M. B lymphocytes may escape tolerance by revising their antigen receptors. J Exp Med 1993;177:1165–73.
4. Tiegs SL, Russell DM, Nemazee D. Receptor editing in self-reactive bone marrow B cells. J Exp Med 1993;177:1009–20.
5. Hartley SB, Crosbie J, Brink R, Kantor AB, Basten A, Goodnow CC. Elimination from peripheral lymphoid tissues of self-reactive B lymphocytes recognizing membrane-bound antigens. Nature 1991;353:765–9.
6. Hartley SB, Cooke MP, Fulcher DA, et al. Elimination of self-reactive B lymphocytes proceeds in two stages: arrested development and cell death. Cell 1993;72:325–35.
7. Allman D, Lindsley RC, DeMuth W, Rudd K, Shinton SA, Hardy RR. Resolution of three nonproliferative immature splenic B cell subsets reveals multiple selection points during peripheral B cell maturation. J Immunol 2001;167:6834–40.
8. Forster I, Rajewsky K. The bulk of the peripheral B-cell pool in mice is stable and not rapidly renewed from the bone marrow. Proc Natl Acad Sci USA 1990;87:4781–4.
9. Fulcher DA, Basten A. Influences on the lifespan of B cell subpopulations defined by different phenotypes. Eur J Immunol 1997;27:1188–99.
10. Goodnow CC, Crosbie J, Adelstein S, et al. Altered immunoglobulin expression and functional silencing of self-reactive B lymphocytes in transgenic mice. Nature 1988;334:676–82.
11. Phan TG, Amesbury M, Gardam S, et al. B cell receptor-independent stimuli trigger immunoglobulin (Ig) class switch recombination and production of IgG autoantibodies by anergic self-reactive B cells. J Exp Med 2003;197:845–60.
12. Cyster JG, Hartley SB, Goodnow CC. Competition for follicular niches excludes self-reactive cells from the recirculating B-cell repertoire. Nature 1994;371:389–95.
13. Cyster JG, Goodnow CC. Antigen-induced exclusion from follicles and anergy are separate and complementary processes that influence peripheral B cell fate. Immunity 1995;3: 691–701.
14. Merrell KT, Benschop RJ, Gauld SB, et al. Identification of anergic B cells within a wild-type repertoire. Immunity 2006;25:953–62.
15. Brink R. Regulation of B cell self-tolerance by BAFF. Semin Immunol 2006;18:276–83.
16. Vinuesa CG, Sze DM, Cook MC, et al. Recirculating and germinal center B cells differentiate into cells responsive to polysaccharide antigens. Eur J Immunol 2003;33:297–305.
17. Srivastava B, Quinn WJ, III, Hazard K, Erikson J, Allman D. Characterization of marginal zone B cell precursors. J Exp Med 2005;202:1225–34.
18. Lopes-Carvalho T, Kearney JF. Development and selection of marginal zone B cells. Immunol Rev 2004;197:192–205.
19. Mason DY, Jones M, Goodnow CC. Development and follicular localization of tolerant B lymphocytes in lysozyme/anti-lysozyme IgM/IgD transgenic mice. Int Immunol 1992;4:163–75.
20. Thien M, Phan TG, Gardam S, et al. Excess BAFF rescues self-reactive B cells from peripheral deletion and allows them to enter forbidden follicular and marginal zone niches. Immunity 2004;20:785–98.

21. Schneider P, Takatsuka H, Wilson A, et al. Maturation of marginal zone and follicular B cells requires B cell activating factor of the tumor necrosis factor family and is independent of B cell maturation antigen. J Exp Med 2001;194:1691–7.
22. Gross JA, Dillon SR, Mudri S, et al. TACI-Ig neutralizes molecules critical for B cell development and autoimmune disease: impaired B cell maturation in mice lacking BLyS. Immunity 2001;15:289–302.
23. Schiemann B, Gommerman JL, Vora K, et al. An essential role for BAFF in the normal development of B cells through a BCMA-independent pathway. Science 2001;293:2111–4.
24. Mackay F, Woodcock SA, Lawton P, et al. Mice transgenic for BAFF develop lymphocytic disorders along with autoimmune manifestations. J Exp Med 1999;190:1697–710.
25. Khare SD, Sarosi I, Xia XZ, et al. Severe B cell hyperplasia and autoimmune disease in TALL-1 transgenic mice. Proc Natl Acad Sci USA 2000;97:3370–5.
26. Mackay F, Silveira PA, Brink R. B cells and the BAFF/APRIL axis: fast-forward on autoimmunity and signaling. Curr Opin Immunol 2007;19:327–36.
27. Thompson JS, Bixler SA, Qian F, et al. BAFF-R, a newly identified TNF receptor that specifically interacts with BAFF. Science 2001;293:2108–11.
28. Shulga-Morskaya S, Dobles M, Walsh ME, et al. B cell-activating factor belonging to the TNF family acts through separate receptors to support B cell survival and T cell-independent antibody formation. J Immunol 2004;173:2331–41.
29. Sasaki Y, Casola S, Kutok JL, Rajewsky K, Schmidt-Supprian M. TNF family member B cell-activating factor (BAFF) receptor-dependent and -independent roles for BAFF in B cell physiology. J Immunol 2004;173:2245–52.
30. Xu S, Lam KP. B-cell maturation protein, which binds the tumor necrosis factor family members BAFF and APRIL, is dispensable for humoral immune responses. Mol Cell Biol 2001;21:4067–74.
31. von Bulow GU, van Deursen JM, Bram RJ. Regulation of the T-independent humoral response by TACI. Immunity 2001;14:573–82.
32. Yan M, Wang H, Chan B, et al. Activation and accumulation of B cells in TACI-deficient mice. Nature Immunol 2001;2:638–43.
33. Hsu BL, Harless SM, Lindsley RC, Hilbert DM, Cancro MP. Cutting edge: BLyS enables survival of transitional and mature B cells through distinct mediators. J Immunol 2002;168:5993–6.
34. Crowley JE, Treml LS, Stadanlick JE, Carpenter E, Cancro MP. Homeostatic niche specification among naive and activated B cells: a growing role for the BLyS family of receptors and ligands. Semin Immunol 2005;17:193–9.
35. Groom J, Kalled SL, Cutler AH, et al. Association of BAFF/BLyS overexpression and altered B cell differentiation with Sjogren"s syndrome. J Clin Invest 2002;109:59–68.
36. Pers JO, Daridon C, Devauchelle V, et al. BAFF overexpression is associated with autoantibody production in autoimmune diseases. Ann NY Acad Sci 2005;1050:34–9.
37. Lesley R, Xu Y, Kalled SL, et al. Reduced competitiveness of autoantigen-engaged B cells due to increased dependence on BAFF. Immunity 2004;20:441–53.
38. Zhang X, Park CS, Yoon SO, et al. BAFF supports human B cell differentiation in the lymphoid follicles through distinct receptors. Int Immunol 2005;17:779–88.
39. Ekland EH, Forster R, Lipp M, Cyster JG. Requirements for follicular exclusion and competitive elimination of autoantigen-binding B cells. J Immunol 2004;172:4700–8.
40. Cyster JG. B cells on the front line. Nature Immunol 2000;1:9–10.
41. Kayagaki N, Yan M, Seshasayee D, et al. BAFF/BLyS receptor 3 binds the B cell survival factor BAFF ligand through a discrete surface loop and promotes processing of NF-kappaB2. Immunity 2002;17:515–24.
42. Claudio E, Brown K, Park S, Wang H, Siebenlist U. BAFF-induced NEMO-independent processing of NF-kappa B2 in maturing B cells. Nat Immunol 2002;3:958–65.
43. Gardam S, Sierro F, Basten A, Mackay F, Brink R. TRAF2 and TRAF3 signal adapters act cooperatively to control the maturation and survival signals delivered to B cells by the BAFF receptor. Immunity 2008;28:391–401.

44. Franzoso G, Carlson L, Xing L, et al. Requirement for NF-kappaB in osteoclast and B-cell development. Genes Dev 1997;11:3482–96.
45. Sha WC, Liou HC, Tuomanen EI, Baltimore D. Targeted disruption of the p50 subunit of NF-kappa B leads to multifocal defects in immune responses. Cell 1995;80:321–30.
46. Grumont RJ, Rourke IJ, O"Reilly LA, et al. B lymphocytes differentially use the Rel and nuclear factor kappaB1 (NF-kappaB1) transcription factors to regulate cell cycle progression and apoptosis in quiescent and mitogen-activated cells. J Exp Med 1998;187:663–74.
47. Caamano JH, Rizzo CA, Durham SK, et al. Nuclear factor (NF)-kappa B2 (p100/p52) is required for normal splenic microarchitecture and B cell-mediated immune responses. J Exp Med 1998;187:185–96.
48. Beinke S, Ley SC. Functions of NF-kappaB1 and NF-kappaB2 in immune cell biology. Biochem J 2004;382:393–409.
49. Liang C, Zhang M, Sun SC. Beta-TrCP binding and processing of NF-kappaB2/p100 involve its phosphorylation at serines 866 and 870. Cell Signal 2006;18:1309–17.
50. Enzler T, Bonizzi G, Silverman GJ, et al. Alternative and classical NF-kappa B signaling retain autoreactive B cells in the splenic marginal zone and result in lupus-like disease. Immunity 2006;25:403–15.
51. Sasaki Y, Derudder E, Hobeika E, et al. Canonical NF-kappaB activity, dispensable for B cell development, replaces BAFF-receptor signals and promotes B cell proliferation upon activation. Immunity 2006;24:729–39.
52. Ramakrishnan P, Wang W, Wallach D. Receptor-specific signaling for both the alternative and the canonical NF-kappaB activation pathways by NF-kappaB-inducing kinase. Immunity 2004;21:477–89.
53. Bren GD, Solan NJ, Miyoshi H, Pennington KN, Pobst LJ, Paya CV. Transcription of the RelB gene is regulated by NF-kappaB. Oncogene 2001;20:7722–33.
54. Yilmaz ZB, Weih DS, Sivakumar V, Weih F. RelB is required for Peyer's patch development: differential regulation of p52-RelB by lymphotoxin and TNF. EMBO J 2003; 22:121–30.
55. Derudder E, Dejardin E, Pritchard LL, Green DR, Korner M, Baud V. RelB/p50 dimers are differentially regulated by tumor necrosis factor-alpha and lymphotoxin-beta receptor activation: critical roles for p100. J Biol Chem 2003;278:23278–84.
56. Lam KP, Kuhn R, Rajewsky K. In vivo ablation of surface immunoglobulin on mature B cells by inducible gene targeting results in rapid cell death. Cell 1997;90:1073–83.
57. Vigorito E, Gambardella L, Colucci F, McAdam S, Turner M. Vav proteins regulate peripheral B-cell survival. Blood 2005;106:2391–8.
58. Franzoso G, Carlson L, Poljak L, et al. Mice deficient in nuclear factor (NF)-kappa B/p52 present with defects in humoral responses, germinal center reactions, and splenic microarchitecture. J Exp Med 1998;187:147–59.
59. Miosge LA, Blasioli J, Blery M, Goodnow CC. Analysis of an ethylnitrosourea-generated mouse mutation defines a cell intrinsic role of nuclear factor kappaB2 in regulating circulating B cell numbers. J Exp Med 2002;196:1113–9.
60. Xie P, Stunz LL, Larison KD, Yang B, Bishop GA. Tumor necrosis factor receptor-associated factor 3 is a critical regulator of B cell homeostasis in secondary lymphoid organs. Immunity 2007;27:253–67.
61. Grech AP, Amesbury M, Chan T, Gardam S, Basten A, Brink R. TRAF2 differentially regulates the canonical and noncanonical pathways of NF-kappaB activation in mature B cells. Immunity 2004;21:629–42.
62. Xu LG, Shu HB. TNFR-associated factor-3 is associated with BAFF-R and negatively regulates BAFF-R-mediated NF-kappa B activation and IL-10 production. J Immunol 2002;169:6883–9.
63. Hauer J, Puschner S, Ramakrishnan P, et al. TNF receptor (TNFR)-associated factor (TRAF) 3 serves as an inhibitor of TRAF2/5-mediated activation of the noncanonical NF-kappaB pathway by TRAF-binding TNFRs. Proc Natl Acad Sci USA 2005;102:2874–9.

64. Liao G, Zhang M, Harhaj EW, Sun SC. Regulation of the NF-kappaB-inducing kinase by tumor necrosis factor receptor-associated factor 3-induced degradation. J Biol Chem 2004;279:26243–50.
65. Xu Y, Cheng G, Baltimore D. Targeted disruption of TRAF3 leads to postnatal lethality and defective T-dependent immune responses. Immunity 1996;5:407–15.
66. He JQ, Zarnegar B, Oganesyan G, et al. Rescue of TRAF3-null mice by p100 NF-kappa B deficiency. J Exp Med 2006;203:2413–8.
67. Morrison MD, Reiley W, Zhang M, Sun SC. An atypical tumor necrosis factor (TNF) receptor-associated factor-binding motif of B cell-activating factor belonging to the TNF family (BAFF) receptor mediates induction of the noncanonical NF-kappaB signaling pathway. J Biol Chem 2005;280:10018–24.
68. Rothe M, Sarma V, Dixit VM, Goeddel DV. TRAF2-mediated activation of NF-kappa B by TNF receptor 2 and CD40. Science 1995;269:1424–7.
69. Song HY, Regnier CH, Kirschning CJ, Goeddel DV, Rothe M. Tumor necrosis factor (TNF)-mediated kinase cascades: bifurcation of nuclear factor-kappaB and c-jun N-terminal kinase (JNK/SAPK) pathways at TNF receptor-associated factor 2. Proc Natl Acad Sci USA 1997;94:9792–6.
70. Malinin NL, Boldin MP, Kovalenko AV, Wallach D. MAP3K-related kinase involved in NF-kappaB induction by TNF, CD95 and IL-1. Nature 1997;385:540–4.
71. Liu YC. Ubiquitin ligases and the immune response. Annu Rev Immunol 2004;22:81–127.
72. Xia ZP, Chen ZJ. TRAF2: a double-edged sword? Sci STKE 2005;pe7 (http://stke.sciencemag.org/cgi/content/full/OC_sigtrans;stke.2722005pe7)
73. Shi CS, Kehrl JH. Tumor necrosis factor (TNF)-induced germinal center kinase-related (GCKR) and stress-activated protein kinase (SAPK) activation depends upon the E2/E3 complex Ubc13-Uev1A/TNF receptor-associated factor 2 (TRAF2). J Biol Chem 2003;278:15429–34.
74. Habelhah H, Takahashi S, Cho SG, Kadoya T, Watanabe T, Ronai Z. Ubiquitination and translocation of TRAF2 is required for activation of JNK but not of p38 or NF-kappaB. Embo J 2004;23:322–32.
75. Rothe M, Pan MG, Henzel WJ, Ayres TM, Goeddel DV. The TNFR2-TRAF signaling complex contains two novel proteins related to baculoviral inhibitor of apoptosis proteins. Cell 1995;83:1243–52.
76. Vince JE, Wong WW, Khan N, et al. IAP antagonists target cIAP1 to induce TNFalpha-dependent apoptosis. Cell 2007;131:682–93.
77. Varfolomeev E, Blankenship JW, Wayson SM, et al. IAP antagonists induce autoubiquitination of c-IAPs, NF-kappaB activation, and TNFalpha-dependent apoptosis. Cell 2007;131(4):669–81.
78. Annunziata CM, Davis RE, Demchenko Y, et al. Frequent engagement of the classical and alternative NF-kappaB pathways by diverse genetic abnormalities in multiple myeloma. Cancer Cell 2007;12:115–30.
79. Keats JJ, Fonseca R, Chesi M, et al. Promiscuous mutations activate the noncanonical NF-kappaB pathway in multiple myeloma. Cancer Cell 2007;12:131–44.
80. Qian Y, Qin J, Cui G, et al. Act1, a negative regulator in CD40- and BAFF-mediated B cell survival. Immunity 2004;21:575–87.
81. Patke A, Mecklenbrauker I, Erdjument-Bromage H, Tempst P, Tarakhovsky A. BAFF controls B cell metabolic fitness through a PKC beta- and Akt-dependent mechanism. J Exp Med 2006;203:2551–62.
82. Woodland RT, Schmidt MR, Thompson CB. BLyS and B cell homeostasis. Semin Immunol 2006;18(5):318–26.
83. Plas DR, Rathmell JC, Thompson CB. Homeostatic control of lymphocyte survival: potential origins and implications. Nat Immunol 2002;3:515–21.
84. Batten M, Groom J, Cachero TG, et al. BAFF mediates survival of peripheral immature B lymphocytes. J Exp Med 2000;192:1453–66.

85. Do RK, Hatada E, Lee H, Tourigny MR, Hilbert D, Chen-Kiang S. Attenuation of apoptosis underlies B lymphocyte stimulator enhancement of humoral immune response. J Exp Med 2000;192:953–64.
86. Bouillet P, Metcalf D, Huang DC, et al. Proapoptotic Bcl-2 relative Bim required for certain apoptotic responses, leukocyte homeostasis, and to preclude autoimmunity. Science 1999;286:1735–8.
87. Enders A, Bouillet P, Puthalakath H, Xu Y, Tarlinton DM, Strasser A. Loss of the pro-apoptotic BH3-only Bcl-2 family member Bim inhibits BCR stimulation-induced apoptosis and deletion of autoreactive B cells. J Exp Med 2003;198:1119–26.
88. Zong WX, Edelstein LC, Chen C, Bash J, Gelinas C. The prosurvival Bcl-2 homolog Bfl-1/A1 is a direct transcriptional target of NF-kappaB that blocks TNFalpha-induced apoptosis. Genes Dev 1999;13:382–7.
89. Lee HH, Dadgostar H, Cheng Q, Shu J, Cheng G. NF-kappaB-mediated up-regulation of Bcl-x and Bfl-1/A1 is required for CD40 survival signaling in B lymphocytes. Proc Natl Acad Sci USA 1999;96:9136–41.
90. Miyamoto A, Nakayama K, Imaki H, et al. Increased proliferation of B cells and auto-immunity in mice lacking protein kinase Cdelta. Nature 2002;416:865–9.
91. Mecklenbrauker I, Saijo K, Zheng NY, Leitges M, Tarakhovsky A. Protein kinase Cdelta controls self-antigen-induced B-cell tolerance. Nature 2002;416:860–5.
92. Mecklenbrauker I, Kalled SL, Leitges M, Mackay F, Tarakhovsky A. Regulation of B-cell survival by BAFF-dependent PKCdelta-mediated nuclear signalling. Nature 2004;431:456–61.

Chapter 4
Role of BAFF and APRIL in Antibody Production and Diversification

Andrea Cerutti and Kang Chen

Abstract BAFF and APRIL are two TNF family members that have emerged as important regulators of antibody production and diversification. They are expressed by innate immune cells and stromal cells and are upregulated in response to various danger signals. By bindings to their receptors and signaling for the activation of NF-κB, they regulate multiple aspects of T cell-dependent and -independent antibody responses including B-cell survival, activation, class switching, somatic hypermutation, plasmacytoid differentitation and tolerance. The understanding of their roles in these processes will offer novel insights in the therapeutic intervention of antibody-related disorders such as immunodeficiency, autoimmunity, infection and inflammation.

Keywords BAFF · APRIL · T cell-independent antibody response · Class switching

4.1 Introduction

Since their discovery in 1998, the TNF family members, BAFF and APRIL, have received increasing attention. In addition to regulating the homeostasis of B cells and plasma cells, BAFF and APRIL participate in the initiation and regulation of antibody responses. Here we focus on the cellular and signaling pathways utilized by BAFF and APRIL to induce antibody production and diversification. We also discuss the immune disorders resulting from a dysregulation of these pathways.

A. Cerutti (✉)
Department of Pathology and Laboratory Medicine, Graduate Program of Immunology and Microbial Pathogenesis, Weill Medical College of Cornell University, New York, USA

M.P. Cancro (ed.), *BLyS Ligands and Receptors,* Contemporary Immunology,
DOI 10.1007/978-1-60327-013-7_4,

4.2 Role of BAFF and APRIL in B-Cell Survival and Activation

4.2.1 Structure of BAFF and APRIL

B-cell activation factor of the tumor necrosis factor family (BAFF, also known as BLyS, TALL-1, THANK, zTNF4, or TNFSF-13b) and a proliferation-inducing ligand (APRIL, also known as TRDL-1, TALL-2, or TNFSF-13a) play an essential role in the homeostasis of peripheral B cells and contribute to the initiation and diversification of B-cell responses to antigens [1, 2]. BAFF and APRIL were initially identified by homology searches of the nucleic acid databases as two molecules structurally related to tumor necrosis factor (TNF) [3, 4]. Similar to TNF, BAFF and APRIL spontaneously form homotrimers [1], while biologically active heterotrimers of BAFF and APRIL were also found [5]. BAFF and APRIL exist as both membrane-bound and soluble ligands. Soluble ligands originate upon cleavage of the N-terminal side of the TNF homology domain of BAFF and APRIL by a furin-like convertase [3, 6, 7]. Cleavage of BAFF takes place on the membrane, whereas cleavage of APRIL mostly occurs in the cytoplasm [6]. An alternative form of APRIL, called TWE–PRIL, arises from a hybrid mRNA encoding the cytoplasmic and transmembrane portions of the adjacent gene TWEAK and carboxy-terminal TNF domain of APRIL [8, 9]. TWE–PRIL may represent a plasma-membrane-bound form of APRIL [2]. Although both BAFF and APRIL are biologically active in a soluble form, it is not known whether processing of BAFF is absolutely required for its activity or whether the membrane-bound form of this molecule suffices to generate biological functions [1].

4.2.2 Expression of BAFF and APRIL

Constitutive BAFF production occurs in radio-resistant stromal cells [10, 11], which provides a tonic signal thought to be essential for the survival of mature B cells. BAFF is also expressed by many innate immune cell types, including monocytes, macrophages, dendritic cells (DCs), follicular DCs (FDCs), granulocytes, epithelial cells, synoviocytes, and astrocytes [12–14]. Immune stimulation can augment the expression of BAFF through a mechanism involving cytokines, including interferon-γ (IFN-γ), IFN-α, IFN-β, interleukin-10 (IL-10), TNF, or thymic stromal lymphopoietin (TSLP) [12, 13, 15]. BAFF production can also be upregulated by CD40 ligand (CD40L) [16], a TNF family member expressed by antigen-activated $CD4^+$ T cells, and by pathogen-associated molecular patterns (PAMPs) [13, 16–18], which are highly conserved microbial products capable of initiating innate immune responses [19]. PAMPs with BAFF-inducing activity include microbial Toll-like receptor (TLR) ligands, such as peptidoglycan, viral double-strand RNA (dsRNA), lipopolysaccharide (LPS), and unmethylated deoxycytidylate-phosphate-deoxyguanylate (CpG) DNA [13, 20]. BAFF-inducing activity is also associated with mannose-binding C-type lectin receptor (MCLR) ligands, a family of PAMPs that includes the envelope glycoprotein gp120 of the human immunodeficiency virus (HIV) [21].

Together with cytokines and CD40L, PAMPs can induce local concentrations of BAFF sufficient to initiate complex adaptive B-cell responses in a T-cell-independent (TI) fashion, including class switching and antibody production. Like BAFF, APRIL is also produced by various innate immune and stromal cell types, including monocytes, macrophages, DCs, epithelial cells, and osteoclasts [2, 17, 20]. These cells increase the production of APRIL upon exposure to cytokines (e.g., IFN-γ, IFN-α, IL-10, TSLP), CD40L or PAMPs (e.g., ligands of TLR3, TLR4, TLR5, and TLR9) [2, 17, 18]. Finally, some studies showed that activated $CD4^+$ T cells also express BAFF and APRIL, suggesting the involvement of these TNF family ligands in the amplification of T-cell-dependent (TD) B-cell responses [3, 4, 22–24].

4.2.3 BAFF and APRIL Receptors

Ligand–receptor interactions within the BAFF–APRIL subfamily of TNF ligands are both redundant and specific. Indeed, BAFF and APRIL share two receptors on B cells that are structurally related to TNF receptors and are known as transmembrane activator and calcium modulator and cyclophylin ligand interactor (TACI) and B-cell maturation antigen (BCMA) [1]. BAFF binds TACI with higher affinity than BCMA, whereas APRIL binds BCMA with higher affinity than TACI [2]. In addition to sharing TACI and BCMA receptors with APRIL, BAFF has a private receptor known as BAFF receptor (BAFF-R/BR3) [25]. APRIL also has a private receptor expressed on both B cells and non-B cells. This receptor has been identified as heparan-sulfate proteoglycans (HSPGs), which bind basic amino acid residues on APRIL through sulfated glycosaminoglycan side chains [26, 27]. Recent evidence indicates that TACI, BCMA, BAFF-R, and HSPGs have a specific expression pattern on peripheral B cells [12]. Indeed, BAFF-R is expressed by all peripheral immature and mature B-cell subsets, including transitional type-1 (T1), T2, marginal zone (MZ), follicular, germinal center (GC), and some plasmacytoid B cells [25, 26, 28–32]. In contrast, TACI is expressed by activated follicular and MZ B cells, whereas BCMA is expressed by activated memory and plasmacytoid B cells [28, 29, 32]. The latter can also express HSPGs [26]. Of note, TACI, BCMA, and BAFF-R appear to deliver non-redundant signals to different B-cell subsets [12]. Thus, while TACI stimulates immunoglobulin (Ig) class switching and production in extrafollicular (e.g., MZ) and follicular B cells [33], BCMA delivers survival signals to bone marrow plasma cells [34]. Finally, BAFF-R delivers class switch- and proliferation-inducing signals to extrafollicular (e.g., MZ) and follicular B cells, and survival signals to all peripheral B cells [25, 33].

4.2.4 BAFF and APRIL Signaling

The signal transduction pathway emanating from TACI, BCMA, and BAFF-R is only partially characterized [35], but resembles that associated with CD40, the receptor of CD40L on B cells [36]. The cytoplasmic tail of TACI has a canonical consensus TNF receptor-associated factor (TRAF)-binding site that recruits TRAF2 and TRAF5 and a minimal consensus TRAF-binding site that recruits TRAF6 [37]. TRAF recruitment by TACI activates the IκB kinase (IKK) complex, a signaling structure that encompasses two catalytic α and β subunits and a regulatory γ subunit [38]. The IKK complex mediates phosphorylation of inhibitors of NF-κB (IκB), which usually retain nuclear factor-κB (NF-κB) p50, p65, and c-Rel transcription factors in a cytoplasmic inactive form [38]. Phosphorylation of IκBα by the IKK complex is followed by ubiquitination and proteasome-dependent degradation of IκBα and nuclear translocation of p50, p65, and c-Rel transcription factors. In addition to recruiting TRAFs, TACI interacts with calcium modulator and cyclophylin ligand interactor (CAML), a positive regulator of the calcium-dependent phosphatase calcineurin, which mediates dephosphorylation and subsequent nuclear translocation of nuclear factor of activated T cells (NF-AT) [39]. Finally, TACI activates activator protein-1 (AP-1) [39]. Together with p50, p65, and c-Rel, NF-AT and AP-1 proteins transactivate genes involved in TACI-mediated B-cell proliferation, Ig class switching, and production.

The cytoplasmic tail of BCMA contains two canonical TRAF-binding sites that recruit TRAF1, TRAF2, and TRAF3 [40]. Recruitment of TRAFs results in the activation of NF-κB, p38 mitogen-activated protein kinase (MAPK), and c-Jun N-terminal kinase (JNK). These signaling pathways would enable BCMA to deliver plasma cell survival signals [34]. Although lacking classical consensus TRAF-binding sites, the cytoplasmic tail of BAFF-R has an atypical TRAF-binding motif, which recruits TRAF3 [41]. In addition to activating the canonical NF-κB pathway via the IKK complex, BAFF-R preferentially activates an alternative NF-κB pathway via IKKα [38]. A recent study showed that the atypical TRAF-binding motif of BAFF-R could explain its preferential activation of the alternative NF-κB pathway [42]. BAFF-R activates IKKα through NF-κB-inducing kinase (NIK), a TRAF3-interacting protein [43–45]. Phosphorylation of p100, a cytoplasmic NF-κB precursor, by IKKα is followed by ubiquitination and proteasome-mediated processing of p100 to p52, which then translocates to the nucleus in association with RelB, another NF-κB protein [38]. Once in the nucleus, p52 and RelB transactivate genes involved in BAFF-R-mediated B-cell survival. Of note, this alternative NF-κB pathway is negatively regulated by the TRAF3-binding protein Act1 [46].

4.2.5 Role of BAFF and APRIL in B-Cell Survival

In mice, the size of the peripheral B-cell compartment is controlled by a constitutive pool of BAFF that is produced in fixed amounts by radio-resistant stromal cells [12]. By engaging BAFF-R, this constitutive pool of BAFF delivers a tonic survival signal to all peripheral B cells positioned beyond the T1 stage, including T2, MZ, and follicular B cells [1, 2]. Consistent with this possibility, mice lacking either BAFF or BAFF-R and mice expressing a mutant BAFF-R exhibit a profound loss of peripheral B cells [10, 25]. A similar loss can be induced by exposing mice to soluble TACI-Ig or BCMA-Ig decoy receptors, which prevent the binding of BAFF to BAFF-R on the surface of B cells [47, 48]. Conversely, mice expressing transgenic

BAFF show a prominent B-cell hyperplasia [49]. BAFF-R delivers survival signals by activating both canonical and alternative NF-κB pathways, although the alternative NF-κB pathway seems to play a more important role [43, 44, 50, 51]. Activation of NF-κB by BAFF is associated with upregulation of anti-apoptotic Bcl-2 family proteins, including Bcl-2, Bcl-xL, and Mcl-1, and with downregulation of pro-apoptotic Bcl-2 family proteins, such as Bax, Bid, and Bad [35]. This BAFF-R-dependent pathway may contribute to the pro-survival activity of the B-cell antigen receptor (BCR, or surface Ig), because BCR signaling induces upregulation of BAFF-R expression on B cells [31]. Interestingly, in some B cells the BAFF–BAFF-R system is complemented by an APRIL–BCMA system that provides additional pro-survival signals. This is the case of long-lived murine plasma cells, which depend on BCMA for their survival [34]. Generally, compared to BAFF, APRIL has a marginal effect on the survival of peripheral B cells. Indeed, mice lacking APRIL do not show any major alteration of peripheral B-cell subsets [52]. A similar phenotype is associated with mice expressing an APRIL transgene, except that these mice develop a clonal expansion of peritoneal B-1 cells as they age [53].

4.2.6 Role of BAFF and APRIL in B-Cell Activation and Differentiation

While the constitutive pool of BAFF delivers a tonic survival signal to peripheral B cells, an inducible pool of BAFF, mostly derived from activated myeloid cells, seems to control the local accumulation, activation, and differentiation of B cells in the context of an immune response [12]. In this regard, it must be noted that the prominent B-cell survival activity of BAFF has complicated the analysis of other functions associated with this ligand. However, in vitro experiments with purified populations of B cells have clearly shown that BAFF can trigger functions distinct from survival. For example, BAFF can upregulate CD21 and CD23 expression and enhance T1-to-T2 B-cell differentiation via both BAFF-R-dependent and -independent mechanisms [54, 55]. In addition, BAFF can enhance the entry of B cells into G1 phase of the cell cycle via BAFF-R, an event that may be implicated in the formation and maintenance of the GC during a TD antibody response [56]. Consistent with this evidence, mice expressing a mutant BAFF-R or lacking BAFF cannot induce a sustained GC reaction in response to a TD antigen [57]. Additional findings show that BAFF or APRIL can upregulate the expression of T cell co-stimulatory molecules on B cells through a pathway involving BCMA [58].

Last but not least, BAFF and APRIL can trigger IgG, IgA, and IgE class switching and production in pre-switched IgD$^+$ B cells without the help from CD4$^+$ T cells via CD40L [13, 16, 17]. This CD40-independent response is mediated by TACI and, to a lesser extent, by BAFF-R, as pre-switched IgD$^+$ B cells lacking either of these receptors do not produce IgG or IgA upon exposure to BAFF or APRIL [33]. In agreement with these in vitro data, TACI-deficient mice exhibit impaired IgG and IgA responses to TI antigens in spite of having an uncontrolled B-cell expansion [59, 60]. Similarly, humans with a congenital defect of TACI signaling develop selective

IgA deficiency or common variable immunodeficiency, a disorder associated with severely reduced IgM, IgG, and IgA production [61, 62]. Furthermore, mice lacking APRIL exhibit less IgA class switching and production in the intestinal mucosa [63]. Taken together, these evidences indicate that BAFF and APRIL possess antibody-inducing and antibody-diversifying functions that become particularly important in the context of B-cell responses to TI antigens.

4.3 Role of BAFF and APRIL in Antibody Production

4.3.1 Regulation of Adaptive Immune Responses by DCs

BAFF and APRIL production involves DCs, a specialized innate immune cell subset that plays a key role in adaptive immune responses, including B-cell responses [64]. To better understand the specific contribution of BAFF and APRIL to adaptive B-cell responses, we will first discuss the general pathways followed by DCs to orchestrate protective immunity. Immune protection requires the interplay between non-antigen-specific innate immunity and antigen-specific adaptive immunity. Distributed as sentinels throughout the body, DCs collect and transfer information on the composition of invading microbes from the external environment to the T and B cells of the adaptive immune system [65]. This process requires the recognition of microbes by DCs through pattern-recognition receptors, including TLRs [66]. After sensing PAMPs through TLRs, DCs become activated and initiate a maturation program that involves upregulation of molecules mediating antigen processing and presentation [67]. This maturation program also involves modulation of chemokine receptors that promote the migration of DCs to draining lymphoid organs [68]. At these sites, DCs function as professional antigen-presenting cells (APCs), in that they orchestrate clonal adaptive responses by instructing T cells on the nature and composition of antigen. In particular, DCs process antigen into immunogenic peptides that are presented to T cells in the context of major histocompatibility (MHC) molecules.

In the presence of co-stimulatory signals provided by DC-derived cytokines such as IL-12, IL-6, and transforming growth factor-β (TGF-β1), engagement of T-cell receptor (TCR) by the peptide–MHC complex triggers the differentiation of antigen-specific T cells into specialized effector subsets. IL-12 drives the differentiation of naive $CD4^{+}$ T helper (Th) cells into Th1 effector cells, which typically produce IFN-γ. In the presence of IL-12-inhibiting signals, DCs stimulate the differentiation of naive $CD4^{+}$ Th cells into Th2 effector cells, which produce IL-4 and IL-10. Finally, IL-6 and TGF-β1 drive the differentiation of mouse naive $CD4^{+}$ Th cells into Th17 cells, and in human, IL-6 and IL-1β have been shown to do this [69]. By activating B cells through CD40L expressed by antigen-activated $CD4^{+}$ T cells, Th1, Th2, and Th17 subsets orchestrate antibody responses to antigen [70].

In addition to activating B cells indirectly through Th cells, DCs can express B-cell-stimulating signals autonomously. Indeed, engagement of CD40 by CD40L augments DC secretion of numerous cytokines with B-cell-stimulating function.

Among them, IL-6 cooperates with CD40L to induce the differentiation of B cells into antibody-secreting plasma cells [71]. In addition to IL-6, DCs release BAFF and APRIL which cooperate with CD40L to amplify the survival, proliferation, and differentiation of B cells [1, 12]. Furthermore, BAFF and APRIL cooperate with Th2 cytokines, such as IL-4 or IL-10, to trigger switching from IgM to IgG, IgA, or IgE [72]. Finally, BAFF and APRIL production can also be effectively induced by the Th1 cytokine IFN-γ [2]. These data imply that BAFF and APRIL are utilized by multiple Th-cell subsets to initiate or amplify TD antibody production.

4.3.2 Role of Specific DC Subsets

Due to their key function of linking innate with adaptive immunity, DCs have evolved to have distinct subsets capable of initiating various immune functions. It is now accepted that DCs comprise at least two subpopulations: one in the lymphoid lineage, known as plasmacytoid DCs (pDCs) and one in the myeloid lineage, known as myeloid DCs (mDCs) [73]. The relative contribution of pDCs and mDCs to BAFF and APRIL production is poorly understood.

pDCs are immature circulating DCs that release massive amounts of IFN-α and IFN-β (IFN-α/β) upon encountering microbial products [74, 75]. In addition to eliciting antiviral and cytotoxic responses, IFN-α/β stimulates the maturation of pDCs, which then become capable of eliciting Th1-like responses yielding both IL-10 and IFN-γ [76]. When incubated with CD40-activated B cells, these cytokines trigger production of IgM as well as switching from IgM to IgG and IgA, indicating the possible participation of Th1-like cells in TD antibody responses [72]. Yet, it must be noted that IFN-γ has an activating effect in mouse but not human B cells [77]. pDCs can also undergo maturation upon exposure to IL-3 released by parasite-activated eosinophils, basophils, and mast cells [78]. Mature pDCs originating from this alternative pathway promote Th2 responses yielding both IL-4 and IL-10 [79]. When incubated with CD40-activated B cells, these cytokines trigger switching from IgM to IgG, IgA or IgE [72].

In addition to generating Th cells with B-cell-helper activity, mature pDCs can directly deliver B-cell-activating signals through IFN-α/β and IL-6. These cytokines augment TD antibody responses by promoting the differentiation of CD40-activated B cells into antibody-secreting plasmablasts and plasma cells in a sequential fashion [71]. IFN-α/β can further amplify antibody responses by enhancing the production of BAFF and APRIL by mDCs [16], suggesting that humoral immunity involves a complex crosstalk between distinct DC subsets. It remains unclear whether pDCs produce BAFF and APRIL in an autonomous fashion. Some data from our laboratory indicate that circulating pDCs may produce BAFF after migrating into secondary lymphoid tissues.

mDCs include circulating immature and interstitial mature $CD11c^+$ DCs that release IL-12 in response to microbial products [73]. IL-12 promotes Th1 responses characterized by a prominent production of IFN-γ. However, in the presence of IL-12-inhibiting signals, mDCs promote Th2 responses characterized by a prevalent production of IL-4 and IL-10 [67]. The role of these cytokines in TD (i.e., CD40-dependent) antibody responses has already been discussed. In addition to directly delivering activating signals to B cells, IFN-γ and IL-10 can stimulate mDCs to release more BAFF and APRIL [7]. Importantly, mDCs augment the release of BAFF and APRIL also in response to CD40L, a TNF family member transiently expressed by all Th-cell subsets [16]. PAMPs, including TLR ligands, have a similar BAFF- and APRIL-inducing effect [13, 16–18]. These data imply that mDCs deploy multiple pathways to amplify TD antibody responses via BAFF and APRIL.

4.3.3 DC Regulation of TD Antibody Responses

Canonical antigens, including complex microbial proteins, initiate antibody responses by inducing the activation of $CD4^+$ T cells with B-cell helper activity via DCs. After capturing antigen, peripheral DCs migrate to the extrafollicular area of draining lymph nodes, where they present peptide-MHC-II complex to $CD4^+$ T cells [80]. Signals generated by the immunogenic peptide and co-stimulatory molecules, including CD80 (B7.1) and CD86 (B7.2), stimulate $CD4^+$ T cells to upregulate the expression of CD40L, a powerful activator of B cells. The subsequent differentiation of $CD4^+$ T cells into either Th1 or Th2 effector cells generates additional B-cell-stimulating factors, including IL-4, IL-10, and IFN-γ. The combination of these factors is thought to trigger the initial activation of antigen-specific follicular B cells.

Follicular B cells interact with antigen-specific Th cells after internalizing and processing antigen through the BCR [81]. The combination of signals emanating from BCR, CD40, and cytokine receptors induces massive proliferation as well as migration of follicular B cells into the germinal center (GC) of secondary lymphoid organs, a specialized microenvironment that fosters antibody selection and diversification through two processes known as Ig V(D)J gene somatic hypermutation and Ig class switching [82]. Somatic hypermutation introduces point mutations within the genes that encode the antigen-binding variable (V) region of the Ig, thereby providing the structural correlate for selection by antigen of higher-affinity Ig mutants [83]. Class switching substitutes the $C\mu$ gene encoding the heavy chain constant (C_H) region of IgM with the $C\gamma$, $C\alpha$, or $C\varepsilon$ gene encoding the C_H region of IgG, IgA, or IgE, thereby providing antibodies with novel effector functions without changing their specificity for antigen [84]. Ultimately, the GC reaction generates memory B cells or antibody-secreting plasma cells expressing IgG, IgA, or IgE antibodies with high affinity for antigen. While memory B cells function as sentinels ready to respond to secondary antigenic challenges either in the circulation or in the MZ of the spleen, plasma cells migrate to the bone marrow to become long-lived antibody-secreting cells.

4.3.4 Function of BAFF and APRIL in TD Antibody Responses

BAFF production can be detected in the GC, the key site for TD antibody responses [85]. At this site, BAFF is mainly produced by FDCs and macrophages [85]. This circumstance strongly suggests the involvement of BAFF in the antibody response to TD antigens, including the GC reaction. Such involvement would require induction of both survival and activation signals in GC B cells. Consistent with this possibility, BAFF-deficient mice cannot mount a sustained GC reaction although having preserved GCs [57]. One possibility is that BAFF helps the development of an optimal GC reaction due to its ability to support the survival of follicular B cells producing lymphotoxin-β (LT-β), a cytokine involved in the maturation of the FDC network.

Although developing a normal FDC network, BAFF-R mutant mice also fail to mount a sustained GC reaction, indicating that BAFF exerts part of its effect on the GC independently of FDCs. Indeed, BAFF can deliver pro-survival co-signals to B cells exposed to CD40L, the main driver of the GC reaction [86]. Therefore, BAFF may optimize the proliferation, survival, differentiation, diversification, and selection of antigen-specific GC B cells by delivering either direct signals via BAFF receptors or indirect signals via BCR as a result of its ability to enhance antigen exposure by LT-β-activated FDCs. This possibility would be consistent with prior findings showing that BAFF enhances the proliferation and differentiation of B cells stimulated via BCR [3, 87].

The involvement of BAFF in the GC reaction is further suggested by evidence indicating that BAFF enhances the survival of B cells exposed to CD40L [86]. Engagement of CD40 by CD40L is central to the GC reaction as a result of its ability to deliver powerful activation, proliferation, survival, and class switch-inducing signals to B cells. In addition to enhancing CD40L-induced B-cell proliferation and survival [56], BAFF may augment CD40L-induced class switching (Fig. 4.1). This hypothesis would be in agreement with the observation that humans expressing a mutant form of TACI develop pan-hypogammaglobulinemia and their B cells undergo less IgA class switching in response to CD40L in vitro [61]. However, other findings suggest that TACI negatively modulates the activation of B cells via CD40L [59, 88, 89]. Clearly, more studies are needed to reconcile these discrepancies.

Additional evidence points to a role of APRIL in the enhancement of TD antibody responses (Fig. 4.1). Indeed, APRIL-transgenic mice exhibit enhanced IgM responses to TD antigens [22]. This alteration could be due to the ability of APRIL to induce proliferation of B cells via TACI, although the percentage of major peritoneal and splenic B-cell subsets is mostly normal in APRIL-transgenic mice [22]. An alternative possibility is that APRIL provides a co-stimulatory signal that enhances antibody production in CD40-activated B cells. A third possibility is that APRIL enhances CD40 signaling in B cells by delivering activating signals to $CD4^+$ T cells. Indeed, $CD4^+$ T cells can express TACI and undergo a significant increase in their number in mice expressing an APRIL transgene [89]. Finally, APRIL may further enhance TD antibody responses by enhancing the survival of antibody-secreting plasmablasts originating from memory B cells through BCMA [34]. This pathway would also account for the survival of long-lived antibody-secreting plasma cells lodged in the bone marrow.

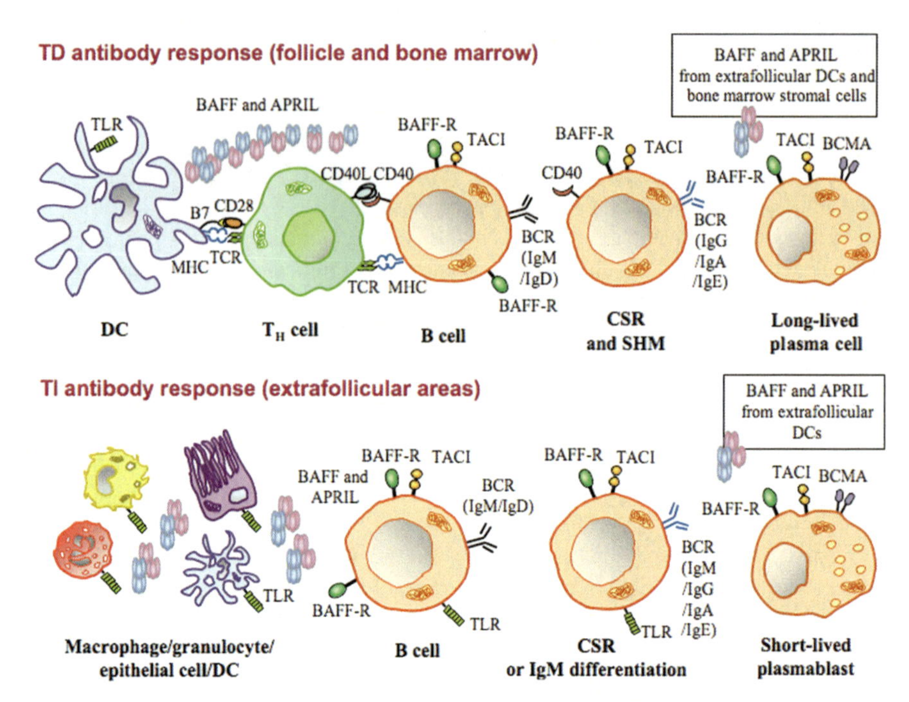

Fig. 4.1 Role of BAFF and APRIL in TD and TI antibody production and diversification. BAFF and APRIL are implicated in both T-cell-dependent and T-cell-independent antibody responses. In the TD antibody response, CD4$^+$ Th cells activated by DCs upregulate CD40L. Engagement of CD40 on BCR-activated follicular B cells by CD40L on Th cells triggers B-cell proliferation, selection, SHM, and IgG/A/E CSR. The concomitant engagement of TACI and BAFF-R by APRIL and/or BAFF from TLR-activated DCs optimizes the proliferation, survival, and diversification of follicular B cells. Ultimately, these follicular B cells differentiate into either memory B cells or antibody-secreting plasma cells, which migrate to extrafollicular areas and bone marrow, respectively. Engagement of TACI, BCMA, and BAFF-R by APRIL or BAFF from extrafollicular DCs elicits the differentiation of memory B cells into plasma cells, whereas engagement of BCMA by APRIL from stromal cells promotes the long-term survival of bone marrow plasma cells. In TI antibody responses, engagement of TACI and BAFF-R on BCR- and TLR-activated extrafollicular B cells by BAFF and APRIL from TLR-activated DCs, monocytes, macrophages, epithelial cells, or granulocytes initiates IgM differentiation or IgG/A CSR. Then, extrafollicular B cells rapidly differentiate into antibody-secreting plasmablasts, which upregulate BCMA. Engagement of TACI or BCMA by APRIL and/or BAFF from extrafollicular DCs regulates the survival and differentiation of these extrafollicular plasmablasts

4.3.5 Role of TI Antibody Responses in Protective Immunity

TD B-cell responses do not provide immune protection for at least 5–7 days, which is too much of a delay to control quickly replicating viruses and bacteria either systemically or at mucosal sites of entry [90]. To compensate for this limitation, specialized extrafollicular B cells, including splenic marginal zone (MZ) B cells and mucosal B cells, rapidly undergo production of IgM as well as class-switched IgG and IgA antibodies in a TI fashion [72]. This frontline antibody response usually targets highly conserved molecular patterns embedded within a large family of microbial products comprising TI type-1 (TI-1) and TI type-2 (TI-2) antigens. TI-1 antigens include viral and bacterial TLR ligands, such as hypomethylated CpG-rich DNA and lipopolysaccharide (LPS), whereas TI-2 antigens include bacterial polysaccharides with repetitive structure [91].

TI-1 antigens can activate B cells by engaging multiple innate antigen receptors encoded by germline genes, including TLRs. Extrafollicular B cells express TLRs at higher density than conventional B cells [32, 92, 93], a property that may account for the ability of extrafollicular B cells to respond to TI-1 antigens with an unmatched speed [94, 95]. Extrafollicular B cells express additional semi-invariant and invariant innate antigen receptors, including BCR, complement receptors (e.g., CD21 and CD35), and MHC-like CD1 molecules that contribute to the recognition of free or complement-opsonized TI-1 antigens and to the initiation of TI class switching and antibody production [96, 97].

TI-1 antigens can also activate B cells through an indirect mechanism involving induction of BAFF and APRIL by innate cell types. Indeed, DCs, macrophages, and epithelial cells increase the release of BAFF and APRIL after sensing microbes or their products through TLRs [13, 16]. Then, BAFF and APRIL would activate extrafollicular B cells by engaging TACI and BAFF-R. In this regard, recent findings indicate that extrafollicular B cells express more TACI than conventional B cells, perhaps as a result of their continuous exposure to antigenic BCR and TLR ligands [32, 55, 98]. Of note, these ligands synergize with BAFF and APRIL to activate B cells [55, 99], establishing a further link between the innate and adaptive immune systems.

As for TI-2 antigens, these compounds activate extrafollicular B cells through the BCR. In mice, splenic MZ B cells and peritoneal B-1 B cells sense bacterial polysaccharides through a low-affinity recognition system involving a semi-conserved BCR, i.e., a surface Ig receptor encoded by unmutated V(D)J genes [95, 96, 100]. In contrast, human splenic MZ B cells utilize a high-affinity recognition system involving a mutated BCR [101]. In spite of these differences, polysaccharide-specific B cells from both mice and humans undergo class switching and antibody production in a TI fashion. It remains unknown whether TI-2 antigens utilize additional recognition systems capable of upregulating the production of BAFF and APRIL in DCs.

4.3.6 Dendritic Cell Regulation of TI Antibody Responses

TI antibody responses appear to require the interaction of B cells with DCs, including mDCs [16, 64, 102, 103]. Indeed, mDCs are capable of sampling and internalizing TI antigens through a non-degradative endocytic pathway [104]. Subsequent recycling of antigen-containing vesicles to the cell surface would enable mDCs to present intact TI antigens to the BCR and TLRs on B cells. BCR-mediated internalization of antigen could also allow the activation of cytosolic TLRs in B cells [105, 106].

In addition to presenting TI antigens to B cells, mDCs would upregulate the release of BAFF and APRIL. Indeed, these innate B-cell-stimulating factors are released by mDCs upon exposure to TLR ligands or to cytokines generated by TLR-activated pDCs (i.e., IFN-α/β) [16]. Engagement of TACI and BAFF-R by BAFF and APRIL delivers powerful CD40-independent signals to B cells [16, 33] (Fig. 4.1). These signals cooperate not only with signals originating from DC-derived cytokines, such as IL-10, IL-15, and TGF-β1, but also with signals originating from the presentation of TI antigens to B cells by mDCs. Indeed, BAFF and APRIL can enhance the activation of extrafollicular B cells exposed to BCR or TLR ligands [16, 32, 55, 98, 99], possibly because the signaling pathways emanating from TACI receptor, BAFF-R, BCR, and TLRs converge at the level of NF-κB.

Thus, DCs would initiate TI antibody responses by activating B cells through multiple, intertwined CD40-independent pathways involving both somatically recombined (i.e., BCR) and germline gene-encoded (i.e., TLRs) antigen receptors as well as TACI, BAFF-R, and cytokine receptors. In addition to a prominent and well-known IgM component, these "innate" antibody responses include a class-switched IgG and IgA component that may be crucial to protect mucosal sites of entry, including the respiratory, intestinal, and urogenital mucosae. Indeed, growing evidence shows that IgG and IgA antibodies generated in a TI fashion help the mucosal immune systems to prevent the systemic penetration of both commensal and pathogenic microorganisms, including viruses [107–109]. In mice, B-1 cells expressing low-affinity and polyreactive antibodies are the major player of mucosal TI antibody responses [100]. The human equivalent of mouse B-1 cells remains to be identified. Nonetheless, the human mucosa can mount antibody responses, including IgA class switching, in a TI fashion [17].

DCs producing BAFF and APRIL may also activate splenic MZ B cells and initiate rapid TI IgM as well as class-switched IgG and IgA responses against quickly replicating blood-borne pathogens, including viruses and encapsulated bacteria [110]. In mice, these responses are characterized by the production of low-affinity and polyreactive antibodies encoded by semi-conserved V(D)J genes [95]. In humans, splenic MZ B cells produce IgM, IgG and IgA antibodies encoded by mutated V(D)J genes in a TI fashion and independently of any GC reaction [101]. These antibodies appear to have a high affinity for bacterial polysaccharides and play a critical protective role. Indeed, children below the age of 5 are more susceptible to invasive infections by encapsulated bacteria due to the immaturity of their splenic MZ [111]. Similarly, splenectomized adults are more susceptible to sepsis by encapsulated bacteria [112]. When, where, and how human splenic MZ B cells acquire somatic hypermutations remains unclear, but innate signals generated by BCR, TLR, TACI, and BAFF-R may be playing a key role.

4.3.7 Function of BAFF and APRIL in TI Antibody Responses

Although severely reduced, IgG and IgA responses are not abrogated in humans and mice with congenital CD40-signaling defects [113–116]. While impairing the

GC reaction, CD40-signaling defects spare the production of IgG and IgA antibodies in the splenic MZ and the intestinal lamina propria, two extrafollicular districts exposed to mucosal and blood-borne microorganisms, respectively [17, 117]. These in vivo observations imply that extrafollicular B cells can undergo class switching and antibody production through an alternative pathway that does not require help from CD4^{+} T cells through CD40L. Of note, extrafollicular districts are heavily populated by DCs and macrophages that produce large amounts of BAFF and APRIL, particularly after activation by microbial or immune stimuli [64]. In addition to innate immune cells, the intestinal mucosa includes epithelial cells, which also produce large amounts of BAFF and APRIL [13, 17]. Considering their ability to induce CD40-independent class switching and antibody production in vitro [16, 33] it is likely that BAFF and APRIL are central to the initiation of TI antibody responses taking place in extrafollicular areas, including subepithelial mucosal districts.

In agreement with this possibility, mice lacking TACI, the main class switch-inducing receptor in the BAFF–APRIL system, have defective IgG and IgA responses to TI antigens [59]. Interestingly, these mice exhibit conserved IgG and IgA responses to TD antigens and do not develop B-cell lymphopenia as mice lacking BAFF or BAFF-R do [59]. On the contrary, TACI-deficient mice develop B-cell hyperplasia [59], strongly indicating that the IgG and IgA defects arising in these mice cannot originate from the reduced survival of peripheral B cells. Indeed, in vitro studies show that engagement of TACI by BAFF or APRIL delivers IgG and IgA class switch-inducing signals, at least in mouse B cells [33]. TACI may deliver similar signals to human B cells, because loss-of-function mutations of the gene encoding TACI are associated with common variable immunodeficiency and selective IgA deficiency, two disorders characterized by decreased IgG and IgA (and IgM) production [61].

As for the B-cell hyperplasia developing in TACI-deficient mice, this finding is interpreted by some as a sign of TACI's ability to negatively modulate B-cell survival signals emanating from BAFF-R [89]. However, in vitro data do not support this possibility in a univocal fashion [118]. An alternative explanation is that the B-cell hyperplasia developing in TACI-deficient mice may derive from the chronic activation of B cells by commensal bacteria or bacterial products "leaking" from the intestinal mucosa as a result of a local IgA deficiency. Consistent with this possibility, mice lacking activation-induced cytidine deaminase (AID), an enzyme essential for class switching, develop not only IgA deficiency, but also a B-cell hyperplasia that is likely due to profound alterations of the intestinal microbiota [119].

The key role of the BAFF–APRIL system in TI antibody responses is further documented by data derived from APRIL-deficient and APRIL-transgenic mice. Indeed, APRIL-deficient mice develop intestinal IgA deficiency, suggesting that APRIL is the major partner of TACI for the induction of IgA responses in the intestinal mucosa [63]. In agreement with this interpretation, the human intestinal mucosa is extremely reach in APRIL, particularly at the level of epithelial cells, and fully supports IgA production in a TI fashion [17]. As for APRIL-transgenic mice, these animals show an enhancement of TI IgG and IgA responses, further pointing to APRIL as a major factor in TI class switching and antibody production [22].

Additional studies performed with BAFF-deficient, BAFF-R-mutant (i.e., A/WySnJ), and BAFF-R-deficient mice show that BAFF not only supports the survival of peripheral B cells, but also plays an important role in TI antibody formation [1]. Indeed, BAFF-R-deficient mice exhibit a loss of mature B cells similar to that observed in BAFF-deficient and BAFF-R-mutant mice [120]. However, while BAFF-R-deficient mice cannot carry out TD antibody formation, they differ from BAFF-deficient mice in generating normal levels of antibodies to at least some TI antigens [121]. These studies clearly indicate that BAFF regulates antibody responses in vivo through receptors other than BAFF-R. In agreement with this interpretation, in vitro data show that TACI is the main receptor utilized by BAFF to trigger CD40-independent class switching and antibody production [33].

4.4 Role of BAFF and APRIL in Antibody Diversification

4.4.1 Processes Mediating Antibody Diversification

Antibody diversification is essential for the immune system to mount protective responses against invading pathogens. The great diversity of antigen receptors in the mammalian immune system depends on the unique ability of B and T cells to somatically alter their genomes. B cells are particularly remarkable because they can undergo three distinct types of genetic alterations in two distinct phases of their development.

In the antigen-independent phase of B-cell development, B cell precursors lodged in the bone marrow generate antigen-recognition diversity by assembling the exons that encode Ig heavy (IgH) and light (IgL) chain variable regions from individual variable (V), diversity (D), and joining (J) gene segments through a process known as V(D)J gene recombination [122]. This site-directed, antigen-independent recombination event is initiated by a lymphoid-specific and sequence-specific recombination-activating gene (RAG)-1 and RAG-2 endonuclease complex and subsequently completed by components of the non-homologous end-joining machinery. Productive assembly of V_HDJ_H and V_LJ_L exons allows the expression of IgH and IgL chains as cell-surface IgM by newly generated B cells. After further maturation and acquisition of IgD, these cells migrate to secondary lymphoid organs, including the spleen and lymph nodes, where they can undergo further antigen-driven Ig gene diversification.

In the antigen-dependent phase of B-cell development, mature B cells diversify their antibody repertoire through somatic hypermutation (SHM) and class switching. SHM introduces point mutations at high rates into V_HDJ_H and V_LJ_L exons, thereby providing the structural correlate for selection by antigen of higher-affinity Ig mutants [83]. By contrast, class switching substitutes the IgH constant region μ (Cμ) and Cδ genes initially expressed by the primary IgM and IgD isotypes with an alternative set of downstream Cγ, Cα, or Cε genes through a process known as class switch DNA recombination (CSR) [84]. In this fashion, class switching generates secondary IgG, IgA, and IgE isotypes that have the same antigen specificity but different effector functions. Unlike primary IgM and IgD isotypes, secondary IgG, IgA, and IgE isotypes are capable of enhancing the pro-inflammatory, phagocytic, and microbicidal functions of various innate immune cell subsets, including neutrophils, eosinophils, basophils, monocytes, macrophages, mast cells, and natural killer (NK) cells [77]. Indeed, these cells augment their ability to capture, internalize, remove, and/or kill invading pathogens upon binding the Cγ, Cα, or Cε region of secondary IgG, IgA, and IgE antibodies through specific Fc receptors. An additional structure known as polymeric Ig receptor enables mucosal epithelial cells to secrete IgA antibodies into body fluids. Thus, by linking the innate and adaptive immune systems, CSR critically enhances the clearance of antigen.

4.4.2 Mechanism and Requirements of Class Switching

Despite being distinct processes targeted to distinct Ig regions, class switching and somatic hypermutation are similar in that they both occur in antigen-stimulated B cells, require transcription through the target Ig gene region, and require the DNA-editing enzyme activation-induced cytidine deaminase (AID) [123]. Here we will focus on class switching because this is the antibody diversification process that is specifically induced by BAFF and APRIL, although the additional involvement of these TNF ligands in somatic hypermutation cannot be excluded (Fig. 4.2).

The mouse and human IgH loci contain eight (Cμ, Cδ, Cγ1, Cγ3, Cγ2b, Cγ2a, Cε, Cα) and nine (Cμ, Cδ, Cγ1, Cγ3, Cα1, Cγ2, Cγ4, Cε, Cα2) genes encoding proteins with different effector functions, respectively [77]. Class switching involves an exchange of upstream Cμ and Cδ genes with downstream Cγ, Cα, or Cε genes through CSR [84]. This process is mediated by specialized, highly repetitive intronic sequences known as switch (S) regions, which precede each C_H gene, except Cδ. In CSR, Sμ recombines with a downstream Sγ, Sα, or Sε region, leading to the deletion of the intervening IgH DNA. As a result of this process, the V_HDJ_H exon is juxtaposed with a different C_H gene. Similar to V(D)J recombination, CSR is a deletional recombination reaction in which double-strand breaks are introduced into the DNA of the two participating S regions followed by their fusion. However, unlike V(D)J recombination, CSR does not involve RAG proteins and does not target a consensus sequence within the S regions. A currently accepted working model for the initiation of CSR involves transcription through the targeted donor and acceptor S regions, which renders these regions substrates for subsequent modification by AID [84]. Then, several DNA repair pathways complete CSR by mediating the processing and joining of AID-initiated S-region breaks.

In general, B cells need two signals to initiate CSR, one provided by a TNF (e.g., CD40L, BAFF, APRIL) or a TLR (e.g., LPS, CpG DNA) ligand, and the other provided by a cytokine (e.g., IL-4, IL-5, IL-10, IL-21, IFN-γ or TGF-β1). The combination of these two signals initiates germline transcription of the I_H-S-C_H DNA

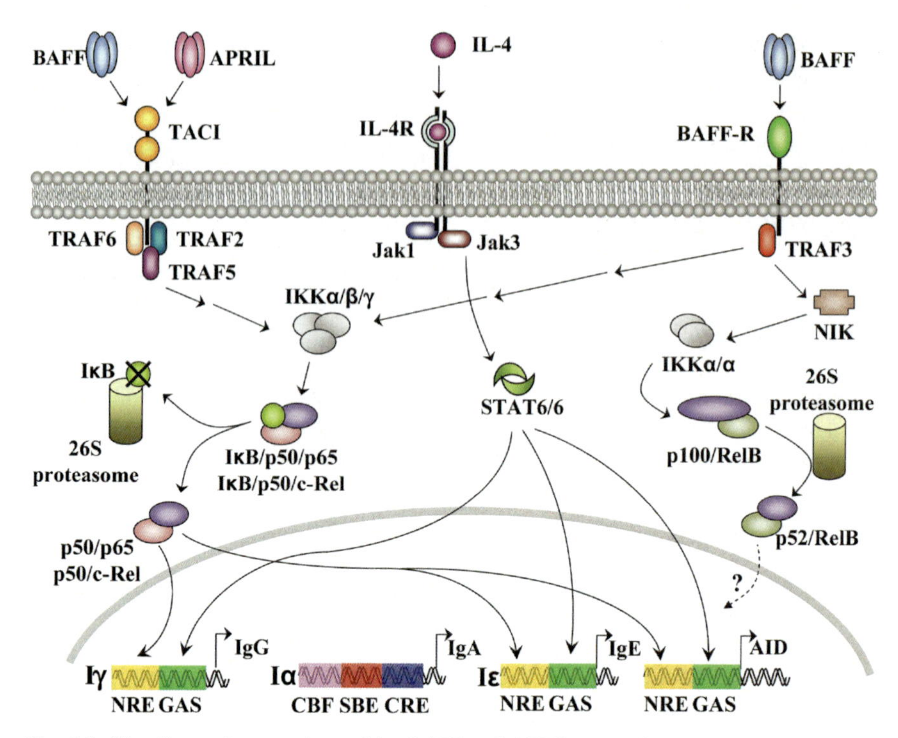

Fig. 4.2 Signaling pathways triggered by BAFF and APRIL to mediate class switching. TACI and BAFF-R are the major receptors involved in CSR triggered by BAFF and APRIL. TACI activates the classical NF-κB pathway and BAFF-R activates both the classical and the alternative NF-κB pathway. The classical NF-κB pathway has been shown to be important for CSR. The role of the alternative NF-κB pathway in CSR is unknown. NF-κB signals emanating from TACI and BAFF-R cooperate with Jak-STAT signals emanating from cytokine receptors, such as the IL-4R, to activate intronic promoters upstream of Cγ, Cε, as well as AID genes. NF-κB proteins initiate transcription by binding *cis*-regulatory NF-κB response elements (NRE), whereas STAT proteins initiate transcription by binding *cis*-regulatory γ-interferon gamma activated (GAS) elements adjacent to NREs. Transcription at Ig intronic regions allows AID to gain access to the DNA at these regions and mediate CSR. The intronic promoter in front of Cα genes has *cis*-regulatory elements distinct from those in front of the Cγ or Cε genes, and may bind transcription factors such as Smad proteins, cAMP response element (CRE)-binding protein, and core-binding factor (CBF) α3. These transcription factors are usually activated by the cytokine TGF-β, which does not seem to be required for the induction of IgA CSR by BAFF and APRIL. NRE: NF-κB response elements; GAS: interferon gamma activated site; SRE: Smad-responsive element

region by synergistically activating an intronic (I_H) promoter located upstream of the targeted S region [84]. By facilitating the recruitment of AID to the two participating S regions, germline transcription initiates CSR. In addition to initiating germline transcription of the targeted S regions, CSR-inducing signals induce transcriptional upregulation and subsequent expression of AID.

Interestingly, cytokines can direct switching to specific isotypes as a result of their ability to activate distinct I_H promoters through specific transcription factors. In mouse, B cells exposed to LPS, CD40L and/or a dextran-conjugated anti-IgD antibody, IL-4 directs switching to IgG1 and IgE, IFN-γ switching to IgG2a and IgG3, IL-5 switching to IgA, and TGF-β1 switching to IgA and IgG2b. In human B cells exposed to CD40L, IL-4 triggers switching to IgG (all subclasses), IgA and IgE, IL-10 switching to IgG (IgG1, IgG2, IgG3) and IgA, IL-13 switching to IgG and IgE, IL-21 switching to IgG, and TGF-β1 switching to IgA [124]. It can be noted that human cytokines seem to have a more promiscuous effect than their mouse counterparts. Nonetheless, IL-4 appears to be absolutely required for induction of IgG4 and IgE and TGF-β1 for the induction of IgA. An additional difference between mice and humans is that IL-10 plays a marginal role in the induction of class switching in mouse B cells [77]. Finally, IL-5 and IFN-γ do not induce class switching in human B cells [77].

4.4.3 CD40L Signaling in TD Class Switching

CD40L is the main inducer of IgG, IgA, and IgE class switching during TD antibody responses [77]. Engagement of CD40 on B cells by CD40L on $CD4^+$ T cells initiates class switching through a pathway that is largely dependent on NF-κB. In resting B cells, p65–p50, p50–c-Rel, and p50–p50 NF-κB dimers are kept in an inactive cytoplasmic form by inhibitor of NF-κB (IκB) proteins, including IκBα [38]. By eliciting CD40 oligomerization and recruitment of multiple TNF receptor-associated factor (TRAF) adapter proteins to the CD40 cytoplasmic tail, engagement of CD40 by CD40L triggers activation of an IκB kinase (IKK) complex encompassing two catalytic α and β subunits and one structural γ subunit [36]. Phosphorylation of IκBα by this IKK complex leads to ubiquitination and proteasome-mediated degradation of IκBα, with subsequent translocation of IκBα-free NF-κB dimers from the cytoplasm to the nucleus [38]. Once in the nucleus, NF-κB dimers initiate germline trascription of Cγ and Cε genes by binding to multiple, tandemly arrayed κB sites located within a *cis* CD40-regulatory element associated to both Iγ and Iε promoters [77, 125].

CD40L needs to cooperate with IL-4, a $CD4^+$ T-cell-derived cytokine, to complete switching to IgG and IgE. IL-4 enhances CD40-mediated IgG and IgE class switching by signaling through signal transducer and activator of transcription-6 (STAT-6) [77, 125]. This transcription factor becomes phosphorylated as a result of the activation of IL-4-receptor-associated Jak1, Jak3, and Tyk1 tyrosine kinases by IL-4 [125]. After undergoing dimerization, phosphorylated STAT6 translocates from the cytoplasm to the nucleus, where it initiates germline trascription of Cγ and Cε genes upon binding to a γ-IFN-activated sequence (GAS) located within a relatively conserved region of Iγ and Iε promoters [77, 125, 126]. This IL-4-responsive element is adjacent to the CD40-responsive element, which may explain the synergistic effect of NF-κB and STAT6 on the activation of Iγ and Iε promoters.

Finally, CD40L requires TGF-β1, another $CD4^+$ T cell-derived cytokine, to complete switching to IgA [77]. In this case, induction of NF-κB by CD40 does not play a direct role in the activation of the Iα promoter, which is rather activated by the TGF-β receptor type 2 (TGFβRII) through Smad3 and Smad4 proteins [127]. By cooperating with CREB-ATF, AML, Ets, CBFα, PEBPα, and Runx3 transcription factors, Smad3 and Smad4 activate multiple TGF-β1-responsive (or Smad-responsive) elements within the Iα promoter [127]. Although playing a marginal role in the activation of the Iα promoter, induction of NF-κB by CD40 is critical for IgA class switching in vivo [128, 129]. This observation may reflect the requirement of NF-κB for the expression of AID and for the expansion and survival of IgA class-switched B cells [130, 131]. Indeed, CD40L cooperates with cytokines not only to elicit germline transcription of Cγ, Cα and Cε genes, but also to induce expression of AID.

4.4.4 BAFF and APRIL Signaling in TI Class Switching

BAFF and APRIL appear to be the major inducers of class switching during TI antibody responses. Indeed, in vitro experiments showed that BAFF and APRIL trigger CD40-independent class switching by activating B cells through TACI and, to a lesser extent, BAFF-R [33]. Conversely, BAFF and APRIL do not seem to require BCMA to exert their class switch-inducing activity [33]. The in vivo evidence documenting the key role of TACI and BAFF-R in TI class switching and antibody production has been discussed in a previous paragraph. The signaling events underlying the induction of class switching by TACI and BAFF-R remain largely unclear. However, TACI and BAFF-R may utilize a pathway largely dependent on NF-κB as CD40 does [35].

As already discussed, TACI activates a canonical TRAF-IKK-NF-κB signaling pathway leading to the nuclear translocation of p50, c-Rel, and p65 [35]. In addition to the classical NF-κB pathway, BAFF-R activates an alternative TRAF-NIK-IKKα-NF-κB signaling pathway involving nuclear translocation of p52 and RelB [35]. While p50, p65, and c-Rel are critical for the induction of IgG, IgA, and IgE class switching, p52 and RelB are not, suggesting that BAFF-R induces antibody diversification through the classical pathway as TACI does [33, 37, 39, 131–133]. Consistent with this possibility, BAFF- and APRIL-induced class switching is associated with nuclear translocation of p50, p65, and c-Rel proteins and with subsequent binding of these proteins to a key κB site on the human Iγ3 promoter [16]. Yet, an additional contribution of p52 and RelB proteins cannot be excluded as these proteins also can bind the Iγ3 as well as other Iγ promoters.

TACI and BAFF-R cooperate with IL-4 and/or IL-10 to induce IgG, IgA, and IgE class switching [16, 33]. The molecular basis for this cooperation is likely similar to that underpinning the cooperation between CD40 and cytokines [77]. Interestingly, TGF-β1 does not seem to be required for the induction of IgA by BAFF or APRIL, at least in mouse B cells [33]. However, B cells could release TGF-β1 or another IgA-inducing TGF-β family member in an autocrine fashion upon exposure to BAFF or APRIL [134]. Of note, IL-4, IL-10, and TGF-β1 can be produced not only by $CD4^+$ T cells, but also by innate immune cells. Indeed, IL-4 can be released by basophils and mast cells, whereas DCs, macrophages, and epithelial cells can release IL-10 and TGF-β [135–138]. DCs, macrophages, and epithelial cells can also produce BAFF and APRIL, suggesting that TI IgG, IgA, and IgE responses require a complex crosstalk between B cells and multiple innate immune cell subsets [13, 16, 17].

In human B cells, the combination of BAFF or APRIL and cytokines is sufficient to induce CSR, but not antibody secretion [16]. The latter requires additional signals provided by either BCR or TLRs, indicating that antigen plays an important role in the formation of TI antibody responses [16, 99]. Indeed, signals emanating from BCR and/or TLRs may be necessary for the optimal activation, expansion, and differentiation of BAFF- and APRIL-stimulated B cells lodged within front-line districts. Consistent with this possibility, BCR and TLR ligands are capable of enhancing IgM, IgG, and IgA production as induced by BAFF or APRIL [13, 17].

4.5 Role of BAFF and APRIL in Antibody Disorders

4.5.1 Autoimmune Disorders

BAFF and APRIL are thought to be involved in the establishment and/or maintenance of systemic autoimmune diseases associated with abnormal activation, expansion, and differentiation of self-reactive B cells, including systemic lupus erythematosus (SLE), rheumatoid arthritis, Sjögren syndrome, and multiple sclerosis [139]. In these diseases, there is an abnormally increased production of self-reactive IgG and IgA antibodies, which, in general, are more pathogenic than their IgM counterpart. Indeed, immune complexes containing IgG and IgA, but not IgM autoantibodies are capable of activating powerful pro-inflammatory Fcγ and Fcα receptors expressed on the surface of phagocytic cells, including macrophages and granulocytes. This process would promote tissue damage as a result of a massive release of pro-inflammatory mediators by Fc-activated phagocytes.

In general, BAFF and APRIL are thought to promote autoimmunity due to their ability to increase the survival of autoreactive B cells and plasma cells. Consistent with this possibility, BAFF-transgenic animals develop a lupus-like disease characterized by B-cell hyperplasia and tissue deposition of IgG and IgA autoantibodies [140]. Conversely, autoimmune mice treated with TACI-Ig, a soluble decoy receptor blocking both BAFF and APRIL, undergo peripheral B-cell depletion in concomitance with a remarkable attenuation of disease signs and symptoms [48]. The involvement of BAFF and APRIL in autoimmunity arising in humans is indicated by the presence of elevated levels of BAFF and APRIL in the serum and/or synovial fluids of patients with SLE, rheumatoid arthritis, and Sjögren syndrome [139]. In some of these patients, the serum and/or synovial levels of BAFF and APRIL positively correlate with the serum levels of autoantibodies and/or with the disease activity [141].

Although the pro-survival activity of BAFF and APRIL on autoreactive B cells is certainly central to the establishment, maintenance and/or progression of autoimmunity, additional mechanisms may also be important. Indeed, autoreactive B cells may undergo class switching and subsequent production of highly pathogenic IgG and IgA autoantibodies upon exposure to BAFF and/or APRIL. In agreement with this possibility, BAFF-transgenic animals can develop pathogenic IgG and IgA autoantibodies in the absence of $CD4^+$ T-cell help to B cells via CD40L [140]. The importance of class switching in autoimmunity is further documented by studies showing that AID deficiency ameliorates autoimmune signs and symptoms in a subset of autoimmune-prone mice [142]. Clearly, further studies are warranted to better understand the relative contribution of TD and TI pathways to the autoantibody response developing in different animal models and human disease subsets.

In addition to producing and secreting autoantibodies, pathogenic B cells activate and differentiate autoreactive $CD4^+$ T cells by presenting self-antigen and releasing cytokines. In some animal models, these antibody-independent effects of autoreactive B cells seem to be dominant over the effects triggered by antibody-dependent processes [143]. Interestingly, some studies indicate that BAFF and APRIL might be implicated also in these antibody-independent effects. Indeed, by engaging BCMA on the surface of autoreactive B cells, BAFF and APRIL could upregulate the expression of costimulatory molecules on $CD4^+$ T cells, thereafter enhancing the presentation of self-antigen to autoreactive $CD4^+$ T cells [58].

4.5.2 Immunodeficiencies

Recent studies show that mutations affecting the *TNFRSF13B* gene encoding TACI are associated with antibody-deficiency syndromes, including common variable immunodeficiency (CVID) and selective IgA deficiency (sIgAD) [61, 62, 144]. While CVID causes pan-hypogammaglobulinemia, sIgAD causes a humoral defect restricted to IgA that, in some patients, may later develop into a full-blown CVID. Interestingly, both homozygous and heterozygous mutations of the *TNFRSF13B* gene are associated with disease, implying that TACI deficiency can occur as either an autosomal recessive trait or as an autosomal dominant trait. Heterozygous mutations would impair TACI functions through a dominant-negative mode of action, in which ligand-induced recruitment of a single TACI mutant into the trimeric receptor is sufficient to perturb the signaling complex.

Interestingly, the phenotype of patients with TACI mutations differs from that of genetically engineered mice lacking TACI. These mice develop a massive hyperplasia of peripheral B cells as well as SLE-like signs and symptoms [59]. In contrast, TACI-deficient patients develop neither SLE nor an overt peripheral B-cell hyperplasia. In fact, some of these patients show a reduction of peripheral B cells, particularly memory B cells [62]. Although autoimmune phenomena and to a lesser extent B-cell tumors are rather frequent complications of TACI deficiency, similar complications are common in CVID patients in general and therefore cannot be directly ascribed to *TNFRSF13B* mutations. One possibility is that TACI-deficient mice develop peripheral B-cell hyperplasia as a result of a systemic leakage of the gut microflora secondary to defective TI production of intestinal IgM and IgA antibodies. This bacterial leakage may be less pronounced in TACI-deficient patients.

Similar to TACI-deficient mice, TACI-deficient patients respond poorly to TI antigens, including microbial polysaccharides, thus supporting prior evidence that BAFF and APRIL induce CD40-independent antibody production in human B cells [16]. The association of TACI deficiency with pan-hypogammaglobulinemia may be due to either a role of human TI antibody responses more pervasive than originally anticipated or, more likely, the fact that TACI regulates TD in addition to TI antibody responses. Indeed, engagement of TACI and BAFF-R by BAFF may be important to stimulate the plasmacytoid differentiation of memory B cells, a canonical product of the GC reaction [29, 145]. Alternatively, TACI may regulate essential aspects of CD40 signaling in B cells.

4.6 Conclusions

The discovery of BAFF and APRIL has provided a wealth of new information about the immune system. It is now recognized that BAFF and APRIL play an important role not only in B-cell survival and tolerance, but also in B-cell maturation and diversification, including IgG and IgA class switching. The association of BAFF and APRIL with autoimmune diseases associated with pathogenic IgG and IgA production has led to widespread efforts to develop agents capable of neutralizing these innate B-cell-stimulating ligands. TACI-Ig and a human monoclonal antibody specific for BAFF are currently being tested in clinical trials with SLE patients. These and other therapeutic agents may achieve clinically significant results not only by reducing the survival of autoreactive B cells, but also by dampening autoreactive class switching. Finally, targeting the BAFF–APRIL system may also represent a valuable option for the treatment of humoral immunodeficiencies. Although initial studies suggested the use of recombinant BAFF and APRIL to increase antibody production in CVID patients, this strategy may actually involve a significant risk of developing autoimmunity. In addition, CVID and sIgAD patients with TACI mutations may be poorly responsive to BAFF and/or APRIL. Clearly, further efforts are needed to circumvent these limitations and devise better therapeutic options.

References

1. Mackay F, Schneider P, Rennert P, Browning J. BAFF and APRIL: a tutorial on B cell survival. Annu Rev Immunol 2003;21:231–64.
2. Dillon SR, Gross JA, Ansell SM, Novak AJ. An APRIL to remember: novel TNF ligands as therapeutic targets. Nat Rev Drug Discov 2006;5:235–46.
3. Schneider P, MacKay F, Steiner V, Hofmann K, Bodmer JL, Holler N, Ambrose C, Lawton P, Bixler S, Acha-Orbea H, et al. BAFF, a novel ligand of the tumor necrosis factor family, stimulates B cell growth. J Exp Med 1999;189:1747–56.

4. Hahne M, Kataoka T, Schroter M, Hofmann K, Irmler M, Bodmer JL, Schneider P, Bornand T, Holler N, French LE, et al. APRIL, a new ligand of the tumor necrosis factor family, stimulates tumor cell growth. J Exp Med 1998;188:1185–90.
5. Roschke V, Sosnovtseva S, Ward CD, Hong JS, Smith R, Albert V, Stohl W, Baker KP, Ullrich S, Nardelli B, et al. BLyS and APRIL form biologically active heterotrimers that are expressed in patients with systemic immune-based rheumatic diseases. J Immunol 2002;169:4314–21.
6. Lopez-Fraga M, Fernandez R, Albar JP, Hahne M. Biologically active APRIL is secreted following intracellular processing in the Golgi apparatus by furin convertase. EMBO Rep 2001;2:945–51.
7. Nardelli B, Belvedere O, Roschke V, Moore PA, Olsen HS, Migone TS, Sosnovtseva S, Carrell JA, Feng P, Giri JG, et al. Synthesis and release of B-lymphocyte stimulator from myeloid cells. Blood 2001;97:198–204.
8. Pradet-Balade B, Medema JP, Lopez-Fraga M, Lozano JC, Kolfschoten GM, Picard A, Martinez AC, Garcia-Sanz JA, Hahne M. An endogenous hybrid mRNA encodes TWE-PRIL, a functional cell surface TWEAK-APRIL fusion protein. Embo J 2002;21:5711–20.
9. Kolfschoten GM, Pradet-Balade B, Hahne M, Medema JP. TWE-PRIL; a fusion protein of TWEAK and APRIL. Biochem Pharmacol 2003;66:1427–32.
10. Gorelik L, Gilbride K, Dobles M, Kalled SL, Zandman D, Scott ML. Normal B cell homeostasis requires B cell activation factor production by radiation-resistant cells. J Exp Med 2003;198:937–45.
11. Lesley R, Xu Y, Kalled SL, Hess DM, Schwab SR, Shu HB, Cyster JG. Reduced competitiveness of autoantigen-engaged B cells due to increased dependence on BAFF. Immunity 2004;20:441–53.
12. Schneider P. The role of APRIL and BAFF in lymphocyte activation. Curr Opin Immunol 2005;17:282–9.
13. Xu W, He B, Chiu A, Chadburn A, Shan M, Buldys M, Ding A, Knowles DM, Santini PA, Cerutti A. Epithelial cells trigger frontline immunoglobulin class switching through a pathway regulated by the inhibitor SLPI. Nat Immunol 2007;8:294–303.
14. Krumbholz M, Theil D, Derfuss T, Rosenwald A, Schrader F, Monoranu CM, Kalled SL, Hess DM, Serafini B, Aloisi F, et al. BAFF is produced by astrocytes and up-regulated in multiple sclerosis lesions and primary central nervous system lymphoma. J Exp Med 2005;201:195–200.
15. Mackay F, Ambrose C. The TNF family members BAFF and APRIL: the growing complexity. Cytokine Growth Factor Rev 2003;14:311–24.
16. Litinskiy MB, Nardelli B, Hilbert DM, He B, Schaffer A, Casali P, Cerutti A. DCs induce CD40-independent immunoglobulin class switching through BLyS and APRIL. Nat Immunol 2002;3:822–9.
17. He B, Xu W, Santini PA, Polydorides AD, Chiu A, Estrella J, Shan M, Chadburn A, Villanacci V, Plebani A, et al. Intestinal bacteria trigger T cell-independent immunoglobulin A(2) class switching by inducing epithelial-cell secretion of the cytokine APRIL. Immunity 2007;26:812–26.
18. Hardenberg G, Planelles L, Schwarte CM, van Bostelen L, Le Huong T, Hahne M, Medema JP. Specific TLR ligands regulate APRIL secretion by dendritic cells in a PKR-dependent manner. Eur J Immunol 2007;37:2900–11.
19. Pasare C, Medzhitov R. Toll-like receptors: linking innate and adaptive immunity. Microbes Infect 2004;6:1382–7.
20. Kato A, Truong-Tran AQ, Scott AL, Matsumoto K, Schleimer RP. Airway epithelial cells produce B cell-activating factor of TNF family by an IFN-beta-dependent mechanism. J Immunol 2006;177:7164–72.
21. He B, Qiao X, Klasse PJ, Chiu A, Chadburn A, Knowles DM, Moore JP, Cerutti A. HIV-1 envelope triggers polyclonal Ig class switch recombination through a CD40-independent mechanism involving BAFF and C-type lectin receptors. J Immunol 2006;176: 3931–41.

22. Stein JV, Lopez-Fraga M, Elustondo FA, Carvalho-Pinto CE, Rodriguez D, Gomez-Caro R, De Jong J, Martinez AC, Medema JP, Hahne M. APRIL modulates B and T cell immunity. J Clin Invest 2002;109:1587–98.
23. Ng LG, Sutherland AP, Newton R, Qian F, Cachero TG, Scott ML, Thompson JS, Wheway J, Chtanova T, Groom J, et al. B cell-activating factor belonging to the TNF family (BAFF)-R is the principal BAFF receptor facilitating BAFF costimulation of circulating T and B cells. J Immunol 2004;173:807–17.
24. Huard B, Arlettaz L, Ambrose C, Kindler V, Mauri D, Roosnek E, Tschopp J, Schneider P, French LE. BAFF production by antigen-presenting cells provides T cell co-stimulation. Int Immunol 2004;16:467–75.
25. Schiemann B, Gommerman JL, Vora K, Cachero TG, Shulga-Morskaya S, Dobles M, Frew E, Scott ML. An essential role for BAFF in the normal development of B cells through a BCMA-independent pathway. Science 2001;293:2111–4.
26. Ingold K, Zumsteg A, Tardivel A, Huard B, Steiner QG, Cachero TG, Qiang F, Gorelik L, Kalled SL, Acha-Orbea H, et al. Identification of proteoglycans as the APRIL-specific binding partners. J Exp Med 2005;201:1375–83.
27. Hendriks J, Planelles L, de Jong-Odding J, Hardenberg G, Pals ST, Hahne M, Spaargaren M, Medema JP. Heparan sulfate proteoglycan binding promotes APRIL-induced tumor cell proliferation. Cell Death Differ 2005;12:637–48.
28. Chiu A, Xu W, He B, Dillon SR, Gross JA, Sievers E, Qiao X, Santini P, Hyjek E, Lee JW, et al. Hodgkin lymphoma cells express TACI and BCMA receptors and generate survival and proliferation signals in response to BAFF and APRIL. Blood 2007;109:729–39.
29. Darce JR, Arendt BK, Wu X, Jelinek DF. Regulated expression of BAFF-binding receptors during human B cell differentiation. J Immunol 2007;179:7276–86.
30. Hsu BL, Harless SM, Lindsley RC, Hilbert DM, Cancro MP. Cutting edge: BLyS enables survival of transitional and mature B cells through distinct mediators. J Immunol 2002;168:5993–6.
31. Smith SH, Cancro MP. Cutting edge: B cell receptor signals regulate BLyS receptor levels in mature B cells and their immediate progenitors. J Immunol 2003;170:5820–3.
32. Treml LS, Carlesso G, Hoek KL, Stadanlick JE, Kambayashi T, Bram RJ, Cancro MP, Khan WN. TLR stimulation modifies BLyS receptor expression in follicular and marginal zone B cells. J Immunol 2007;178:7531–9.
33. Castigli E, Wilson SA, Scott S, Dedeoglu F, Xu S, Lam KP, Bram RJ, Jabara H, Geha RS. TACI and BAFF-R mediate isotype switching in B cells. J Exp Med 2005;201:35–9.
34. O'Connor BP, Raman VS, Erickson LD, Cook WJ, Weaver LK, Ahonen C, Lin LL, Mantchev GT, Bram RJ, Noelle RJ. BCMA is essential for the survival of long-lived bone marrow plasma cells. J Exp Med 2004;199:91–8.
35. Bossen C, Schneider P. BAFF, APRIL and their receptors: structure, function and signaling. Semin Immunol 2006;18:263–75.
36. Bishop GA, Moore CR, Xie P, Stunz LL, Kraus ZJ. TRAF proteins in CD40 signaling. Adv Exp Med Biol 2007;597:131–51.
37. Xia XZ, Treanor J, Senaldi G, Khare SD, Boone T, Kelley M, Theill LE, Colombero A, Solovyev I, Lee F, et al. TACI is a TRAF-interacting receptor for TALL-1, a tumor necrosis factor family member involved in B cell regulation. J Exp Med 2000;192:137–43.
38. Karin M, Greten FR. NF-kappaB: linking inflammation and immunity to cancer development and progression. Nat Rev Immunol 2005;5:749–59.
39. von Bulow GU, Bram RJ. NF-AT activation induced by a CAML-interacting member of the tumor necrosis factor receptor superfamily. Science 1997;278:138–41.
40. Hatzoglou A, Roussel J, Bourgeade MF, Rogier E, Madry C, Inoue J, Devergne O, Tsapis A. TNF receptor family member BCMA (B cell maturation) associates with TNF receptor-associated factor (TRAF) 1, TRAF2, and TRAF3 and activates NF-kappa B, elk-1, c-Jun N-terminal kinase, and p38 mitogen-activated protein kinase. J Immunol 2000;165:1322–30.
41. Ni CZ, Oganesyan G, Welsh K, Zhu X, Reed JC, Satterthwait AC, Cheng G, Ely KR. Key molecular contacts promote recognition of the BAFF receptor by TNF receptor-

associated factor 3: implications for intracellular signaling regulation. J Immunol 2004;173: 7394–400.
42. Morrison MD, Reiley W, Zhang M, Sun SC. An atypical tumor necrosis factor (TNF) receptor-associated factor-binding motif of B cell-activating factor belonging to the TNF family (BAFF) receptor mediates induction of the noncanonical NF-kappaB signaling pathway. J Biol Chem 2005;280:10018–24.
43. Kayagaki N, Yan M, Seshasayee D, Wang H, Lee W, French DM, Grewal IS, Cochran AG, Gordon NC, Yin J, et al. BAFF/BLyS receptor 3 binds the B cell survival factor BAFF ligand through a discrete surface loop and promotes processing of NF-kappaB2. Immunity 2002;17:515–24.
44. Claudio E, Brown K, Park S, Wang H, Siebenlist U. BAFF-induced NEMO-independent processing of NF-kappa B2 in maturing B cells. Nat Immunol 2002;3:958–65.
45. He JQ, Saha SK, Kang JR, Zarnegar B, Cheng G. Specificity of TRAF3 in its negative regulation of the noncanonical NF-kappa B pathway. J Biol Chem 2007;282:3688–94.
46. Qian Y, Qin J, Cui G, Naramura M, Snow EC, Ware CF, Fairchild RL, Omori SA, Rickert RC, Scott M, et al. Act1, a negative regulator in CD40- and BAFF-mediated B cell survival. Immunity 2004;21:575–87.
47. Thompson JS, Schneider P, Kalled SL, Wang L, Lefevre EA, Cachero TG, MacKay F, Bixler SA, Zafari M, Liu ZY, et al. BAFF binds to the tumor necrosis factor receptor-like molecule B cell maturation antigen and is important for maintaining the peripheral B cell population. J Exp Med 2000;192:129–35.
48. Gross JA, Dillon SR, Mudri S, Johnston J, Littau A, Roque R, Rixon M, Schou O, Foley KP, Haugen H, et al. TACI-Ig neutralizes molecules critical for B cell development and autoimmune disease. impaired B cell maturation in mice lacking BLyS. Immunity 2001;15:289–302.
49. Mackay F, Woodcock SA, Lawton P, Ambrose C, Baetscher M, Schneider P, Tschopp J, Browning JL. Mice transgenic for BAFF develop lymphocytic disorders along with autoimmune manifestations. J Exp Med 1999;190:1697–710.
50. He B, Chadburn A, Jou E, Schattner EJ, Knowles DM, Cerutti A. Lymphoma B cells evade apoptosis through the TNF family members BAFF/BLyS and APRIL. J Immunol 2004;172:3268–79.
51. Hatada EN, Do RK, Orlofsky A, Liou HC, Prystowsky M, MacLennan IC, Caamano J, Chen-Kiang S. NF-kappa B1 p50 is required for BLyS attenuation of apoptosis but dispensable for processing of NF-kappa B2 p100 to p52 in quiescent mature B cells. J Immunol 2003;171:761–8.
52. Varfolomeev E, Kischkel F, Martin F, Seshasayee D, Wang H, Lawrence D, Olsson C, Tom L, Erickson S, French D, et al. APRIL-deficient mice have normal immune system development. Mol Cell Biol 2004;24:997–1006.
53. Planelles L, Carvalho-Pinto CE, Hardenberg G, Smaniotto S, Savino W, Gomez-Caro R, Alvarez-Mon M, de Jong J, Eldering E, Martinez AC, et al. APRIL promotes B-1 cell-associated neoplasm. Cancer Cell 2004;6:399–408.
54. Gorelik L, Cutler AH, Thill G, Miklasz SD, Shea DE, Ambrose C, Bixler SA, Su L, Scott ML, Kalled SL. Cutting edge: BAFF regulates CD21/35 and CD23 expression independent of its B cell survival function. J Immunol 2004;172:762–6.
55. Ng LG, Ng CH, Woehl B, Sutherland AP, Huo J, Xu S, Mackay F, Lam KP. BAFF costimulation of Toll-like receptor-activated B-1 cells. Eur J Immunol 2006;36: 1837–46.
56. Huang X, Di Liberto M, Cunningham AF, Kang L, Cheng S, Ely S, Liou HC, Maclennan IC, Chen-Kiang S. Homeostatic cell-cycle control by BLyS: induction of cell-cycle entry but not G1/S transition in opposition to p18INK4c and p27Kip1. Proc Natl Acad Sci USA 2004;101:17789–94.
57. Rahman ZS, Rao SP, Kalled SL, Manser T. Normal induction but attenuated progression of germinal center responses in BAFF and BAFF-R signaling-deficient mice. J Exp Med 2003;198:1157–69.

58. Yang M, Hase H, Legarda-Addison D, Varughese L, Seed B, Ting AT. B cell maturation antigen, the receptor for a proliferation-inducing ligand and B cell-activating factor of the TNF family, induces antigen presentation in B cells. J Immunol 2005;175: 2814–24.
59. Yan M, Wang H, Chan B, Roose-Girma M, Erickson S, Baker T, Tumas D, Grewal IS, Dixit VM. Activation and accumulation of B cells in TACI-deficient mice. Nat Immunol 2001;2:638–43.
60. von Bulow GU, van Deursen JM, Bram RJ. Regulation of the T-independent humoral response by TACI. Immunity 2001;14:573–82.
61. Castigli E, Wilson SA, Garibyan L, Rachid R, Bonilla F, Schneider L, Geha RS. TACI is mutant in common variable immunodeficiency and IgA deficiency. Nat Genet 2005;37:829–34.
62. Salzer U, Chapel HM, Webster AD, Pan-Hammarstrom Q, Schmitt-Graeff A, Schlesier M, Peter HH, Rockstroh JK, Schneider P, Schaffer AA, et al. Mutations in TNFRSF13B encoding TACI are associated with common variable immunodeficiency in humans. Nat Genet 2005;37:820–8.
63. Castigli E, Scott S, Dedeoglu F, Bryce P, Jabara H, Bhan AK, Mizoguchi E, Geha RS. Impaired IgA class switching in APRIL-deficient mice. Proc Natl Acad Sci USA 2004;101:3903–8.
64. MacLennan I, Vinuesa C. Dendritic cells, BAFF, and APRIL: innate players in adaptive antibody responses. Immunity 2002;17:235–8.
65. Banchereau J, Briere F, Caux C, Davoust J, Lebecque S, Liu YJ, Pulendran B, Palucka K. Immunobiology of dendritic cells. Annu Rev Immunol 2000;18:767–811.
66. Janeway CA, Jr, Medzhitov R. Innate immune recognition. Annu Rev Immunol 2002;20:197–216.
67. Guermonprez P, Valladeau J, Zitvogel L, Thery C, Amigorena S. Antigen presentation and T cell stimulation by dendritic cells. Annu Rev Immunol 2002;20:621–67.
68. Pulendran B, Palucka K, Banchereau J. Sensing pathogens and tuning immune responses. Science 2001;293:253–6.
69. Laurence A, O'Shea JJ. T(H)-17 differentiation of mice and men. Nat Immunol 2007;8:903–5.
70. Quezada SA, Jarvinen LZ, Lind EF, Noelle RJ. CD40/CD154 interactions at the interface of tolerance and immunity. Annu Rev Immunol 2004;22:307–28.
71. Jego G, Palucka AK, Blanck JP, Chalouni C, Pascual V, Banchereau J. Plasmacytoid dendritic cells induce plasma cell differentiation through type I interferon and interleukin 6. Immunity 2003;19:225–34.
72. Cerutti A, Qiao X, He B. Plasmacytoid dendritic cells and the regulation of immunoglobulin heavy chain class switching. Immunol Cell Biol 2005;83:554–62.
73. Liu YJ. Dendritic cell subsets and lineages, and their functions in innate and adaptive immunity. Cell 2001;106:259–62.
74. Siegal FP, Kadowaki N, Shodell M, Fitzgerald-Bocarsly PA, Shah K, Ho S, Antonenko S, Liu YJ. The nature of the principal type 1 interferon-producing cells in human blood. Science 1999;284:1835–7.
75. Asselin-Paturel C, Boonstra A, Dalod M, Durand I, Yessaad N, Dezutter-Dambuyant C, Vicari A, O'Garra A, Biron C, Briere F, et al. Mouse type I IFN-producing cells are immature APCs with plasmacytoid morphology. Nat Immunol 2001;2:1144–50.
76. Cella M, Jarrossay D, Facchetti F, Alebardi O, Nakajima H, Lanzavecchia A, Colonna M. Plasmacytoid monocytes migrate to inflamed lymph nodes and produce large amounts of type I interferon. Nat Med 1999;5:919–23.
77. Stavnezer J. Antibody class switching. Adv Immunol 1996;61:79–146.
78. Liu YJ. IPC: professional type 1 interferon-producing cells and plasmacytoid dendritic cell precursors. Annu Rev Immunol 2005;23:275–306.
79. Kadowaki N, Liu YJ. Natural type I interferon-producing cells as a link between innate and adaptive immunity. Hum Immunol 2002;63:1126–32.

80. Cyster JG. Leukocyte migration: scent of the T zone. Curr Biol 2000;10:R30–3.
81. Lanzavecchia A. Antigen-specific interaction between T and B cells. Nature 1985;314:537–9.
82. Allen CD, Okada T, Cyster JG. Germinal-center organization and cellular dynamics. Immunity 2007;27:190–202.
83. Odegard VH, Schatz DG. Targeting of somatic hypermutation. Nat Rev Immunol 2006;6:573–83.
84. Chaudhuri J, Alt FW. Class-switch recombination: interplay of transcription, DNA deamination and DNA repair. Nat Rev Immunol 2004; 4:541–52.
85. Zhang X, Park CS, Yoon SO, Li L, Hsu YM, Ambrose C, Choi YS. BAFF supports human B cell differentiation in the lymphoid follicles through distinct receptors. Int Immunol 2005;17:779–88.
86. Do RK, Hatada E, Lee H, Tourigny MR, Hilbert D, Chen-Kiang S. Attenuation of apoptosis underlies B lymphocyte stimulator enhancement of humoral immune response. J Exp Med 2000;192:953–64.
87. Moore PA, Belvedere O, Orr A, Pieri K, LaFleur DW, Feng P, Soppet D, Charters M, Gentz R, Parmelee D, et al. BLyS: member of the tumor necrosis factor family and B lymphocyte stimulator. Science 1999;285:260–3.
88. Sakurai D, Kanno Y, Hase H, Kojima H, Okumura K, Kobata T. TACI attenuates antibody production costimulated by BAFF-R and CD40. Eur J Immunol 2007;37:110–8.
89. Seshasayee D, Valdez P, Yan M, Dixit VM, Tumas D, Grewal IS. Loss of TACI causes fatal lymphoproliferation and autoimmunity, establishing TACI as an inhibitory BLyS receptor. Immunity 2003;18:279–88.
90. MacLennan IC. Germinal centers. Annu Rev Immunol 1994;12:117–39.
91. Mond JJ, Vos Q, Lees A, Snapper CM. T cell independent antigens. Curr Opin Immunol 1995;7:349–54.
92. Lenert P, Brummel R, Field EH, Ashman RF. TLR-9 activation of marginal zone B cells in lupus mice regulates immunity through increased IL-10 production. J Clin Immunol 2005;25:29–40.
93. Genestier L, Taillardet M, Mondiere P, Gheit H, Bella C, Defrance T. TLR agonists selectively promote terminal plasma cell differentiation of B cell subsets specialized in thymus-independent responses. J Immunol 2007;178:7779–86.
94. Oliver AM, Martin F, Kearney JF. IgMhighCD21high lymphocytes enriched in the splenic marginal zone generate effector cells more rapidly than the bulk of follicular B cells. J Immunol 1999;162:7198–207.
95. Martin F, Kearney JF. Marginal-zone B cells. Nat Rev Immunol 2002;2:323–35.
96. Bendelac A, Bonneville M, Kearney JF. Autoreactivity by design: innate B and T lymphocytes. Nat Rev Immunol 2001;1:177–86.
97. Fagarasan S, Honjo T. Intestinal IgA synthesis: regulation of front-line body defences. Nat Rev Immunol 2003;3:63–72.
98. Katsenelson N, Kanswal S, Puig M, Mostowski H, Verthelyi D, Akkoyunlu M. Synthetic CpG oligodeoxynucleotides augment BAFF- and APRIL-mediated immunoglobulin secretion. Eur J Immunol 2007;37:1785–95.
99. He B, Qiao X, Cerutti A. CpG DNA induces IgG class switch DNA recombination by activating human B cells through an innate pathway that requires TLR9 and cooperates with IL-10. J Immunol 2004;173:4479–91.
100. Hardy RR. B-1 B cells: development, selection, natural autoantibody and leukemia. Curr Opin Immunol 2006;18:547–55.
101. Weller S, Braun MC, Tan BK, Rosenwald A, Cordier C, Conley ME, Plebani A, Kumararatne DS, Bonnet D, Tournilhac O, et al. Human blood IgM "memory" B cells are circulating splenic marginal zone B cells harboring a prediversified immunoglobulin repertoire. Blood 2004;104:3647–54.
102. Le Bon A, Schiavoni G, D'Agostino G, Gresser I, Belardelli F, Tough DF. Type i interferons potently enhance humoral immunity and can promote isotype switching by stimulating dendritic cells in vivo. Immunity 2001;14:461–70.

103. Balazs M, Martin F, Zhou T, Kearney J. Blood dendritic cells interact with splenic marginal zone B cells to initiate T-independent immune responses. Immunity 2002;17:341–52.
104. Bergtold A, Desai DD, Gavhane A, Clynes R. Cell surface recycling of internalized antigen permits dendritic cell priming of B cells. Immunity 2005;23:503–14.
105. Vinuesa CG, Goodnow CC. Immunology: DNA drives autoimmunity. Nature 2002;416:595–8.
106. Peng SL. Signaling in B cells via Toll-like receptors. Curr Opin Immunol 2005;17:230–6.
107. Szomolanyi-Tsuda E, Welsh RM. T-cell-independent antiviral antibody responses. Curr Opin Immunol 1998;10:431–5.
108. Macpherson AJ. IgA adaptation to the presence of commensal bacteria in the intestine. Curr Top Microbiol Immunol 2006;308:117–36.
109. Lee BO, Rangel-Moreno J, Moyron-Quiroz JE, Hartson L, Makris M, Sprague F, Lund FE, Randall TD. CD4 T cell-independent antibody response promotes resolution of primary influenza infection and helps to prevent reinfection. J Immunol 2005;175:5827–38.
110. Lopes-Carvalho T, Foote J, Kearney JF. Marginal zone B cells in lymphocyte activation and regulation. Curr Opin Immunol 2005;17:244–50.
111. Zandvoort A, Timens W. The dual function of the splenic marginal zone: essential for initiation of anti-TI-2 responses but also vital in the general first-line defense against blood-borne antigens. Clin Exp Immunol 2002;130:4–11.
112. Sumaraju V, Smith LG, Smith SM. Infectious complications in asplenic hosts. Infect Dis Clin North Am 2001;15:551–65, x.
113. Xu J, Foy TM, Laman JD, Elliott EA, Dunn JJ, Waldschmidt TJ, Elsemore J, Noelle RJ, Flavell RA. Mice deficient for the CD40 ligand. Immunity 1994;1:423–31.
114. Renshaw BR, Fanslow WC, III, Armitage RJ, Campbell KA, Liggitt D, Wright B, Davison BL, Maliszewski CR. Humoral immune responses in CD40 ligand-deficient mice. J Exp Med 1994;180:1889–900.
115. Ferrari S, Giliani S, Insalaco A, Al-Ghonaium A, Soresina AR, Loubser M, Avanzini MA, Marconi M, Badolato R, Ugazio AG, et al. Mutations of CD40 gene cause an autosomal recessive form of immunodeficiency with hyper IgM. Proc Natl Acad Sci USA 2001;98:12614–9.
116. Jain A, Ma CA, Lopez-Granados E, Means G, Brady W, Orange JS, Liu S, Holland S, Derry JM. Specific NEMO mutations impair CD40-mediated c-Rel activation and B cell terminal differentiation. J Clin Invest 2004;114:1593–602.
117. Oliver AM, Martin F, Gartland GL, Carter RH, Kearney JF. Marginal zone B cells exhibit unique activation, proliferative and immunoglobulin secretory responses. Eur J Immunol 1997;27:2366–74.
118. Woodland RT, Schmidt MR, Thompson CB. BLyS and B cell homeostasis. Semin Immunol 2006;18:318–26.
119. Muramatsu M, Kinoshita K, Fagarasan S, Yamada S, Shinkai Y, Honjo T. Class switch recombination and hypermutation require activation-induced cytidine deaminase (AID), a potential RNA editing enzyme. Cell 2000;102:553–63.
120. Thompson JS, Bixler SA, Qian F, Vora K, Scott ML, Cachero TG, Hession C, Schneider P, Sizing ID, Mullen C, et al. BAFF-R, a newly identified TNF receptor that specifically interacts with BAFF. Science 2001;293:2108–11.
121. Miller DJ, Hanson KD, Carman JA, Hayes CE. A single autosomal gene defect severely limits IgG but not IgM responses in B lymphocyte-deficient A/WySnJ mice. Eur J Immunol 1992;22:373–9.
122. Hardy RR. B-Lymphocyte Development and Biology. *In Fundamental Immunology*, WE. Paul, editor. Philadelphia Lippincott Williams & Wilkins, 2003, pp. 159–94.
123. Honjo T, Kinoshita K, Muramatsu M. Molecular mechanism of class switch recombination: linkage with somatic hypermutation. Annu Rev Immunol 2002;20:165–96.
124. Pan-Hammarstrom Q, Zhao Y, Hammarstrom L. Class switch recombination: a comparison between mouse and human. Adv Immunol 2007;93:1–61.
125. Geha RS, Jabara HH, Brodeur SR. The regulation of immunoglobulin E class-switch recombination. Nat Rev Immunol 2003;3:721–32.

126. Schaffer A, Cerutti A, Shah S, Zan H, Casali P. The evolutionarily conserved sequence upstream of the human Ig heavy chain S gamma 3 region is an inducible promoter: synergistic activation by CD40 ligand and IL-4 via cooperative NF-kappa B and STAT-6 binding sites. J Immunol 1999;162:5327–36.
127. Johansen FE, Brandtzaeg P. Transcriptional regulation of the mucosal IgA system. Trends Immunol 2004;25:150–7.
128. Doi TS, Takahashi T, Taguchi O, Azuma T, Obata Y. NF-kappa B RelA-deficient lymphocytes: normal development of T cells and B cells, impaired production of IgA and IgG1 and reduced proliferative responses. J Exp Med 1997;185:953–61.
129. Snapper CM, Zelazowski P, Rosas FR, Kehry MR, Tian M, Baltimore D, Sha WC. B cells from p50/NF-kappa B knockout mice have selective defects in proliferation, differentiation, germ-line CH transcription, and Ig class switching. J Immunol 1996;156:183–91.
130. Dedeoglu F, Horwitz B, Chaudhuri J, Alt FW, Geha RS. Induction of activation-induced cytidine deaminase gene expression by IL-4 and CD40 ligation is dependent on STAT6 and NFkappaB. Int Immunol 2004;16:395–404.
131. Sha WC, Liou HC, Tuomanen EI, Baltimore D. Targeted disruption of the p50 subunit of NF-kappa B leads to multifocal defects in immune responses. Cell 1995;80: 321–30.
132. Enzler T, Bonizzi G, Silverman GJ, Otero DC, Widhopf GF, Anzelon-Mills A, Rickert RC, Karin M. Alternative and classical NF-kappa B signaling retain autoreactive B cells in the splenic marginal zone and result in lupus-like disease. Immunity 2006;25:403–15.
133. Yu G, Boone T, Delaney J, Hawkins N, Kelley M, Ramakrishnan M, McCabe S, Qiu WR, Kornuc M, Xia XZ, et al. APRIL and TALL-I and receptors BCMA and TACI: system for regulating humoral immunity. Nat Immunol 2000;1:252–6.
134. Zan H, Cerutti A, Dramitinos P, Schaffer A, Casali P. CD40 engagement triggers switching to IgA1 and IgA2 in human B cells through induction of endogenous TGF-beta: evidence for TGF-beta but not IL-10-dependent direct S mu→S alpha and sequential S mu→S gamma, S gamma→S alpha DNA recombination. J Immunol 1998;161:5217–25.
135. Kawakami T, Galli SJ. Regulation of mast-cell and basophil function and survival by IgE. Nat Rev Immunol 2002;2:773–86.
136. Reis e Sousa C. Dendritic cells in a mature age. Nat Rev Immunol 2006;6:476–83.
137. Gordon S, Taylor PR. Monocyte and macrophage heterogeneity. Nat Rev Immunol 2005;5:953–64.
138. Sansonetti PJ. War and peace at mucosal surfaces. Nat Rev Immunol 2004;4:953–64.
139. Mackay F, Silveira PA, Brink R. B cells and the BAFF/APRIL axis: fast-forward on autoimmunity and signaling. Curr Opin Immunol 2007;19:327–36.
140. Groom JR, Fletcher CA, Walters SN, Grey ST, Watt SV, Sweet MJ, Smyth MJ, Mackay CR, Mackay F. BAFF and MyD88 signals promote a lupuslike disease independent of T cells. J Exp Med 2007;204:1959–71.
141. Cancro MP. The BLyS/BAFF family of ligands and receptors: key targets in the therapy and understanding of autoimmunity. Ann Rheum Dis 2006;65(Suppl 3):iii34–6.
142. Jiang C, Foley J, Clayton N, Kissling G, Jokinen M, Herbert R, Diaz M. Abrogation of lupus nephritis in activation-induced deaminase-deficient MRL/lpr mice. J Immunol 2007;178:7422–31.
143. Chan OT, Hannum LG, Haberman AM, Madaio MP, Shlomchik MJ. A novel mouse with B cells but lacking serum antibody reveals an antibody-independent role for B cells in murine lupus. J Exp Med 1999;189:1639–48.
144. Castigli E, Geha RS. TACI, isotype switching, CVID and IgAD. Immunol Res 2007;38:102–11.
145. Avery DT, Kalled SL, Ellyard JI, Ambrose C, Bixler SA, Thien M, Brink R, Mackay F, Hodgkin PD, Tangye SG. BAFF selectively enhances the survival of plasmablasts generated from human memory B cells. J Clin Invest 2003;112:286–97.

Chapter 5
Signal Transduction by Receptors for BAFF and APRIL

Joanne M. Hildebrand, Ping Xie, and Gail A. Bishop

Abstract The cytokines B-cell-activating factor of the TNF family (BAFF) and A proliferation-inducing ligand (APRIL) stimulate several of the major signaling cascades responsible for B-cell homeostasis. Transmitting these stimuli to the intracellular milieu are the receptors B-cell maturation antigen (BCMA), transmembrane activator and CAML interactor (TACI), and BAFF receptor (BAFFR). While evidence for the activation of several nuclear-independent pathways is accumulating, the most consistent, reproducibly detectable molecular effect of stimulating cells with BAFF or APRIL is activation of the NF-κB family of transcription factors. Here, we will summarize the current state of knowledge about BAFF and APRIL signaling; from the membrane-proximal events that are distinct to BCMA, TACI, and BAFFR to the differences in the downstream pathway types and kinetics that are manifest in the phenotypic extremes of their respective knockout mouse models.

Keywords BAFF · APRIL · BAFFR · TACI · BCMA · TRAF · NF-κB · B lymphocyte · Signal transduction

5.1 Proximal Signaling by BAFF and APRIL Receptors

5.1.1 Recruitment of Adaptor Molecules

Like other members of the tumor necrosis factor receptor (TNFR) superfamily, BCMA, TACI, and BAFFR lack intrinsic enzymatic activity and thus initiate signal propagation by recruiting various adaptor molecules to their cytoplasmic domains. While several TNFR-associated factors (TRAFs), CAML, and Act1 have been shown to interact with these receptors, there is a paucity of information regarding the precise roles played by each adaptor protein in APRIL and BAFF signaling. It

G.A. Bishop (✉)
Department of Microbiology, The University of Iowa, Iowa City, IA 52242, USA

M.P. Cancro (ed.), *BLyS Ligands and Receptors,* Contemporary Immunology,
DOI 10.1007/978-1-60327-013-7_5,

is difficult to ascribe functions to the adaptor proteins recruited by BAFFR, TACI, and BCMA through comparison with more thoroughly studied TNFR family members, given the growing evidence that their functions are very much context dependent [1, 2]. It also remains to be seen whether adaptor protein interaction data to date, obtained mostly through use of tagged receptors and/or adaptor molecules that are overexpressed in transformed epithelial cell lines, are physiologically relevant (Table 5.1).

Table 5.1 Proteins that associate with the cytoplasmic domains of BAFFR, TACI, and BCMA

Receptor	Interacting protein/s	Experimental system	References
BCMA	TRAF6 and TRAF5	HEK 293 cells	[13]
	TRAFs1, 2 and 3, 4 and 5	COS7 cells	[12]
TACI	CAML	Yeast two hybrid and HEK 293 cells	[31]
	TRAFs 2, 5 and 6	Yeast two hybrid	[14]
	CAML	HEK 293 cells	[14]
	TRAF3	Bjab B-cell line	[87]
	TRAF2 and TRAF3	A20 B-cell line	K. Larison and J. Hildebrand, (unpublished observations)
	TRAF3	Murine primary B splenocytes	
BAFFR	TRAF3	HEK 293 cells and Bjab B-cell line	[87]
	TRAF3	M12 B-cell line.	[5]
	TRAF3	A20 B-cell line and murine primary B splenocytes	K. Larison and J. Hildebrand, (unpublished observations)
	Act1	Human IM9 B-cell line and Human primary lymphocytes	[30]

5.1.2 TRAF-Binding Sites in the TNF Receptor Family

TRAFs are cytoplasmic proteins that act as adaptors between TNFR and interleukin-1 receptor/Toll-like receptor (IL-1R/TLR) superfamily members and their downstream signaling pathways. The six known TRAFs (TRAFs 1, 2, 3, 4, 5, and 6) contain a coiled-coil, leucine zipper, or "TRAF-N" domain which mediates their characteristic homo- or heterotrimerization. Zinc finger and zinc ring domains in all but TRAF1 provide the physical link to proteins at the threshold of several signaling pathways. TRAFs are recruited to distinct amino acid motifs in the intracellular portion of TNFR and IL-1R/TLR receptors via their carboxy terminal "TRAF-C" domains (reviewed by [3]). Comparison of the regions of TNFRs utilized for TRAF2 binding led to the compilation of a major TRAF 2-binding consensus motif (P/S/A/T)x(Q/E)E and a minor consensus motif PxQxxD [4].

Further studies show that TRAFs 1, 3, and 5 can associate in a similar way with regions that overlap TRAF2-binding sites by virtue of conserved surface exposed residues within their TRAF-C domains [4–7]. TRAF6 normally binds a distinct, more membrane-proximal region with the consensus sequence PxExx(Aromatic/Acidic) [8]. An additional independent lower-affinity TRAF2-binding motif (SXXE) has been demonstrated in CD40 and CD30 [9]. For each TNFR family molecule, unique aspects of the competition and cooperation between TRAFs at these binding sites determine the final signals that are transduced [1, 10, 11].

5.1.3 TRAF-Binding Sites in BAFFR

For the receptors of interest here (BAFFR, TACI, and BCMA), the TRAF-binding region has been best resolved for BAFFR. The intracellular portion of BAFFR does not contain a region that fits an aforementioned TRAF-binding consensus, but selective studies of truncation mutants and X-ray crystallography have pinpointed the PVPAT sequence as a specific, albeit atypical, TRAF3-binding motif [5, 6] (Fig. 5.1). Replacement of this motif with AVAAA abolishes TRAF3 binding,

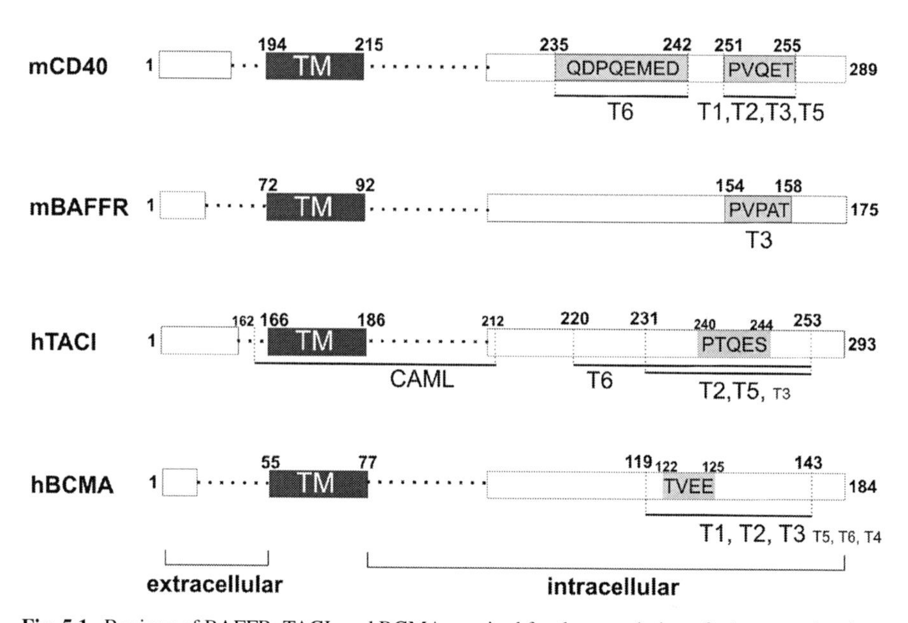

Fig. 5.1 Regions of BAFFR, TACI, and BCMA required for the association of adapter molecules. The association of TRAF3 with mBAFFR has been shown to require the conserved PVPAT motif. The regions required for TRAF binding to hTACI and hBCMA have been delimited to 20–30 amino acid regions (indicated by broken lines). Sequences within these regions that best correspond to known TRAF-binding motifs are shaded in *gray*. TRAFs presented in subscript were shown to associate with these receptors in separate studies, and are expected to utilize overlapping sites based on TRAF association with other TNFRs like CD40. TM; transmembrane segment

TRAF3 degradation, and NF-κB2 signaling in the M12 mouse B-cell line [5]. TRAF3 can also be co-immunoprecipitated alongside BAFFR when both proteins are at endogenous levels in the mouse B-cell line A20, and in mouse primary splenocytes (Fig. 5.2).

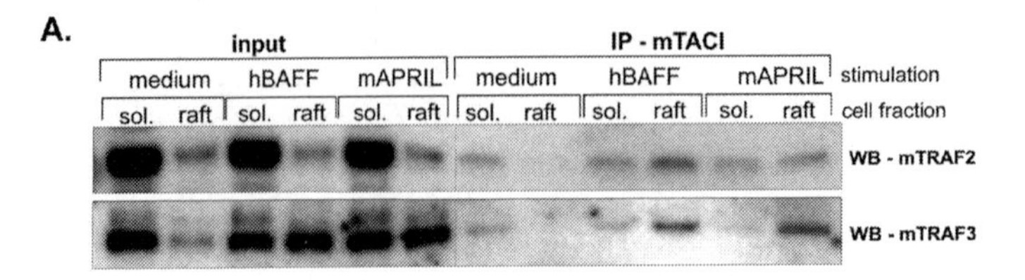

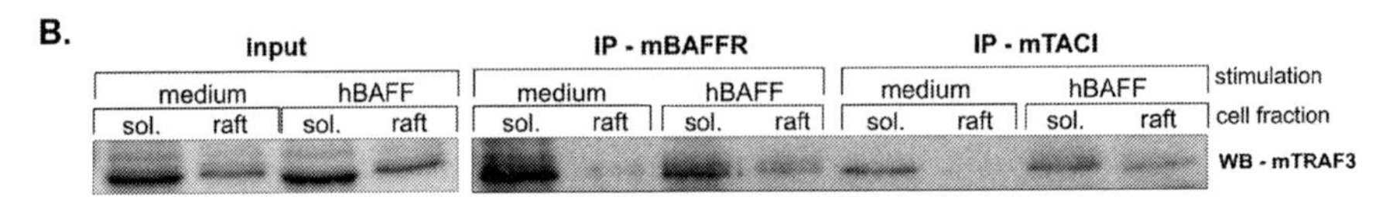

Fig. 5.2 TRAF association with BAFFR and TACI following BAFF and APRIL stimulation of B lymphocytes. A20 murine B cells (**a**) and murine primary B splenocytes (**b**) were stimulated with recombinant human BAFF or recombinant mouse APRIL. Cells were divided into non-membrane raft (sol.) and membrane raft (raft) fractions based on detergent solubility. In A20 B-cell extracts, TRAF2 and TRAF3 can be co-immunoprecipitated with TACI using polyclonal anti-TACI antiserum. TRAF3 can be co-immunoprecipitated from mouse splenic B cells using polyclonal antisera raised against TACI or BAFFR. The association of other TRAFs with TACI are not detected using this particular system, indicating the importance of qualifying protein–protein interaction data using relevant cell types, with candidates expressed at endogenous levels and free of affinity tags

5.1.4 TRAF-Binding Sites in BCMA

The work of various groups has reduced the TRAF-binding portions of TACI and BCMA to 20–30 amino acid regions. Residues 119–143 of human BCMA are required for the co-immunprecipitation of TRAFs 1, 2, and 3, and the activation of NF-κB, Elk1, and JNK in the transformed epithelial cell line HEK 293 [12]. Though not reported to date, the recruitment of TRAFs 1, 2, 3, and possibly 5 may be mediated via the consensus matching TVEE sequence contained within this region. Co-immunoprecipitation of TRAF6 with BCMA has also been reported, though there is no region matching the TRAF6-binding consensus [13]. There is currently no TRAF-binding data available for BCMA in B cells.

5.1.5 TRAF-Binding Sites in TACI

Like BCMA, TACI does not contain a region matching the TRAF6-binding consensus, but can be co-immunoprecipitated with TRAFs 2, 5, and 6 expressed in a transformed epithelial cell line [14]. The closest match to the TRAF6-binding motif is the $_{226}$PVETCSF$_{232}$ sequence that lies upstream and close to the sequence $_{240}$PTQES$_{244}$ in hTACI. The latter motif fits the major TRAF2-binding consensus, and both motifs lie within the regions necessary for positive interactions with TRAF6 (residues 220–253) and TRAFs 2, 5, and 6 (residues 231–253) as assessed by the yeast two-hybrid binding assay [14]. It should also be noted that these motifs lie within the best-conserved regions of the TACI intracellular domain. TRAF3 is the only TRAF reported to interact with TACI in a human B-cell line, Bjab. Recent work in our laboratory demonstrates that endogenous levels of TACI can co-immunoprecipitate TRAFs 2 and 3 from the A20 mouse B-cell line and TRAF3 from primary mouse B cells (Fig. 5.2). In contrast to observations using the yeast two-hybrid system and transformed epithelial cell lines, the association of TRAFs 5 and 6 cannot be detected in transformed or primary B cells using this method.

5.1.6 Lipid Raft Movement of BAFFR and TACI

Many cell surface receptors including TNFR family members have been shown to associate with cholesterol and glycosphingolipid rich, detergent insoluble membrane microdomains or "lipid rafts". These rafts are thought to be a reservoir of various integral membrane and membrane-associated protein components that permit rapid propagation of signals following ligand stimulation. CD40 has been shown to partition into B-cell membrane rafts within minutes of ligand engagement [15] and BAFFR and TACI are no exception (JH and K. Larison, unpublished observation). While detection of untagged endogenous mouse BAFFR or TACI is difficult due to lack of available antibodies suitable for immunoblotting, their movement into membrane rafts is evident in measurements of TRAF movement and recruitment following ligand stimulation. Figure 5.2A shows that in the A20 mouse B-cell line, stimulation with BAFF and APRIL leads to a significant increase in the total amounts of TRAF3 in the membrane raft fraction. While the movement of TRAF3 in primary splenocytes and TRAF2 in A20 B cells is not obvious in total cell extracts, the amount of TRAFs 2 and 3 that co-immunoprecipitate with BAFFR and TACI in membrane raft fractions is significantly increased following ligand stimulation (Fig. 5.2a, b).

5.1.7 TRAF Degradation Induced by BAFF and APRIL Receptors

CD40 and other members of the TNFR family have been shown to induce the degradation of various TRAFs following ligand stimulation [15–20]. BAFF stimulation can induce the degradation of TRAF3 in B-cell lines M12 and CH12.LX ([18], JH and K. Larison, unpublished), but induces little or no detectableTRAF3 degradation in primary resting B cells [21]. BAFFR-mediated degradation of TRAF3 in the B-cell line M12.4.1 requires the direct association of TRAF3 to the PVPAT TRAF-binding motif [5]. TRAF3 degradation is likely also induced by TACI and/or BCMA, as stimulation with APRIL also induces TRAF3 degradation by the CH12.LX B-cell line (JH and K. Larison, data not shown). There has been no report of TACI- or BCMA-mediated degradation of other TRAF molecules. A report by Liao et al. demonstrated that the degradation of TRAF3 may occur not only as a means of attenuating positive signaling through these receptors, but as a means of releasing a TRAF3-dependent signaling blockade [18]. When M12 B cells were stimulated with BAFF, the degradation of TRAF3 was found to coincide with an increase in NF-κB-inducing kinase (NIK) levels and subsequent activation of the NF-κB2 pathway. TRAF3 was found to bind NIK directly and facilitate its ubiquitin-mediated degradation, which could explain why TRAF3 depletion leads to constitutive activation of NF-κB2 in B lymphocytes [22]. However, it should be noted that evidence for TRAF3-NIK association has not been seen in primary splenic B cells [22].

The degradation of TRAF3 by CD40 is dependent upon TRAF2, which may act as an ubiquitin ligase itself or permit the action of an associated ubiquitin ligase [20, 23]. Given that TRAF2 has not been reported to associate with BAFFR, what is mediating the degradation of TRAF3 by CD40-BAFFR chimeras? The only other reported BAFFR interacting protein – Act1 – does not have any reported E3 ligase activity, nor is TRAF3 known to mediate its own degradation [24]. The degradation of TRAF3 may be mediated by a yet unidentified protein that is recruited by BAFFR. Alternatively, BAFFR and TACI may form hetero-oligomers to bind their common ligand BAFF, bringing TACI-associated TRAF2 into the vicinity of BAFFR-associated TRAF3. There is evidence that TNFR1 and TNFR2 (CD120a and CD120b) may cooperate in a similar way on cells that express them both [25].

5.1.8 BAFFR Recruitment of Act1 and TRAF3

BAFF ligand stimulation has been shown to activate both NF-κB1 and NF-κB2 pathways, of which a major outcome is the alteration of Bcl2 family protein levels that favor cell survival [26, 27]. It is widely held that the BAFF-induced survival signal is mediated mainly through BAFFR, as opposed to TACI which also binds BAFF. Like BAFF-deficient mice, BAFFR knockout and mutant mice show a severe reduction in splenic B-cell numbers, and do not show attenuated B-cell apoptosis in response to BAFF administration. In contrast, TACI-deficient mice exhibit significant lymphoproliferation, indicating that this receptor may negate rather than transmit the BAFF survival signal [28]. One of the major questions for investigators of these pathways is how BAFFR can act as the major transducer of this survival signal, when there is only evidence for the recruitment to this receptor of TRAF3 and Act1, both of which are considered negative regulators of B-cell survival pathways.

Act1 was initially cloned using screens designed to recognize new components of NF-κB-dependent signaling pathways [29]. It was found to interact with TNFR family members CD40 and BAFFR in the human B-cell line IM9 following stimulation with CD154 and BAFF, respectively [30]. The region of BAFFR responsible for Act1 association has not been delineated. The significance of this interaction is evident in the enhanced activation of CD40 and BAFFR survival pathways in Act1-deficient cells, indicating a negative regulatory role for Act1 in TNFR family signaling. Act1 contains two putative TRAF-binding sites [29] and was found to co-immunoprecipitate with TRAF3 following BAFF stimulation [30]. Act1 was also found to physically interact with IKKγ/NEMO, IKKα, and IKKβ, and may prove to be an important link between BAFFR recruited TRAF3 and the NF-κB2 pathway in a similar way to NIK, though the exact mechanism remains unclear.

5.1.9 CAML Binding to TACI

TACI was first identified in a yeast two-hybrid screen for interacting partners of CAML (calcium-modulator and cyclophilin ligand) [31]. CAML is an integral membrane protein located in intracellular vesicles and can act as a signaling intermediate for TACI activation of the T-cell-specific transcription factor NF-AT in Jurkat human T cells [31]. TACI interacts with the cytosol-exposed portion of CAML, with studies using the yeast two-hybrid system and transformed epithelial cell lines limiting the TACI-binding region to a 50 amino acid span starting prior to the transmembrane segment. This region is independent of regions shown to be necessary for TRAF binding [14, 31]. Given that TACI bears no sequence similarities to known activators of NF-AT, and levels are negligible on primary resting T lymphocytes, the connection between TACI and CAML pathways remains an intriguing question, but one that should be preceded by confirmation of TACI–CAML interaction in cells known to express TACI naturally.

5.2 Kinase Regulation by BAFF Receptors

Increasing evidence indicates that engagement of BAFF receptors by BAFF may mobilize multiple kinase signaling pathways to regulate B-cell survival. These include inhibition of protein kinase C delta (PKCδ) nuclear translocation, upregulation of the kinase Pim-2, and activation of AKt and ERK kinase signaling pathways [32–38] (Fig. 5.3). Furthermore, recent evidence demonstrates that BAFF stimulation induces the expression of a number of proteins controlling cell cycle progression in B cells [36, 39], suggesting that BAFF may also prime B cells for antigen-induced proliferation.

5.2.1 Inhibition of PKCδ Nuclear Translocation by BAFF

In the absence of stimulation, *ex vivo*-cultured resting splenic B cells undergo apoptosis, and this process is coupled with an increase in the levels of nuclear PKCδ [33]. Treatment of B cells with BAFF inhibits nuclear accumulation of PKCδ [33, 34]. It is found that one of the nuclear targets of PKCδ is histone 2B. Phosphorylation of histone 2B at Ser 14 (H2B-S14) is significantly reduced in PKCδ$^{-/-}$ B cells,

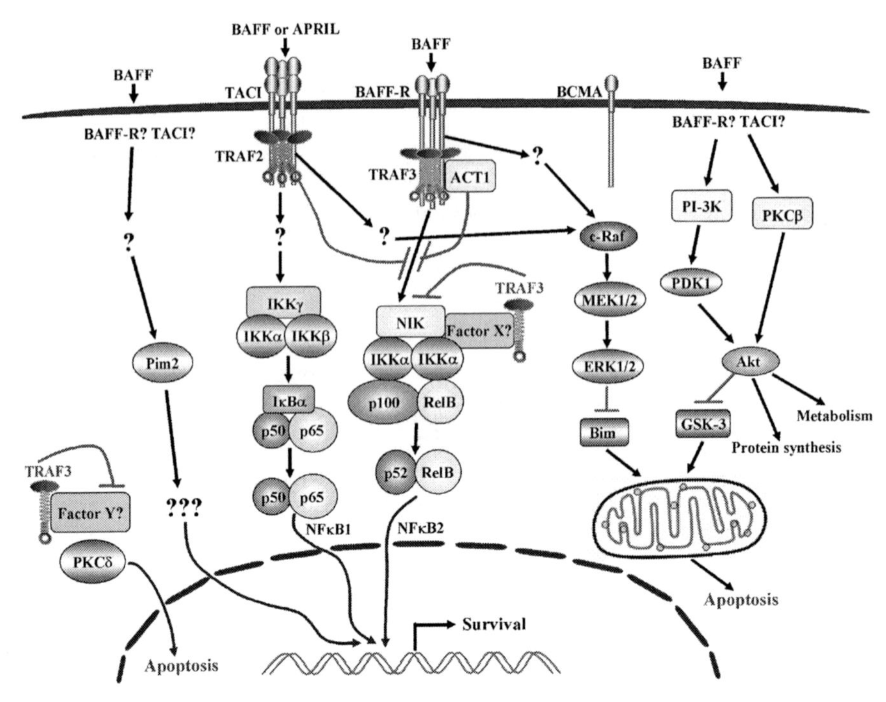

Fig. 5.3 Signaling pathways implicated in BAFF-mediated survival in mouse splenic B cells. A summary scheme depicts BAFF-induced multiple signaling pathways that may participate in the regulation of B-cell survival. These include activation of both classical and alternative NF-κB pathways, inhibition of PKCδ nuclear translocation, upregulation of the kinase Pim-2, and activation of AKt and ERK signaling pathways. A link between BAFFR and the NF-κB2 pathway has been clearly established. In the absence of BAFFR engagement, TRAF3 may bind to NIK directly or indirectly through an unknown factor X, thus inhibiting p100 processing. Meanwhile, TRAF3 may bind to another unknown Factor Y (or a multi-protein complex Y), thus preventing factor Y from binding to PKCδ, so PKCδ is free to enter the nucleus and induce apoptosis. BAFF signals recruit TRAF3 to the BAFFR signaling complex in membrane rafts, and thus remove TRAF3 from its cytoplasmic binding partners, leading to B-cell survival. Unknown links in BAFF-induced activation of Akt, ERK, and Pim2 signaling pathways are indicated by question marks. See text for details

pointing to a major contribution of PKCδ in the regulation of H2B-S14 phosphorylation [33]. Previous evidence has shown that phosphorylation of H2B-S14 is associated with DNA fragmentation and apoptotic cell death [40, 41]. Together, these findings suggest that BAFF may inhibit PKCδ nuclear translocation to prevent H2B-S14 phosphorylation, and thus protect resting splenic B cells from spontaneous apoptosis.

Direct evidence supporting a positive role of PKCδ in B-cell apoptosis is that PKCδ$^{-/-}$ B cells exhibit prolonged survival capacity both in vivo and *ex vivo* [33]. Furthermore, PKCδ$^{-/-}$ mice display expanded B-cell compartments in the spleen and lymph nodes, leading to splenomegaly, lymphadenopathy, and a lupus-like autoimmune disease [42, 43]. The phenotype of PKCδ$^{-/-}$ mice recapitulates that observed in BAFF-transgenic mice [44–46]. However, interpretation of the expanded B-cell compartment observed in PKCδ$^{-/-}$ mice is complicated because proliferation responses to treatment with LPS or engagement of BCR in combination with CD40 are also enhanced in PKCδ$^{-/-}$ B cells *ex vivo* [43]. Thus, severe B-cell hyperplasia and autoimmunity of PKCδ$^{-/-}$ mice may reflect the negative roles of PKCδ in both B-cell survival and proliferation. Understanding of the detailed molecular mechanisms by which PKCδ is involved in the regulation of B-cell proliferation, whether PKCδ nuclear translocation is required, and identity of the downstream targets of PKCδ remains to be determined.

Interestingly, resting splenic TRAF3$^{-/-}$ B cells show markedly reduced constitutive levels of nuclear PKCδ, and treatment with BAFF does not further decrease these levels [22]. Moreover, TRAF3$^{-/-}$ B cells have greatly increased constitutive levels of nuclear p52 and RelB, and BAFF stimulation does not further increase the levels of these NF-κB2 subunits in the nucleus [22]. These observations indicate that TRAF3 functions upstream of PKCδ and the NF-κB2 (noncanonical) pathway. Like PKCδ$^{-/-}$ B cells, TRAF3$^{-/-}$ B cells also exhibit prolonged survival capacity *ex vivo* [22]. However, TRAF3$^{-/-}$ B cells do not show any alterations in proliferation responses following treatment with CD40 ligation alone, or in combination with anti-IgM or IL-4 *ex vivo* [22]. So unlike PKCδ, TRAF3 only inhibits B-cell survival but not proliferation. Nonetheless, B-cell-specific TRAF3$^{-/-}$ mice display drastically expanded peripheral B-cell compartments, splenomegaly, lymphadenopathy, and autoimmune manifestations, all of which closely mirror those observed in PKCδ$^{-/-}$ mice [22, 33].

It should be noted that in vivo administration of TACI-Ig, a soluble protein that blocks BAFF and APRIL from binding to their receptors, is unable to deplete peripheral B cells or reduce B-cell numbers in B-cell-specific TRAF3$^{-/-}$ mice [22]. Similarly, injection of BAFFR-Ig, another soluble protein that blocks BAFF from binding to its receptors, has no effect on the B-cell compartments or B-cell survival in PKCδ$^{-/-}$ mice [33]. The independence of PKCδ$^{-/-}$ B cells from BAFF is further confirmed by the generation and characterization of PKCδ$^{-/-}$ A/WySnJ mice, which still exhibit severe peripheral B-cell hyperplasia [33]. Taken together, the above evidence suggests that TRAF3 may constitutively promote nuclear translocation of PKCδ to induce spontaneous apoptosis in resting splenic B cells, and that BAFF stimulation may disrupt the TRAF3/PKCδ pathway by recruiting TRAF3 from the cytoplasm to membrane rafts. However, further study is needed to address how TRAF3 directly or indirectly regulates the movement of PKCδ from the cytoplasm to the nucleus in B cells.

5.2.2 Activation of Phosphatidyl Inositol-3 Kinase (PI-3K)/Akt and PKCβ/Akt Signaling Pathways by BAFF

BAFF induces phosphorylation and activation of Akt simultaneously in B cells through at least two distinct kinase pathways, PI-3 K and PKCβ. Activation of Akt requires its recruitment to the plasma membrane via its pleckstrin homology domain. Such translocation induces a conformational change of Akt, allowing for Akt phosphorylation at Thr 308 and Ser 473. BAFF stimulates Akt phosphorylation at Thr 308 by inducing the tyrosine phosphorylation of p110δ, the catalytic subunit of PI-3 K [36]. Specific inhibition of PI-3 K activity diminishes BAFF-induced Akt phosphorylation at Thr 308, and compromises BAFF-induced B-cell survival [36]. Consistent with a role for PI-3 K in BAFF-induced Akt phosphorylation and B-cell survival, it has been shown that B cells from mice deficient in the p85 regulatory subunit of PI-3 K show poor survival capacity [47, 48]. Similarly, p110$\delta^{-/-}$ mice exhibit decreased numbers of mature B cells [49, 50]. However, PI-3 K has also been shown to be involved in the signal transduction cascades of a variety of receptors in B cells, including BCR, CD40, and IL-4 [49–51]. p85$^{-/-}$ and p110$\delta^{-/-}$ mice have more complex defects than impaired B-cell survival, and decreased mature B-cell numbers observed in these mice may be due to combined effects of PI-3 K deficiency on both BAFF- and BCR-mediated B-cell survival and proliferation.

Interestingly, BAFF promotes the association of PKCβ with Akt independent of PI-3 K activity, and thus allows PKCβ to phosphorylate Akt at Ser 473 [36]. BAFF-induced Akt phosphorylation at Ser 473 is greatly reduced, but not completely abolished in PKC$\beta^{-/-}$ B cells, suggesting that additional unidentified kinase(s) may partially compensate for this specific function. In contrast, BAFF-mediated NF-κB2 activation and inhibition of PKCδ nuclear translocation are not altered by PKCβ deficiency. PKC$\beta^{-/-}$ B cells also exhibit poor survival [36, 52]. Although PKC$\beta^{-/-}$ B cells are partially responsive to BAFF, its ability to support B-cell survival is dampened by the absence of PKCβ. Furthermore, PKC$\beta^{-/-}$ mice exhibit impaired B-cell maturation and a reduced peripheral B-cell compartment. The size of the peripheral B-cell compartment in PKC$\beta^{-/-}$ mice is about 70% of wild-type levels, and the frequency of T1 transitional B cells is increased 2–3-fold in the spleen in PKC$\beta^{-/-}$ mice [36, 52, 53]. These observations mirror those obtained from BAFF$^{-/-}$, A/WySnJ, or BAFF-R$^{-/-}$ mice, albeit at a lower magnitude [54–56]. Taken together, the above evidence indicates that activation of Akt by both PI-3 K and PKCβ contributes to BAFF-mediated B-cell survival and maturation. However, there is not any evidence regarding whether BAFF receptors directly bind to p85/p110δ subunits of PI-3 K or PKCβ. Which and how BAFF receptor(s) stimulate tyrosine phosphorylation of p110δ and promote association between PKCβ and Akt await further investigation.

BAFF-induced Akt phosphorylation and activation result in the phosphorylation of a number of Akt targets, proteins known to control cell survival and metabolism. Akt targets regulated by BAFF signals in B cells include glycogen synthase kinase-3 (GSK-3), eukaryotic translation initiation factor 4E (eIF4E), eIF4E-binding protein 1 (4E-BP1), S6 ribosomal protein, tuberous sclerosis complex 2 (TSC2), and FoxO1 [36]. (1) BAFF induces the phosphorylation, and thus inactivation of glycogen synthase kinase-3 (GSK-3). GSK-3 has been shown to induce apoptosis by inducing MCL1 degradation and damaging mitochondrial membrane integrity [57]. (2) BAFF treatment causes phosphorylation of eIF4E and its inhibitor 4E-BP1, which are required for active protein synthesis [58, 59]. Phosphorylation of eIF4E promotes its binding to capped mRNAs, while 4E-BP1 phosphorylation disrupts its binding to eIF4E [58]. (3) BAFF stimulates phosphorylation of S6 ribosomal protein, a hallmark of active protein synthesis [59]. (4) Treatment of BAFF leads to phosphorylation of the GTPase-activating protein TSC2, and thus may promote protein synthesis through activation of mammalian target of rapamycin (mTOR) [60–62]. (5) BAFF induces the phosphorylation of FoxO1 and its subsequent degradation. It is known that the FoxO family of transcription factors plays important roles in regulating genes required for cell survival, cell cycle progression, and glycolytic metabolism [63, 64]. Thus, phosphorylation of these Akt target proteins by BAFF helps to inhibit B-cell apoptosis and promotes B-cell protein synthesis and metabolism.

5.2.3 Regulation of the MEK/ERK/Bim Signaling Pathway by BAFF

Craxton et al. demonstrated that BAFF stimulates the delayed but sustained phosphorylation and activation of extracellular signal-regulated kinase (ERK) through phosphorylation of its upstream kinases mitogen-activated protein kinase/ERK kinase 1/2 (MEK1/2) and c-Raf in B cells [37]. One downstream effect of BAFF-induced activation of the c-Raf/MEK/ERK pathway is the phosphorylation and downregulation of Bim, a BH3 only proapoptotic protein of the Bcl-2 family [37, 65, 66]. Specific inhibition of the MEK/ERK pathway decreases the phosphorylation of Bim and increases the levels of BimEL protein in BAFF-treated B cells [37]. Thus, BAFF-induced Bim phosphorylation and degradation may also contribute to the inhibitory effects of BAFF on B-cell apoptosis, leading to B-cell survival. To determine which BAFF receptor(s) are involved in BAFF-induced Bim phosphorylation, Craxon et al. produced chimeric BAFF receptors, which are composed of the extracellular and transmembrane domains of mouse CD8α and the intracellular domains of each BAFF receptor. Using WEHI-231 B cells retrovirally transduced with these chimeric receptors, Craxon et al. found that ligation of chimeric CD8α-BAFF-R or CD8α-TACI results in phosphorylation and decreased levels of Bim [37]. These findings indicate that both BAFF-R and TACI may participate in the modulation of the phosphorylation as well as protein levels of Bim in B cells. However, how BAFF-R and TACI induce the phosphorylation of c-Raf remains to be determined.

In support of an important function of Bim in the regulation of B-cell apoptosis/survival, $Bim^{-/-}$ mice exhibit increased splenic B-cell numbers, including elevated numbers of T2 transitional and follicular mature B cells [37, 67]. This phenotype of $Bim^{-/-}$ mice correlates with that of BAFF transgenic mice [44–46]. It is perplexing, however, that $Bim^{-/-}$ mice display reduced marginal zone B cells in the spleen, whereas BAFF transgenic mice show increased numbers of splenic marginal zone B cells [37, 44]. Interestingly, similar to $Bim^{-/-}$ splenic T2 and follicular B cells, $Bim^{-/-}$ marginal zone B cells also show prolonged survival *ex vivo* [37]. Thus, these findings together suggest that Bim-induced B-cell apoptosis may be counteracted by BAFF signals, and that Bim may also regulate marginal zone B-cell development through yet unknown BAFF-independent mechanisms.

5.2.4 Upregulation of Pim2 by BAFF

An additional protein that may also participate in BAFF-induced B-cell survival is Pim2, a member of the Pim family of Ser/Thr kinases [35, 36, 68, 69]. The Pim family genes were originally identified as common sites of retroviral insertion in lymphoid tumors arising in animals after Moloney virus infection [68, 69]. BAFF stimulates the mRNA expression of Pim2 and increases the protein levels of all three isoforms of Pim2 in B cells [35, 36]. It has been previously shown that upregulation of Pim2 promotes cell survival in response to a wide variety of growth factors and mitogens in hematopoietic cells, including B cells [69, 70]. Pim2 appears to protect cells from apoptosis by maintaining mitochondrial inner membrane potential via a PI3K/Akt/mTOR-independent pathway [71]. Available evidence indicates that ectopic expression of Pim2 promotes NF-κB1 activation and the phosphorylation of 4E-BP1, Cot (a Ser/Thr kinase) and BAD (a proapopototic protein of the Bcl2 family) [69, 71–73]. Dysregulation of Pim2 expression has been documented in several human malignancies, including leukemia, lymphoma, and multiple myeloma [69]. About 10–20% of Eμ-Pim2 transgenic mice develop lymphoma at 6 months of age, similar to that observed in Eμ-Myc transgenic mice albeit at a lower incidence [74, 75]. Notably, transgenic mice expressing both Eμ-Pim2 and Eμ-Myc display remarkably accelerated leukemia/lymphoma development, with all mice dying before or shortly after birth [74]. Thus, it is very likely that BAFF-induced Pim2 expression may be involved in the regulation of B-cell survival. Paradoxically, $Pim2^{-/-}$ mice do not show obvious defects in B-cell survival [70], probably due to overlapping or redundant functions with other BAFF-induced signaling pathways. Detailed molecular mechanisms about how Pim2 fits in BAFF signaling pathways to promote B-cell survival is unclear, and is one important area for future research.

5.2.5 Induction of Proteins Controlling Cell Cycle by BAFF

Interestingly, although BAFF does not stimulate B-cell proliferation, it has been reported that BAFF induces the expression of a number of proteins regulating cell cycle progression and mitosis at both the mRNA and protein levels [36, 39]. BAFF-upregulated cell cycle progression proteins include cyclin D, Cdk4, Mcm2, and Mcm3 [36, 39]. BAFF also induces the expression of the proliferation maker Ki67 and Survivin [36, 39]. Furthermore, BAFF stimulates the phosphorylation of the key cell cycle regulating protein Rb, which is a prerequisite for the release of E2F and cell cycle entry into S phase [36, 39, 76]. However, many studies demonstrate that BAFF does not induce DNA replication or proliferation in B cells. The failure of BAFF to drive B cells past G1/S transition may be due to its inability to reduce the levels of the Cdk inhibitor protein p27 [39]. Nevertheless, it is found that pre-treatment of B cells with BAFF accelerates B-cell proliferation in response to B-cell antigen receptor (BCR) stimulation [36]. Thus, BAFF-induced expression/activation of these cell cycle regulating proteins may "prime" B cells, so they are more quickly able to respond to mitogenic stimuli.

5.3 NF-κB Activation by BAFF and APRIL

5.3.1 *Two Pathways of NF-κB Activation*

The most consistent, reproducibly detectable molecular effect of stimulating cells with BAFF or APRIL is activation of the NF-κB family of transcription factors. Two major NF-κB-activating pathways have now been extensively characterized. The first to be investigated, referred to as the NF-κB1/classical/canonical pathway, involves the activation of kinases which phosphorylate the inhibitor of NF-κB (IκB) family of proteins, and are hence called IκB kinases (IKK). The phosphorylated IκB proteins are subsequently polyubiquitinated and targeted for proteasome-mediated degradation. This in turn unmasks the nuclear localization sites of p50, p65, and cRel subunits, which can form homo and heterodimers, and translocate into the nucleus (reviewed in [77, 78]. This pathway is typically (but not always) rapidly and transiently induced, and leads to the transcriptional regulation of a large variety of rapid response genes in B lymphocytes and other cell types [79]. The second (NF-κB2/alternative/non-canonical) pathway involves the activation of NF-κB-inducing kinase (NIK), processing of the NF-κB family member p100–p52, and the movement of p52 and RelB into the nucleus. This pathway is typically induced more slowly, and its more sustained activation plays an important role in cell survival [79].

5.3.2 *Challenges to Determination of Which Pathways Are Activated by BAFF and APRIL*

It is not surprising that BAFF and APRIL induce these pathways via their receptors, as virtually all members of the TNFR superfamily induce NF-κB activation when engaged by ligand [80]. However, it is proving challenging to determine which specific receptors for these TNF family ligands activate which specific NF-κB pathways. Results of numerous experiments designed to investigate the ability of receptors for BAFF and APRIL to activate NF-κB have been published by multiple investigators for over the past 10 years, and many overlapping as well as divergent conclusions have been reached. As is often true when this is seen in biological research, many of these apparent differences may be largely or completely due to important differences in details of model systems and experimental design.

One of the most persistent complications to data interpretation in studies of BAFF and/or APRIL signaling is that both of these ligands can bind to more than a single receptor, with BAFF binding to both BAFFR and TACI, and APRIL binding to TACI, BCMA, and proteoglycan (or an additional unknown receptor). While BCMA is not expressed until the Ig-secreting stage of B-cell differentiation, mature B cells express both BAFFR and TACI, so effects of treating mature or unseparated B-cell populations with BAFF, often attributed solely to BAFFR signaling, may also be the result of TACI signals to the B cells. Early in the study of these receptors, mice lacking TACI were shown to have a marked autoimmune and lymphoproliferative phenotype [28]. This clearly demonstrated that TACI has important negative regulatory roles in vivo, but may have also led some to conclude that the only functions of TACI in B cells are inhibitory, and therefore any activating signals seen following BAFF treatment must be attributable to BAFFR. However, subsequent studies have revealed that TACI also plays important positive roles in mucosal immunity and isotype-switching events [81–84], as well as in the survival of transformed B cells derived from tumors [85, 86]. Additionally, various types of B-cell tumors have been found to express all three receptors for BAFF and APRIL [85, 86], so multiple receptors could transmit signals from these ligands.

By transfecting non-B-cell lines with plasmids that express only one specific receptor for BAFF and/or APRIL, the above complication can be circumvented. However, signaling by members of the TNFR superfamily has been shown in a considerable number of instances to have strongly cell-type-specific characteristics. This is particularly true when comparing B lymphocytes, the cells that principally express BAFFR, TACI, and BCMA, to other cell types. TRAF3 appears to bind more readily to TNFR superfamily members in B cells compared to other cell types [1], and TRAF3 is the only TRAF demonstrated to bind to BAFFR (see above). In particular, transfection of the popular model adenocarcinoma cell line, HEK293, with various receptors and signaling molecules permits very high levels of protein expression. However, the stoichiometry of receptor/signaling protein complexes is thus quite abnormal, and has led to misleading conclusions about how endogenous B-cell TNFR superfamily receptors actually transmit signals [1]. Given all these caveats, what is the present state of understanding of NF-κB activation by BAFFR, TACI, and BCMA?

Early studies using transiently expressed NF-κB reporter genes found that the human B-cell line Bjab, transfected to overexpress BAFFR, can induce modestly elevated NF-κB activity, which is further increased upon BAFF treatment [87]. Similarly, transfectants of the human T-cell line Jurkat, expressing TACI, activate an NF-κB reporter gene in response to a cross-linking anti-TACI Ab [31], and the human kidney adenocarcinoma cell line 293, transfected with TACI, activates a similar reporter even in the absence of receptor ligation [28]. Finally, 293 cells, transfected with BCMA, activate an NF-κB reporter in response to recombinant APRIL [13]. Thus, all three receptors can activate NF-κB, but reporter gene studies could not distinguish between the NF-κB1 and NF-κB2 pathways, and the true situation with endogenous receptors in B lymphocytes requires studies using primary B cells.

5.3.3 Activation of NF-κB by BAFFR

Subsequently, it has been convincingly shown that BAFF potently activates the NF-κB2/alternative pathway in mouse and human primary B cells, but most studies were not designed to determine whether both BAFFR and TACI make contributions to this activation. B cells lacking either TACI or BCMA were shown to still activate p100 processing in response to ligand, while B cells from mice expressing the A/WySnJ mutant BAFFR [55] cannot activate NF-κB2 following exposure to BAFF [88]. Thus, BAFFR appears to be both necessary and sufficient to activate NF-κB2 in response to BAFF ligand.

However, whether BAFF binding to BAFFR can also activate the NF-κB1 pathway has proven confusing. In normal resting mouse B cells, treatment of BAFF does not result in the rapid, transient phosphorylation and degradation of IκBα, followed by movement of p50/p65 heterodimers to the nucleus, which is typically seen when NF-κB1 is activated. Hatada et al. observed the activation of p50 homodimers in such cells, at late time points of 12–24 hour. This suggests that BAFF-induced NF-κB2 activation, seen within several hours, is not dependent upon prior NF-κB1 activation, an impression reinforced by normal BAFF-dependent p100 processing in B cells from p50-deficient mice. These investigators did see IκBα phosphorylation 10–60 min after BAFF treatment, but no subsequent detectable IκBα degradation [27]. They hypothesize that BAFF induces phosphorylation of a proportion of IκBα molecules, whose degradation is counterbalanced by an increase in synthesis of new IκBα. The authors also conclude that BAFF preferentially activates complexes containing p50 but not c-Rel, and that these complexes are important for BAFF-mediated B-cell survival.

The model system used by Hatada et al. had the advantage of relatively homogenous mature resting B cells, but the BAFF used in the experiments could have exerted its effects via BAFFR, TACI, or both. A more recent study in human B-CLL cells used a blocking Ab to BAFFR to distinguish BAFFR-specific effects of BAFF, and found that p100 processing to p52 is specific to this receptor, while both TACI and BCMA activate the NF-κB1 pathway [86]. These findings suggest that the BAFF-attributed effects seen in the earlier study were primarily the result of TACI signaling. Morrison et al. reported that a hybrid molecule composed of the external and transmembrane domains of CD40 and the cytoplasmic domain of BAFFR, when cross-linked with anti-CD40 Ab, stimulates p100 processing, but no IκBα degradation [5]. Taken together, these studies paint a picture, in mature resting B cells, of BAFF-mediated NF-κB1 activation that occurs mainly via BAFF binding to TACI, and NF-κB2 activation that does not require prior NF-κB1 activation, and is mediated primarily by BAFFR signaling.

Several recent studies, however, contend that the NF-κB1 pathway does play an important role in BAFFR signaling. Sasaki et al. found that mice expressing constitutively active IKK2 in B cells have the phenotype of B-cell hyperplasia, prolonged B-cell survival, and retention of PKCδ in the cytoplasm that suggests constitutive BAFF signaling, and B cells from these mice were not dependent upon BAFF for in vitro survival [34]. When bred to BAFFR-deficient mice, this constitutive NF-κB1 activation allows B-cell development to proceed, without restoring p100 processing. Similar to earlier work described above [27], the authors report that BAFF treatment leads to a slow increase in phospho-IκBα over 12 or more hours, that is then sustained for at least 24 hours [34]. The rescue of developmental defects in BAFFR$^{-/-}$ mice by constitutive NF-κB1 activation suggests that this pathway may be downstream of essential BAFFR-mediated signals. However, it is also possible that overexpressed IKK2 drastically alters the survival programs of B cells through BAFF-independent pathways, as NF-κB1 activation has been shown to play important roles in B-cell survival mediated by other receptors, such as BCR and CD40 [1, 89–92]. Another important gap in this very interesting study is that it does not specifically indicate that BAFFR itself directly activates NF-κB1, and the authors did not specifically attempt to distinguish effects of TACI from BAFFR in their experiments.

Shinners et al. studied the relationship between BAFFR and the kinase Btk by characterizing the results of BAFF stimulation of Btk-deficient B cells. In littermate controls, IKK activation and IκBα degradation was observed 90 min after BAFF treatment; this was not seen in $Btk^{-/-}$ B cells. Some loss of IκBα was observed after 90 min in $TACI^{-/-}$ B cells, but a concomitant decrease in total cellular actin was also seen [93], and all cells in the experiments were pretreated with a total protein synthesis inhibitor, so differential protein stability could also contribute to the results. The authors conclude that Btk couples the BAFFR directly to the NF-κB1 pathway, which then induces p100 expression. However, several studies described above indicate that BAFF-induced p100 processing is independent of NF-κB1 activation [27, 34]. Additionally, the only experiment in which the authors attempted to distinguish between BAFFR and TACI-mediated BAFF effects was the total amount of IκBα at 90 min, and complications of that experiment are discussed above. Thus, while this study provides important new information on the involvement of Btk in BAFF-mediated effects, the respective roles of BAFFR and TACI are not completely resolved. It is clear that this tantalizing question requires more investigation, with experimental systems carefully designed to provide receptor-specific information in physiologically relevant contexts.

5.3.4 NF-κB Activation by TACI

A related question is whether TACI, in a normal biological context, activates the NF-κB2 pathway. This question is also complex, as TACI can bind either BAFF or APRIL, with some preference noted for APRIL in an exogenous overexpression system in non-B cells [94]. In a variety of human B-cell lines and tumor samples, TACI-Ig inhibits IκBα phosphorylation and Rel protein binding to DNA, again supporting the ability of TACI to activate NF-κB1 [85]. However, convincing evidence that TACI can directly activate the NF-κB2 pathway in B cells is lacking. Although this ability was attributed to TACI several years ago [11], this conclusion was based upon transient overexpression of TACI in 293 cells, after which constitutive nuclear p52 was measured (p100 processing was not shown). The amount of nuclear p52 in the sample blot shown for TACI was barely detectable, although 3–6-fold more DNA was transfected than for other TNFR family members that showed orders of magnitude more nuclear p52. Thus, the overall picture suggests that TACI primarily activates the NF-κB1 pathway, although a thorough examination of this question in APRIL-treated mature B cells (which do not express BCMA) would be welcomed.

5.3.5 *NF-κB Activation by BCMA*

Finally, the least studied of these receptors is BCMA, which binds only to APRIL and is expressed only late in B-cell differentiation. Overexpressed BCMA in 293 cells activates an NF-κB reporter gene; this is inhibited by overexpression of "DN" mutant proteins for both NF-κB1 (IKKα and IκBα) and NF-κB2 (NIK) pathways [13]. However, subsequent studies show only activation of the NF-κB1 pathway by BCMA in human B-cell tumor samples [85, 86]. It is thus likely that, like TACI, BCMA preferentially activates NF-κB1.

5.3.6 *Adaptor Proteins Involved in BAFF-Mediated NF-κB Activation*

Available evidence indicates that BAFFR interacts with TRAF3, but not with other members of the TRAF family in B cells [38, 87]. Recent evidence suggests that BAFFR triggers NF-κB2 activation by recruiting TRAF3, thus inhibiting TRAF3 from binding to NF-κB-inducing kinase (NIK) [18, 95]. Interaction between TRAF3 and NIK induces proteasome-mediated degradation of NIK [18, 95]. Recruitment of TRAF3 by BAFF-R may thus allow NIK protein to accumulate and increase NF-κB processing. Consistent with this model, TRAF3$^{-/-}$ resting splenic B cells exhibit prolonged survival *ex vivo* and increased constitutive levels of nuclear NF-κB2 [22]. Co-culture of TRAF3$^{-/-}$ and littermate control B cells does not promote the survival of the latter, indicating that TRAF3$^{-/-}$ B cells do not spontaneously secrete any survival factors such as BAFF or IL-6. Furthermore, in vivo administration of TACI-Ig, a soluble fusion protein that blocks both BAFF and APRIL from binding to their receptors, does not reverse peripheral B-cell hyperplasia of B-cell-specific TRAF3$^{-/-}$ mice [22]. Taken together, these results suggest that TRAF3 may constitutively inhibit NF-κB2 activation to induce spontaneous apoptosis in peripheral B cells, and BAFF/BAFF-R signaling functions to remove TRAF3 from NIK to promote B-cell survival. However, it should be noted that in splenic B cells from either B-cell-specific TRAF3$^{-/-}$ mice or their littermate controls, TRAF3-NIK interactions could not be detected [22].

Similarly, resting splenic B cells from a conditional TRAF2$^{-/-}$ mouse strain also show improved survival and higher levels of nuclear NF-κB2 *ex vivo* in the absence of stimuli [21]. Further experiments using TACI-Ig or BAFF-Ig, or alternatively, breeding conditional TRAF2$^{-/-}$ mice with BAFF$^{-/-}$ or BAFFR$^{-/-}$ mice are required to determine whether TRAF2-mediated inhibition of B-cell survival is downstream of BAFF/APRIL signaling or a constitutive TRAF2 function. In addition, it is unclear that through which BAFF receptors TRAF2 regulates NF-κB2 activation in B cells.

Interestingly, Act1, another adaptor protein has been shown to interact with BAFF-R, CD40, and TRAF3 in B cells as determined by co-immunoprecipitation and immunoblot analysis [30]. Similar to B-cell-specific TRAF3$^{-/-}$ and conditional

TRAF2$^{-/-}$ mice, Act1$^{-/-}$ mice also exhibit dramatic increase in peripheral B cells, which culminates in lymphadenopathy and splenomegaly, hypergammaglobulinemia, and autoantibodies [30]. However, unlike TRAF3$^{-/-}$ or TRAF2$^{-/-}$ B cells, Act1$^{-/-}$ splenic B cells display normal basal levels of both NF-κB1 and NF-κB2, but show enhanced responses of both pathways following BAFF or CD40 stimulation [30]. Notably, both BAFF$^{-/-}$Act1$^{-/-}$ and CD40$^{-/-}$Act1$^{-/-}$ mice show reduced enlargement of spleen and lymph nodes as compared to Act1$^{-/-}$ mice [30]. Furthermore, the hypergammaglobulinemia observed in Act1$^{-/-}$ mice is abolished in BAFF$^{-/-}$Act1$^{-/-}$ and CD40$^{-/-}$Act1$^{-/-}$ mice [30]. These findings indicate that Act1 does not constitutively regulate B-cell survival, and that Act1 is a genuine downstream negative regulator of BAFF- and CD40-induced NF-κB activation in B cells.

5.4 Conclusions

It is clear from all the combined information presented in this volume that BAFF, APRIL, and their receptors cause a variety of potent and important biological effects in B-cell survival, development, and differentiated functions. Interestingly, some of the currently least well-understood aspects of this complex regulatory system are the early signals initiated by receptors for BAFF and APRIL and the molecular mechanisms by which this is accomplished. In this chapter, we have presented an overview of the current state of knowledge in this area, but it is evident that many interesting and important questions remain.

References

1. Bishop GA. The multifaceted roles of TRAFs in the regulation of B cell function. Nat Rev Immunol 2004;4:775–86.
2. Qian Y, Liu C, Hartupee J, et al. The adaptor Act1 is required for interleukin 17-dependent signaling associated with autoimmune and inflammatory disease. Nat Immunol 2007;8: 247–56.
3. Inoue J, Ishida T, Tsukamoto N, et al. TRAF family: adapter proteins that mediate cytokine signaling. Exp Cell Res 2000;254:14–24.
4. Ye H, Park YC, Kreishman M, Kieff E, Wu H. The structural basis for the recognition of diverse receptor sequences by TRAF2. Mol Cell 1999;4:321–30.
5. Morrison MD, Reiley W, Zhang M, Sun S-C. An atypical TRAF binding motif of BAFFR mediates induction of the noncanonical NF-kB signaling pathway. J Biol Chem 2005;280:10018–24.
6. Ni C-Z, Oganesyan G, Welsh K, et al. Key molecular contacts promote recognition of the BAFF receptor by TRAF3: implications for intracellular signaling regulation. J Immunol 2004;173:7394–400.
7. Park YC, Burkitt V, Villa AR, Tong L, Wu H. Structural basis for self-association and receptor recognition of human TRAF2. Nature 1999;398:533–8.
8. Ye H, Arron JR, Lamothe B, et al. Distinct molecular mechanism for initiating TRAF6 signalling. Nature 2002;418:443–7.
9. Lu L, Cook WJ, Lin L, Noelle RJ. CD40 signaling through a newly identified TRAF2 binding site. J Biol Chem 2003;278:45414–8.
10. Xie P, Hostager BS, Munroe ME, Moore CR, Bishop GA. Cooperation between TRAFs 1 and 2 in CD40 signaling. J Immunol 2006;176:5388–400.

11. Hauer J, Püschner S, Ramakrishnan P, et al. TRAF3 serves as an inhibitor of TRAF2/5-mediated activation of the noncanonical NF-kB pathway by TRAF-binding TNFRs. Proc Natl Acad Sci (USA) 2005;102:2874–9.
12. Hatzoglou A, Roussel J, Bourgeade MF, et al. TNF receptor family member BCMA (B cell maturation) associates with TRAF1, TRAF2, and TRAF3 and activates NF-kB, elk-1, JNK, and p38 MAPK. J Immunol 2000;165:1322–30.
13. Shu HB, Johnson H. B cell maturation protein is a receptor for the TNF family member TALL-1. Proc Natl Acad Sci (USA) 2000;97:9156–61.
14. Xia X, Treanor J, Senaldi G, et al. TACI is a TRAF-interacting receptor for TALL-1, a TNF family member involved in B cell regulation. J Exp Med 2000;192:137–43.
15. Hostager BS, Catlett IM, Bishop GA. Recruitment of CD40, TRAF2 and TRAF3 to membrane microdomains during CD40 signaling. J Biol Chem 2000;275:15392–8.
16. Duckett CS, Thompson CB. CD30-dependent degradation of TRAF2: implications for negative regulation of TRAF signaling and the control of cell survival. Genes Devel 1997;11:2810–21.
17. Li X, Yang Y, Ashwell JD. TNF-RII and c-IAP1 mediate ubiquitination and degradation of TRAF2. Nature 2002;416:344–9.
18. Liao G, Zhang M, Harhaj EW, Sun S-C. Regulation of NIK by TRAF3-induced degradation. J Biol Chem 2004;279:26243–50.
19. Takayanagi H, Ogasawara K, Hida S, et al. T-cell mediated regulation of osteoclastogenesis by signalling cross-talk between RANKL and IFN-g. Nature 2000;408:600–5.
20. Brown KD, Hostager BS, Bishop GA. Regulation of TRAF2 signaling by self-induced degradation. J Biol Chem 2002;277:19433–8.
21. Grech AP, Amesbury M, Chan T, Gardam S, Basten A, Brink R. TRAF2 differentially regulates the canonical and noncanonical pathways of NF-kB activation in mature B cells. Immunity 2004;21:629–42.
22. Xie P, Stunz LL, Larison KD, Yang B, Bishop GA. TRAF3 is a critical regulator of B cell homeostasis in secondary lymphoid organs. Immunity 2007;27:253–67.
23. Hostager BS, Haxhinasto SA, Rowland SR, Bishop GA. TRAF2-deficient B lymphocytes reveal novel roles for TRAF2 in CD40 signaling. J Biol Chem 2003;278:45382–90.
24. Moore CR, Bishop GA. Differential regulation of CD40-mediated TRAF degradation in B lymphocytes. J Immunol 2005;175:3780–9.
25. Pinckard J, Sheehan K, Arthur C, Schreiber R. Constitutive shedding of both p55 and p75 murine TNF receptors in vivo. J Immunol 1997;158:3869–73.
26. Do RKG, Hatada E, Lee H, Tourigny MR, Hilbert D, Chen-Kiang S. Attenuation of apoptosis underlies BLyS enhancement of humoral immune response. J Exp Med 2000;192:953–64.
27. Hatada EN, Do RKG, Orlofsky A, et al. NF-kB1 p50 is required for BLyS attenuation of apoptosis but dispensable for processing of NF-kB2 p100 to p52 in quiescent mature B cells. J Immunol 2003;171:761–8.
28. Seshasayee D, Valdez P, Yan M, Dixit VM, Tumas D, Grewal IS. Loss of TACI causes fatal lymphoproliferation and autoimmunity, establishing TACI as an inhibitory BLyS receptor. Immunity 2003;18:279–88.
29. Qian Y, Zhao Z, Jiang Z, Li X. Role of NF-kB activator Act1 in CD40-mediated signaling in epithelial cells. Proc Natl Acad Sci (USA) 2002;99:9386–91.
30. Qian Y, Qin J, Cui G, et al. Act1, a negative regulator in CD40 and BAFF-mediated B cell survival. Immunity 2004;21:575–87.
31. von Bulow GU, Bram RJ. NF-AT activation induced by a CAML-interacting member of the tumor necrosis factor receptor superfamily. Science 1997;278:138–41.
32. Claudio E, Brown K, Park S, Wang H, Siebenlist U. BAFF-induced NEMO-independent processing of NF-kB2 in maturing B cells. Nature Immunology 2002;3:958–65.
33. Mecklenbräuker I, Kalled SL, Leitges M, Mackay CR, Tarakhovsky A. Regulation of B cell survival by BAFF-dependent PKCd-mediated nuclear signaling. Nature 2004;431:456–61.

34. Sasaki Y, Derudder E, Hobeika E, et al. Canonical NF-kB activity, dispensable for B cell development, replaces BAFFR signals and promotes B cell proliferation upon activation. Immunity 2006;24:729–39.
35. Xu LG, Wu M, Hu J, Zhai Z, Shu HB. Identification of downstream genes up-regulated by the tumor necrosis factor family member TALL-1. J Leukoc Biol 2002;72:410–6.
36. Patke A, Mecklenbrauker I, Erdjument-Bromage H, Tempst P, Tarakhovsky A. BAFF controls B cell metabolic fitness through a PKCb- and Akt-dependent mechanism. J Exp Med 2006;203:2551–62.
37. Craxton A, Draves KE, Gruppi A, Clark EA. BAFF regulates B cell survival by downregulating the BH3-only family member Bim via the ERK pathway. J Exp Med 2005;202:1363–74.
38. Mackay F, Silveira PA, Brink R. B cells and the BAFF/APRIL axis: fast-forward on autoimmunity and signaling. Curr Opin Immunol 2007;19:327–36.
39. Huang X, De Liberto M, Cunningham AF, et al. Homeostatic cell-cycle control by BLyS: Induction of cell-cycle entry but not G1/S transition in opposition to p18INK4c and p27Kip1. Proc Natl Acad Sci (USA) 2004;101:17789–94.
40. Ajiro K. Histone H2B phosphorylation in mammalian apoptotic cells. An association with DNA fragmentation. J Biol Chem 2000;275:439–43.
41. Cheung WL, Ajiro K, Samejima K, et al. Apoptotic phosphorylation of histone H2B is mediated by mammalian sterile twenty kinase. Cell 2003;113:507–17.
42. Mecklenbräuker I, Saijo K, Zheng N, Leitges M, Tarakhovsky A. Protein kinase Cd controls self-antigen-induced tolerance. Nature 2002;416:860–5.
43. Miyamoto A, Nakyama K, Imaki H, et al. Increased proliferation of B cells and autoimmunity in mice lacking PKCd. Nature 2002;416:865–9.
44. Mackay F, Woodcock SA, Lawton P, et al. Mice transgenic for BAFF develop lymphocytic disorders along with autoimmune manifestations. J Exp Med 1999;190:1697–710.
45. Gross JA, Johnston J, Mudri S, et al. TACI and BCMA are receptors for a TNF homologue implicated in B-cell autoimmune disease. Nature 2000;404:995–9.
46. Khare SD, Sarosi I, Xia X, et al. Severe B cell hyperplasia and autoimmune disease in TALL-1 transgenic mice. Proc Natl Acad Sci (USA) 2000;97:3370–5.
47. Fruman DA, Snapper SB, Yballe CM, et al. Impaired B cell development and proliferation in absence of PI3-kinase p85a. Science 1999;283:393–7.
48. Suzuki H, Terauchi Y, Fujiwara M, et al. Xid-like immunodeficiency in mice with disruption of the p85a subunit of PI3-kinase. Science 1999;283:390–2.
49. Clayton E, Bardi G, Bell SE, et al. A crucial role for the p110d subunit of PI3-K in B cell development and activation. J Exp Med 2002;196:753–63.
50. Jou ST, Carpino N, Takahashi Y, et al. Essential, nonredundant role for the PI3-kK p110d in signaling by the B-cell receptor complex. Mol Cell Biol 2002;22:8580–91.
51. Bilancio A, Okkenhaug K, Camps M, et al. Key role of the p110d isoform of PI3K in B-cell antigen and IL-4 receptor signaling: comparative analysis of genetic and pharmacologic interference with p110d function in B cells. Blood 2006;107:642–50.
52. Saijo K, Mecklenbrauker I, Santana A, Leitger M, Schmedt C, Tarakhovsky A. PKCb controls NF-κB activation in B cells through selective regulation of the IκB kinase a. J Exp Med 2002;195:1647–52.
53. Leitges M, Schmedt C, Guinamard R, et al. Immunodeficiency in PKCb-deficient mice. Science 1996;273(5276):788–91.
54. Schiemann B, Gommerman JL, Vora K, et al. An essential role for BAFF in the normal development of B cells through a BCMA-independent pathway. Science 2001(14 Sept);293:2111–4.
55. Thompson JS, Bixler SA, Qian F, et al. BAFF-R, a newly identified TNF receptor that specifically interacts with BAFF. Science 2001;293:2108–11.
56. Shulga-Morskaya S, Dobbles M, Walsh ME, et al. BAFF acts through separate receptors to support B cell survival and T cell-independent antibody formation. J Immunol 2004;173:2331–41.

57. Maurer U, Charvet C, Wagman AS, Dejardin E, Green DR. Glycogen synthase kinase-3 regulates mitochondrial outer membrane permeabilization and apoptosis by destabilization of MCL-1. Mol Cell 2006;21:749–60.
58. Minich WB, Balasta ML, Goss DJ, Rhoads RE. Chromatographic resolution of in vivo phosphorylated and nonphosphorylated eukaryotic translation initiation factor eIF-4E: increased cap affinity of the phosphorylated form. Proc Natl Acad Sci USA 1994;91:7668–72.
59. Ruggero D, Sonenberg N. The Akt of translational control. Oncogene 2005;24:7426–34.
60. Manning BD, Cantley LC. Rheb fills a GAP between TSC and TOR. Trends Biochem Sci 2003;28:573–6.
61. Manning BD, Cantley LC. United at last: the tuberous sclerosis complex gene products connect the PI3-K/Akt pathway to mammalian target of rapamycin (mTOR) signalling. Biochem Soc Trans 2003;31:573–8.
62. Hay N, Sonenberg N. Upstream and downstream of mTOR. Genes Dev 2004;18:1926–45.
63. Burgering BM, Kops GJ. Cell cycle and death control: long live Forkheads. Trends Biochem Sci 2002;27:352–60.
64. Tran H, Brunet A, Griffith EC, Greenberg ME. The many forks in FOXO's road. Sci STKE 2003;2003:RE5.
65. Opferman JT, Korsmeyer SJ. Apoptosis in the development and maintenance of the immune system. Nat Immunol 2003;4:410–5.
66. Strasser A. The role of BH3-only proteins in the immune system. Nat Rev Immunol 2005;5:189–200.
67. Bouillet P, Metcalf D, Huang DC, et al. Proapoptotic Bcl-2 relative Bim required for certain apoptotic responses, leukocyte homeostasis, and to preclude autoimmunity. Science 1999;286:1735–8.
68. Mikkers H, Allen J, Knipscheer P, et al. High-throughput retroviral tagging to identify components of specific signaling pathways in cancer. Nat Genet 2002;32:153–9.
69. White E. The pims and outs of survival signaling: role for the Pim-2 protein kinase in the suppression of apoptosis by cytokines. Genes Dev 2003;17:1813–6.
70. Mikkers H, Nawijn M, Allen J, et al. Mice deficient for all PIM kinases display reduced body size and impaired responses to hematopoietic growth factors. Mol Cell Biol 2004;24:6104–15.
71. Fox CJ, Hammerman PS, Cinalli RM, Master SR, Chodosh LA, Thompson CB. The serine/threonine kinase Pim-2 is a transcriptionally regulated apoptotic inhibitor. Genes Dev 2003;17:1841–54.
72. Hammerman PS, Fox CJ, Cinalli RM, et al. Lymphocyte transformation by Pim-2 is dependent on nuclear factor-kappaB activation. Cancer Res 2004;64:8341–8.
73. Yan B, Zemskova M, Holder S, et al. The PIM-2 kinase phosphorylates BAD on serine 112 and reverses BAD-induced cell death. J Biol Chem 2003;278:45358–67.
74. Allen JD, Verhoeven E, Domen J, van der Valk M, Berns A. Pim-2 transgene induces lymphoid tumors, exhibiting potent synergy with c-myc. Oncogene 1997;15:1133–41.
75. Adams JM, Harris AW, Pinkert CA, et al. The c-myc oncogene driven by immunoglobulin enhancers induces lymphoid malignancy in transgenic mice. Nature 1985;318:533–8.
76. Weinberg RA. The retinoblastoma protein and cell cycle control. Cell 1995;81:323–30.
77. Thanos D, Maniatis T. NF-kB: a lesson in family values. Cell 1995;80:529–32.
78. May MJ, Ghosh S. Signal transduction through NF-kB. Immunol Today 1998;19:80–8.
79. Bonizzi G, Karin M. The two NF-kB activation pathways and their role in innate and adaptive immunity. TRENDS Immunol 2004;25:280–8.
80. Smith CA, Farrah T, Goodwin RG. The TNF receptor superfamily of cellular and viral proteins: activation, costimulation, and death. Cell 1994;76:959–62.
81. Castigli E, Wilson SA, Scott S, et al. TACI and BAFF-R mediate isotype switching in B cells. J Exp Med 2005;201:35–9.
82. Mantchev GT, Cortesao CS, Rebrovich M, Cascalho M, Bram RJ. TACI is required for efficient plasma cell differentiation in response to T-independent type 2 antigens. J Immunol 2007;179:2282–8.

83. Sakurai D, Hase H, Kanno Y, Kojima H, Okumura K, Kobata T. TACI regulates IgA production by APRIL in collaboration with HSPG. Blood 2007;109:2961–7.
84. Salzer U, Jennings S, Grimbacher B. To switch or not to switch – the opposing roles of TACI in terminal B cell differentiation. Eur J Immunol 2007;37:17–20.
85. He B, Chadburn A, Jou E, Schattner EJ, Knowles DM, Cerutti A. Lymphoma B cells evade apoptosis through the TNF family members BAFF/BlyS and APRIL. J Immunol 2004;172:3268–79.
86. Endo T, Nishio M, Enzler T, et al. BAFF and APRIL support CLL B cell survival through activation of the canonical NF-kB pathway. Blood 2007;109:703–10.
87. Xu L, Shu H. TRAF3 is associated with BAFF-R and negatively regulates BAFF-R-mediated NF-kB activation and IL-10 production. J Immunol 2002;169:6883–9.
88. Kayagaki N, Yan M, Seshasayee D, et al. BAFF/BlyS receptor 3 binds the B cell survival factor BAFF ligand through a discrete surface loop and promotes procesing of NF-kB2. Immunity 2002;17:515–24.
89. Patke A, Mecklenbrauker I, Tarakhovsky A. Survival signaling in resting B cells. Curr Opin Immunol 2004;16:251–5.
90. Moreno-Garcia ME, Sommer KM, Bandaranayake AD, Rawlings DJ. Proximal signals controlling B-cell antigen receptor (BCR) mediated NF-kB activation. Adv Exp Med Biol 2006;584:89–106.
91. Younes A, Aggarwall BB. Clinical implications of the tumor necrosis factor family in benign and malignant hematologic disorders. Cancer 2003;98:458–67.
92. Bishop GA, Hostager BS. Molecular mechanisms of CD40 signaling. Arch Immun Ther Exper 2001;49:129–37.
93. Shinners NP, Carlesso G, Castro I, et al. Btk mediates NF-kB activation and B cell survival by BAFF of the TNF-R family. J Immunol 2007;179:1675–80.
94. Marsters SA, Yan M, Pitti RM, Haas PE, Dixit VM, Ashkenazi A. Interaction of the TNF homologues BLyS and APRIL with the TNFR homologues BCMA and TACI. Curr Biol 2000;10:785–8.
95. He JQ, Zarnegar B, Oganesyan G, et al. Rescue of TRAF3-null mice by p100 NF-kB deficiency. J Exp Med 2006;203:2413–8.

Chapter 6
TACI Signaling and Its Role in Immunity

Richard J. Bram

Abstract Transmembrane activator and CAML interactor (TACI) was initially described 10 years ago. Since that time, it has been increasingly recognized as an important regulator of immunity in both mice and humans. Although much has been learned about its signaling and physiological functions, its complexity in terms of ligand binding, cell-type-specific effects, as well as potential roles in human disease dictates that much work remains to be done in this area. In this chapter, the basic signaling and physiologic role of TACI will be discussed.

Keywords TACI · Lymphocyte signaling · Humoral immunity · APRIL · BAFF

6.1 Initial Characterization of TACI

TACI (official symbol TNFRSF13B, a.k.a. CD267) was originally discovered in a yeast two-hybrid search for partners of the intracellular protein CAML (calcium-modulating cyclophilin ligand) [1]. The human TACI gene resides on chromosome 17p11 and encodes a 293-amino acid polypeptide. Although the initial computational analysis of the TACI sequence did not reveal significant homology to known proteins, the presence of a cluster of hydrophobic amino acid residues in the middle of the protein suggested it could be an integral membrane protein. Transient transfection studies revealed that it is indeed able to accumulate at the cell surface, with the N-terminus of the protein oriented toward the extracellular space. Although the N-terminus of TACI is extracellular, it lacks a cleaved signal sequence, making it a type-III transmembrane protein. Two regions in the N-terminus of TACI were observed to share limited sequence similarity to cysteine-rich domains (CRDs)

R.J. Bram (✉)
Departments of Immunology and Pediatric and Adolescent Medicine, Mayo Clinic, Rochester, MN 55905, USA

Financial Disclosure: The author is listed as an inventor on three patents related to the TACI gene and protein held by St. Jude Children's Research Hospital and licensed to Zymogenetics.

M.P. Cancro (ed.), *BLyS Ligands and Receptors*, Contemporary Immunology,
DOI 10.1007/978-1-60327-013-7_6,

characteristic of members of the tumor necrosis factor receptor superfamily [2]. These motifs direct the archetypal folding of the ligand-binding domains of TNFR family members. Although the sequence similarity between TACI and other TNFRs is relatively weak, the solution structure of TACI has been determined and validates its inclusion in this category [3].

6.2 Ligands

Two TNF homologs, BAFF and APRIL, bind with high affinity and specificity to TACI (Fig. 6.1) [4]–[8]. BAFF (also known as BLyS, THANK, TALL-1, and zTNF4) enhances survival and proliferation of transitional T2 and later stage B cells, and is thought to participate in several types of autoimmune disease in mice and humans. APRIL has been shown to enhance antibody class switching. A recent review by Dillon et al. [9] summarizes results from multiple different studies examining the binding affinities of these two ligands for TACI and concludes that they are similar to each other in the nanomolar range. Intriguingly, a study from patients with autoimmune rheumatologic diseases revealed the existence of stable heterotrimers of BAFF with APRIL, which were able to signal through TACI [10].

BAFF also binds two other TNFR family members BCMA and BAFF-R, while APRIL can bind BCMA but not BAFF-R [11], [12]. BAFF–APRIL heterotrimers were shown to bind TACI, but neither BCMA nor BAFF-R [10]. Like TACI, BCMA and BAFF-R are type-III transmembrane proteins, further suggestive of a common evolutionary ancestor. Unlike TACI, BCMA and BAFF-R have a single cysteine-rich ligand-binding domain. However, further studies have since demonstrated

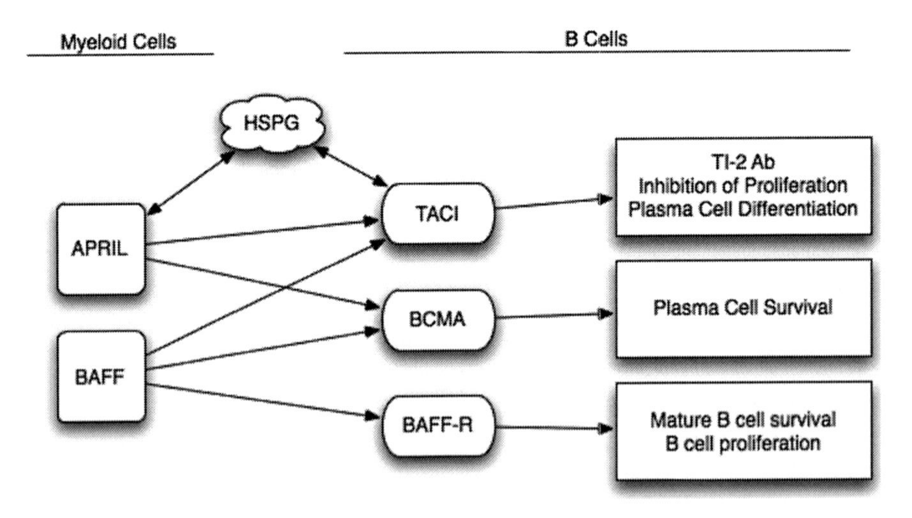

Fig. 6.1 Ligand-receptor interactions in the BAFF/APRIL system. The TNF homologes BAFF and APRIL are expressed primarily by myeloid cells, and bind as indicated to TACI, BCMA, and BAFF-R. Heparan sulfate proteoglycans (HSPG) have been shown to bind to APRIL and to TACI

that only the second CRD of TACI is important for ligand binding [3]. Indeed, alternative splicing of TACI in humans produces a shorter protein ("TACI short") lacking the first CRD, which binds to ligands with similar affinities as full length TACI. Although of unclear significance, there exist potential cleavage sites in the extracellular domain of TACI, and soluble receptor may circulate, thereby reducing the available free BAFF and APRIL.

Recently, the heparan-sulfate proteoglycan (HSPG) syndecan-2 was found to be able to bind to TACI, and not to either BAFF-R or BCMA [13]. This interaction was solely dependent upon the heparan-sulfate side chains of syndecan-2 and suggests that TACI may be triggered by multiple different HSPGs in vivo. APRIL was also shown to bind to HSPGs [14], raising the possibility of ternary complex formation in APRIL–TACI signaling, reminiscent of the bFGF signaling system.

6.3 Patterns of Expression of TACI

TACI is most highly expressed on mature B lymphocytes. It is first detectable at the transitional T1 stage and increases 10-fold in marginal zone B cells [15], and finally downregulates at the plasma cell stage [16]. CpG stimulation through TLR9 in murine splenic B cells induced upregulation of TACI [17]. The induction was rapid and occurred at both the mRNA and protein levels. LPS stimulation through TLR4 had a similar effect, but was primarily observed in marginal zone B cells. Others have reported that cross-linking of IgM, particularly under conditions of CD21 co-stimulation, led to moderate upregulation of TACI in splenic B2 cells. Also, splenic B cells induced TACI to high levels following treatment with CD40 antibody [16]. In human B cells, activation by BCR cross-linking in the presence of multiple cytokines (IL2 plus IL10) lead to robust induction of surface TACI [18].

Although B cells are clearly the highest expressors of TACI, there have been multiple reports of lower level expression by other cells in the immune system. We and others detected low-level TACI expression on activated peripheral T cells from mice [1], [19]. Possibly consistent with this observation was a finding by Huard et al., who demonstrated direct activation of T cells by BAFF [20]. In addition, Sook et al. [21] found evidence for expression of TACI in monocytes, which was induced by BAFF (see below).

6.4 Signaling Events

Elucidation of the signals downstream of TACI has been complicated by its expression pattern and ligand-binding specificity, which overlap with that of BAFF-R and BCMA. Thus, studies on cells exposed to APRIL or BAFF cannot with certainty be attributed to effects via TACI alone. Conversely, antibody cross-linking experiments may not completely accurately reflect the normal signaling consequences of ligand–receptor interaction. Nonetheless, most studies indicate that TACI generates a strong activation of NF-κB, as is the characteristic of TNFR family members

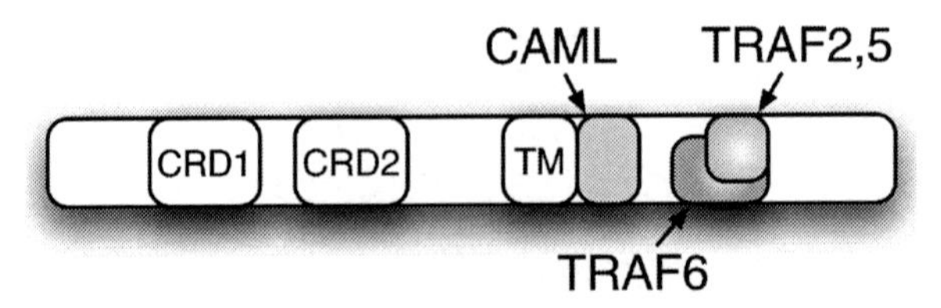

Fig. 6.2 Domains of TACI. The cysteine rich domains (CRD1, CRD2) that characterize TNFR family members lie in the extracellular domain, demarcated by the transmembrane domain (TM). Approximate sites of TRAF and CAML binding are indicated

[7, 19]. Transfection into (TACI negative) Jurkat T cells revealed that TACI can also activate AP-1 and NF-AT transcription factors, suggesting not only TRAF activation, but a potential effect on Ca^{2+} influx[1].

TACI has been shown to bind to TRAF molecules, as is typical for members of the TNFR family. Yeast two-hybrid experiments revealed binding to TRAF 2, 5, and 6 [19]. In addition to NF-κB activation, TACI was also shown to induce a strong JNK signal. Xia et al. [19] also independently reported the interaction of CAML and TACI, and discovered that the TRAF-binding sites of TACI were distinct from each other and the CAML interaction domain (Fig. 6.2). Although CAML has been implicated in Ca^{2+} influx signaling [22] and also in the process of receptor trafficking [23], the role and mechanism of its interaction with TACI are not yet clear.

In an exciting new study, Bossen and colleagues have discovered an unusual twist regarding ligand-induced signaling by TACI [24]. BAFF has been previously shown to stably adopt two alternative higher-order structures in solution: a trimer typical of other TNF family members or a polymeric structure containing 60 copies of the BAFF protein in a virus-like complex [25]. Bossen now demonstrates that although both forms of BAFF can bind with high affinity to TACI, only the 60-mer of BAFF is able to activate intracellular signaling. Since both forms are present in the serum of humans and mice, this provides yet another level of regulation in this system via alteration of the multimeric structure of BAFF. Events that regulate the relative proportions of 3-mer vs. 60-mer are not yet clear. In addition, it was demonstrated that cell-surface-associated BAFF would induce TACI signaling, as expected. This new finding may also help to explain the earlier results from Huard et al., who demonstrated T-cell activation by immobilized BAFF that could not be observed with soluble (presumably trimeric) BAFF [20]. Such T-cell activation may have been mediated through TACI.

6.5 Physiologic Role of TACI

Based on the known effects of BAFF in promoting B-cell survival and development, we and others had initially predicted that TACI knockout mice would demonstrate a defect in B-lymphocyte production or homeostasis. The surprising result, however, was that such mice had double the expected number of peripheral B cells in blood and spleen [26]. B-cell proliferation was enhanced in mutant mice, as evidenced by more rapid incorporation of BrdU in vivo. All of the various B-cell subsets were present in these mice; however, there was a clear deficiency in T-independent type-II antibody production (to NP-ficoll or pneumococcal capsular antigens). T-dependent responses were normal. Independently generated TACI knockout mice had mostly similar phenotypes as young animals [27], although Yan et al. reported moderately enhanced proliferation in response to BAFF. TI-2 responses are an important aspect of immunological protection from rapidly growing encapsulated bacteria; thus, it is of great interest that recent reports indicate that neonates may have significantly reduced levels of TACI on their B cells [28]. Young children are at particularly high risk of overwhelming infection by encapsulated bacteria, so this discovery may lead to improved treatments to help prevent such problems.

An important additional role for TACI in mediating immunoglobulin class switching in mice was demonstrated by Castigli et al. [29]. It was subsequently demonstrated by Sakurai et al. [30] that simultaneous stimulation through APRIL binding to TACI and heparan sulfate proteoglycan provided an optimal signal to induce switching to IgA. This is consistent with the known defect in class switching characteristic of humans with common variable immunodeficiency (CVID), a disorder recently shown to be due to TACI mutations in some cases [31, 32]. Interestingly, at older ages, TACI knockout mice tended to develop autoimmune kidney disease and fatal lymphoproliferation [33]. This could be a direct result of the enhanced proliferation of B cells in these mice. Nonetheless, it is an intriguing finding, because patients with CVID also have frequent development of autoimmunity.

6.6 Negative vs. Positive Signaling by TACI

Although these knockout studies strongly suggested that TACI provides a negative survival or proliferation signal to B cells, the true outcome of TACI signaling in cells is not completely settled. In support of this idea, Seshasayee [33] showed that human B cells stimulated with IL4 and anti-CD40 had less incorporation of tritiated thymidine if co-cultured with agonistic antibodies to TACI. Furthermore, Sakurai et al. [30] reported that agonistic TACI antibodies suppressed human B-cell survival and antibody production following high–level CD40 stimulation of cells in vitro. On the other hand, several studies suggested the opposite that is that the effect of TACI signaling at the individual cellular level is to enhance survival or growth. Craxton et al. [34] found that BAFF signaling through TACI suppressed expression of the pro-apoptotic protein Bim and led to enhanced survival in WEHI-231 B cells. Bischof et al. [13] found that B cells from TACI knockout mice proliferated less than cells from littermate controls when stimulated with BAFF and syndecan-2 in the presence of IL-4 and anti-IgM. Castigli et al. [35] showed that CD40 + IL4-induced antibody secretion by naïve murine B cells was enhanced by stimulation of TACI. These experiments were carefully controlled to ensure the effect was not due to stimulation of BCMA or heparan sulfate proteoglycans. As noted by Castigli, these results appear to directly contradict those of Sakurai et al., and may reflect alternative outcomes of low- vs. high-level CD40 stimulation.

An additional level of complexity that may underlie some of these differences could be due to altered biology of this receptor system in cells at various stages of differentiation. For example, Zhang et al. [36] found that while BAFF was required continuously at multiple time points for optimal B-cell proliferation and differentiation to plasma cells, BAFF-R was only important at the initial point of stimulation. Subsequently, BAFF-R became downregulated and TACI expression increased, as the primary BAFF receptor. Although signaling through BCMA was not ruled out in this study, its expression was very low, and only on a small subset of B cells. Others have also reported that BAFF-R is downregulated upon B-cell activation and is followed by induced expression of TACI, again suggestive of the importance of serial activation of each receptor in turn [37].

An additional factor that may affect the outcome of TACI signaling could be the proliferative state of the cell type. Chiu et al. [38] found that the malignant cells of Hodgkin's lymphoma signal through TACI and BCMA to enhance survival and proliferation. By siRNA-mediated knockdown, the authors showed that TACI provided primarily a proliferation signal, while BCMA was more responsible for the survival promoting effect of APRIL or BAFF.

6.7 Mechanism of Action in T-Independent II Immunity

The defect in TI-2 immune responsiveness in mice lacking TACI has been mysterious, particularly since these mice have significantly increased numbers of mature B-cell subsets. We recently addressed this issue by crossing the TACI knockout gene into mice with a quasi-monoclonal B-cell receptor (BCR) specific for nitrophenyl (QM mice), in order to facilitate identification of antigen-specific events [39]. Importantly, adoptive transfer of a small number of QM TACI$^{-/-}$ B cells into wild-type mice yielded a deficient response to the TI-2 antigen NP-ficoll. This indicated that not only does TACI act in a cell-autonomous fashion in TI-2 responses, but in addition the defect is not related to altered amounts of (decoy) soluble TACI receptor in the blood. Furthermore, the parental QM TACI$^{-/-}$ mice had robust IgM responses to NP-ficoll immunization, similar to littermate QM TACI^{+} mice. Clearly, TACI is not required for the TI-2 response if the precursor frequency of B cells is quite high. Although antigen-induced activation and proliferation were normal in adoptively transferred QM TACI$^{-/-}$ B cells, there was a relative defect in differentiation to the plasma cell stage. We concluded that one potential role of TACI is to block proliferation following B-cell stimulation in order to facilitate differentiation to the plasma cell stage, in which antibody production is much more efficient.

In this sense, TACI could be considered an answer to the biological problem of how an organism can maintain a very high diversity of B cells, which are therefore of necessity present at very low frequencies and yet respond in rapid fashion to an infection. Because TACI appears to be critically important for the T-independent response to bacterial antigens, this may be more of a problem in rapidly growing bacterial infections, rather than viral infections, which allow time for the more controlled T-dependent response to occur.

A role for TACI in plasma cell differentiation was also reported recently by Castigli et al. in the case of B cells stimulated suboptimally with CD40 cross-linking [35]. This would presumably reflect a T-dependent type of antibody production response. Because TACI mutations may contribute to the disease phenotype of CVID, these findings suggest that part of the mechanism may involve a defect in differentiation of B cells to the plasma cell stage, in both T-dependent as well as T-independent immunity.

6.8 Reverse Signaling: TACI as Ligand

A recent paper from the Franco lab [40] described a completely unexpected role for TACI in the priming of naïve cytotoxic T cells (CTLs) in mice. It was discovered that mice lacking B cells (resulting from deletion of the IgM heavy chain) were unable to efficiently generate CTLs in response to immunization with multiple different peptides or intact protein. Reconstitution of these mice with B cells from wild-type or several different mutant mice (beta2-microglobulin, CD40, IL-2, or IL-4 knockout) enabled normal CTL generation. On the other hand, B cells from TACI knockout animals were unable to restore function, suggesting an essential role for TACI in CTL priming in vivo. Remarkably, the requirement for B cells in CTL priming could be bypassed by administration of small amounts of recombinant protein consisting of the extracellular domain of TACI fused to a Fc polypeptide. There was no apparent binding of TACI-Fc to CTLs, as expected, and the presumed target of TACI in this system was dendritic cells, because they were shown to express high levels of surface BAFF and rapidly divide in vitro in the presence of B cells. The requirement for DC maturation via TACI is possibly due to the need of an ongoing expansion of a large number of activated antigen-presenting cells during CTL priming as suggested by the evidence that peptide-pulsed DC terminally differentiated and matured in vitro could not prime CTL in B-cell-deficient and TACI-deficient mice, if not in the presence of TACI administered in vivo. Although the precise mechanism remains to be determined, this new finding clearly indicates that TACI may act as a ligand to facilitate T-cell cytotoxicity, and joins the ranks of other

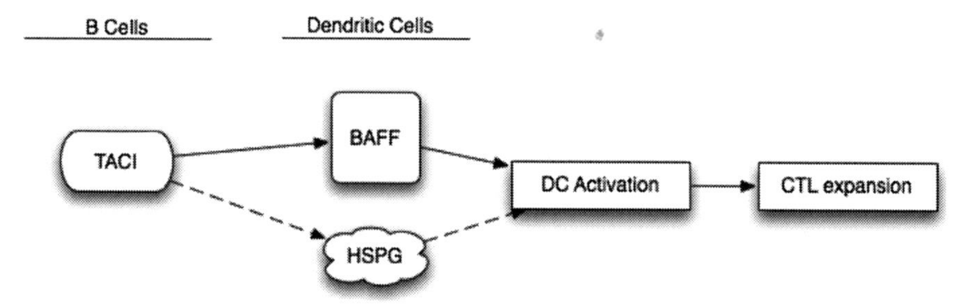

Fig. 6.3 Reverse signaling by TACI. Proposed model by which cell surface TACI might generate a "reverse signal" to dendritic cells through BAFF or HSPG to cause cellular activation, leading ultimately to expansion of cytotoxic lymphocytes

TNFR family members that signal bi-directionally [41]. It is intriguing to consider the possibility that DC-resident heparan sulfate proteoglycans may also participate in this novel signaling event by acting as a receptor for TACI (Franco, personal communication) (Fig. 6.3).

6.9 TACI Expression by Monocytes

An additional alternative role for TACI was suggested by recent work from Chang et al. [21]. It was reported that BAFF provides a potent survival and activation signal in human monocytes, causing them to differentiate to macrophages. TACI was the only known BAFF receptor detectable in these cells, and was proposed by the authors as likely to be mediating this activation event. Interestingly, TACI was solely intracellular in unstimulated cells, but accumulated at higher levels at the cell surface upon stimulation with either BAFF or IL10, suggesting an additional possible mode of regulation.

6.10 Conclusions

In spite of the somewhat conflicting data in the field, it is clear that TACI is an important regulator of multiple functions in mature B cells, including T-independent humoral immunity and immunoglobulin class switching. In light of a possible role for TACI in regulating T-dependent immunity, including aberrant function in patients with CVID, these novel discoveries of possible regulation of antigen-presenting cells by TACI raise interesting new directions that should be pursued. In addition to the improved understanding of immunity that will arise from this field, it has the potential to improve the diagnosis or treatment of autoimmunity, immunodeficiency, and hematologic malignancies.

References

1. von Bulow GU, Bram RJ. NF-AT activation induced by a CAML-interacting member of the tumor necrosis factor receptor superfamily. Science 1997;278(5335):138–41.
2. Ware CF. The TNF superfamily. Cytokine Growth Factor Rev 2003;14(3–4):181–3.
3. Hymowitz SG, Patel DR, Wallweber HJ, et al. Structures of APRIL-receptor complexes: like BCMA, TACI employs only a single cysteine-rich domain for high affinity ligand binding. J Biol Chem 2005;280(8):7218–27.
4. Gross JA, Johnston J, Mudri S, et al. TACI and BCMA are receptors for a TNF homologue implicated in B-cell autoimmune disease. Nature 2000;404(6781):995–9.
5. Marsters SA, Yan M, Pitti RM, Haas PE, Dixit VM, Ashkenazi A. Interaction of the TNF homologues BLyS and APRIL with the TNF receptor homologues BCMA and TACI. Curr Biol 2000;10(13):785–8.
6. Wu Y, Bressette D, Carrell JA, et al. Tumor necrosis factor (TNF) receptor superfamily member TACI is a high affinity receptor for TNF family members APRIL and BLyS. J Biol Chem 2000;275(45):35478–85.

7. Yan M, Marsters SA, Grewal IS, Wang H, Ashkenazi A, Dixit VM. Identification of a receptor for BLyS demonstrates a crucial role in humoral immunity. Nat Immunol 2000;1(1):37–41.
8. Yu G, Boone T, Delaney J, et al. APRIL and TALL-I and receptors BCMA and TACI: system for regulating humoral immunity. Nat Immunol 2000;1(3):252–6.
9. Dillon SR, Gross JA, Ansell SM, Novak AJ. An APRIL to remember: novel TNF ligands as therapeutic targets. Nat Rev Drug Discov 2006;5(3):235–46.
10. Roschke V, Sosnovtseva S, Ward CD, et al. BLyS and APRIL form biologically active heterotrimers that are expressed in patients with systemic immune-based rheumatic diseases. J Immunol 2002;169(8):4314–21.
11. Thompson JS, Bixler SA, Qian F, et al. BAFF-R, a newly identified TNF receptor that specifically interacts with BAFF. Science 2001;293(5537):2108–11.
12. Yan M, Brady JR, Chan B, et al. Identification of a novel receptor for B lymphocyte stimulator that is mutated in a mouse strain with severe B cell deficiency. Curr Biol 2001;11(19): 1547–52.
13. Bischof D, Elsawa SF, Mantchev G, et al. Selective activation of TACI by syndecan-2. Blood 2006;107(8):3235–42.
14. Hendriks J, Planelles L, de Jong-Odding J, et al. Heparan sulfate proteoglycan binding promotes APRIL-induced tumor cell proliferation. Cell Death Differ 2005;12(6):637–48.
15. Hsu BL, Harless SM, Lindsley RC, Hilbert DM, Cancro MP. Cutting edge: BLyS enables survival of transitional and mature B cells through distinct mediators. J Immunol 2002;168(12):5993–6.
16. O'Connor BP, Raman VS, Erickson LD, et al. BCMA is essential for the survival of long-lived bone marrow plasma cells. Journal of Experimental Medicine 2004;199(1):91–8.
17. Treml LS, Carlesso G, Hoek KL, et al. TLR stimulation modifies BLyS receptor expression in follicular and marginal zone B cells. J Immunol 2007;178(12):7531–9.
18. Darce JR, Arendt BK, Wu X, Jelinek DF. Regulated expression of BAFF-binding receptors during human B cell differentiation. J Immunol 2007;179(11):7276–86.
19. Xia XZ, Treanor J, Senaldi G, et al. TACI is a TRAF-interacting receptor for TALL-1, a tumor necrosis factor family member involved in B cell regulation. J Exp Med 2000;192(1):137–43.
20. Huard B, Schneider P, Mauri D, Tschopp J, French LE. T cell costimulation by the TNF ligand BAFF. J Immunol 2001;167:6225–31.
21. Chang SK, Arendt BK, Darce JR, Wu X, Jelinek DF. A role for BLyS in the activation of innate immune cells. Blood 2006;108(8):2687–94.
22. Bram RJ, Crabtree GR. Calcium signalling in T cells stimulated by a cyclophilin B-binding protein. Nature 1994;371(6495):355–8.
23. Tran DD, Russell HR, Sutor SL, van Deursen J, Bram RJ. CAML is required for efficient EGF receptor recycling. Developmental Cell 2003;5:245–56.
24. Bossen C, Cachero TG, Tardivel A, et al. TACI, unlike BAFF-R, is solely activated by oligomeric BAFF and APRIL to support survival of activated B cells and plasmablasts. Blood 2007;179(11): 7276–86.
25. Liu Y, Xu L, Opalka N, Kappler J, Shu HB, Zhang G. Crystal structure of sTALL-1 reveals a virus-like assembly of TNF family ligands. Cell 2002;108(3):383–94.
26. von Bulow GU, van Deursen JM, Bram RJ. Regulation of the T-independent humoral response by TACI. Immunity 2001;14(5):573–82.
27. Yan M, Wang H, Chan B, et al. Activation and accumulation of B cells in TACI-deficient mice. Nat Immunol 2001;2(7):638–43.
28. Kaur K, Chowdhury S, Greenspan NS, Schreiber JR. Decreased expression of tumor necrosis factor family receptors involved in humoral immune responses in preterm neonates. Blood 2007;110(8):2948–54.
29. Castigli E, Scott S, Dedeoglu F, et al. Impaired IgA class switching in APRIL-deficient mice. Proc Natl Acad Sci USA 2004;101:3903.
30. Sakurai D, Hase H, Kanno Y, Kojima H, Okumura K, Kobata T. TACI regulates IgA production by APRIL in collaboration with HSPG. Blood 2007;109(7):2961–7.

31. Castigli E, Wilson SA, Garibyan L, et al. TACI is mutant in common variable immunodeficiency and IgA deficiency. Nat Genet 2005;37(8):829–34.
32. Salzer U, Chapel HM, Webster AD, et al. Mutations in TNFRSF13B encoding TACI are associated with common variable immunodeficiency in humans. Nat Genet 2005;37(8): 820–8.
33. Seshasayee D, Valdez P, Yan M, Dixit V, Tumas D, Grewal IS. Loss of TACI causes fatal lymphoproliferation and autoimmunity, establishing TACI as an inhibitory BLyS receptor. Immunity 2003;18(2):279–88.
34. Craxton A, Draves KE, Gruppi A, Clark EA. BAFF regulates B cell survival by downregulating the BH3-only family member Bim via the ERK pathway. J Exp Med 2005;202(10): 1363–74.
35. Castigli E, Wilson SA, Elkhal A, Ozcan E, Garibyan L, Geha RS. Transmembrane activator and calcium modulator and cyclophilin ligand interactor enhances CD40-driven plasma cell differentiation. J Allergy Clin Immunol 2007;120(4):885–91.
36. Zhang X, Park CS, Yoon SO, et al. BAFF supports human B cell differentiation in the lymphoid follicles through distinct receptors. Int Immunol 2005;17(6):779–88.
37. Mongini PK, Inman JK, Han H, Fattah RJ, Abramson SB, Attur M. APRIL and BAFF promote increased viability of replicating human B2 cells via mechanism involving cyclooxygenase 2. J Immunol 2006;176(11):6736–51.
38. Chiu A, Xu W, He B, et al. Hodgkin lymphoma cells express TACI and BCMA receptors and generate survival and proliferation signals in response to BAFF and APRIL. Blood 2007;109(2):729–39.
39. Mantchev GT, Cortesao CS, Rebrovich M, Cascalho M, Bram RJ. TACI is required for efficient plasma cell differentiation in response to T-independent type 2 antigens. J Immunol 2007;179(4):2282–8.
40. Diaz-de-Durana Y, Mantchev GT, Bram RJ, Franco A. TACI-BLyS signaling via B-cell-dendritic cell cooperation is required for naive CD8+ T-cell priming in vivo. Blood 2006;107(2):594–601.
41. Sun M, Fink PJ. A new class of reverse signaling costimulators belongs to the TNF family. J Immunol 2007;179(7):4307–12.

Chapter 7
The BAFF/APRIL System in Autoimmunity

Fabienne Mackay, William A. Figgett, Pali Verma, and Xavier Mariette

Abstract BAFF, a TNF-like ligand, has emerged as an essential factor for B-cell survival and maturation, and the discovery of this factor has been a major advance for understanding B-cell biology. Excessive BAFF production is associated with the development of autoimmune disorders such as lupus, a problem attributed to the inappropriate survival of self-reactive B cells. These early findings prompted the development of BAFF inhibitors, which are now in human clinical trials. Excessive BAFF production does not cause a total breakdown of B-cell tolerance, but rather the expansion of low/intermediate affinity self-reactive B cells, in particular marginal zone (MZ) B cells. Studies with BAFF transgenic mice have established a new paradigm for autoimmunity, in that BAFF-driven autoimmune disease is entirely T-cell-independent and not linked to affinity maturation of accumulating low/intermediate affinity self-reactive B cells. Interestingly, the expression of the toll-like receptor (TLR) signalling element MyD88 is required for disease, and a reciprocal stimulatory loop exists between TACI and TLR signalling. As TACI is an essential receptor driving T-cell-independent B-cell responses, it raises the possibility of its role in BAFF-mediated autoimmunity, linked to the innate immune system. The BAFF system appears to be truly unique, in that this one factor activates a separate pathogenic autoimmune mechanism not yet represented in any of the current animal models of lupus, and possibly reflecting a mechanism driving disease in a proportion of human autoimmune patients.

Keywords BAFF · APRIL · BAFF-R · TACI · BCMA · Autoantibodies · Systemic

---

F. Mackay (✉)
Autoimmunity Research Unit, The Garvan Institute of Medical Research, Darlinghurst, NSW 2010, Australia

M.P. Cancro (ed.), *BLyS Ligands and Receptors*, Contemporary Immunology,
DOI 10.1007/978-1-60327-013-7_7,

7.1 Introduction

The immune system performs several key functions, other than simply fighting infections and eliminating cancer cells. It produces immune cells from progenitor cells in the bone marrow and it carefully selects protective immune cells while eliminating or neutralising potentially harmful self-reactive cells at several immune checkpoints. A very fine balance between life and death plays an important role in this type of "quality control" that results in immune tolerance. B-lymphocyte responses to B-cell-activating factor from the TNF family (BAFF, also termed TNFSF13b, BLyS) play an important part in this process [1–3] and, as detailed later, failures in the BAFF system can lead to severe immune disorders [1–3].

Failure in B- and/or T-cell tolerance leads to autoimmune disorders. In the case of BAFF, excessive production of this factor leads to systemic lupus erythematosus (SLE)-like symptoms in mice and is associated with SLE and other autoimmune diseases in humans. SLE is a chronic inflammatory autoimmune disorder that affects various tissues, particularly the skin and kidney [4, 5]. Disease pathogenesis is characterised by the production of a range of autoantibodies, in particular anti-nuclear autoantibodies (ANA) [4]. Treatment of SLE still relies on broad immunosuppressants such as corticosteroids and hydroxychloroquine sulphates [6, 7]. While these treatments have been improved, they remain inefficacious in some SLE patients, suggesting that alternative or unappreciated immune mechanisms may be operating.

Autoantibody-producing B cells appear central to the pathogenesis of SLE. Many mechanisms have been proposed to explain their appearance from impaired survival/apoptosis signals preventing negative selection [3], dysfunctional complement or inhibitory Fc receptors [8, 9] to activation of toll-like receptors (TLRs) in response to accumulation of apoptotic bodies [10, 11]. However, substantial evidence indicates that T cells are key players in SLE [5, 12, 13]. Inhibition of T-cell activation or T–B-cell interaction is an effective way to prevent autoimmunity in a number of animal models of SLE [14, 15]. More recently, mutation of the *Roquin* gene, which specifically perturbs the function of follicular helper T cells, induces development of SLE in mice [16]. In humans, various HLA haplotypes have been associated with susceptibility to SLE [5, 12]. T-cell signalling in SLE patients can be abnormal and analysis of anti-DNA autoantibodies has revealed somatic mutations, suggestive of T-dependent affinity maturation [5, 12]. B-cell-depleting reagents, such as Rituximab, are currently tested in SLE clinical trials [17]. It is still unclear whether the efficacy of this treatment is due to reduced autoantibody production or the depletion of B cells with potent antigen-presenting cell (APC) function for T cells, or both. Despite the obvious pathogenic role of autoantibodies and B cells, the corruption of T-cell tolerance is considered the underlying defect in antibody-mediated systemic autoimmune diseases [5, 16].

In conclusion, the field of autoimmunity, in particular rheumatic autoimmune diseases such as not only SLE, but also Sjögren's syndrome (SS) and rheumatoid arthritis (RA), is dominated by the notion of a tight cooperation between dendritic cells (DCs) (e.g. TLR-activated plasmacytoid DC (pDC)), T cells, and B cells as the underlying cause of disease. In this chapter, we will detail the autoimmune mechanisms triggered by dysregulated expression of BAFF, which is very unusual, as it does not require any involvement of T cells. These findings should alert us to the fact that there may be a subset of SLE patients suffering from a subtype of disease, who may benefit from alternative treatments.

7.2 The BAFF/APRIL Ligands/Receptors System

7.2.1 BAFF and APRIL

7.2.1.1 Structure

B-cell-activating factor of the tumour necrosis factor family (known as BAFF, BLyS, TALL-1, THANK, zTNF-4, and TNFSF-13B) and a proliferation-inducing ligand (APRIL, TALL-2, TRDL-1, and TNFSF13a) are members of the TNF superfamily (reviewed in [3, 18, 19, 20]). BAFF and APRIL are produced as type-II transmembrane proteins (like many of the TNF ligands), then proteolytically cleaved at a furin protease site and released in a soluble form [21]. APRIL is cleaved in the Golgi prior to the release and exists in a soluble form only [22]. BAFF is cleaved by furin convertase between amino acids R133 and A134, which are situated at the end of a consensus furin cleavage motif in the stalk region, and BAFF is released as a soluble 152-amino acid 17-kDa protein [23, 24]. Similarly, 250-amino acid APRIL is cleaved between R104 and A105, and is released as a 146-amino acid 17-kDa protein [22] (Fig. 7.1).

The human BAFF gene contains a transmembrane domain and flanking regions in exon 1, a furin-processing site in exon 2, and a TNF homology domain (THD) in exons 3–6 [18]. The THD mediates binding to receptors, which are discussed in the second part of this section. The mouse BAFF gene contains an additional exon

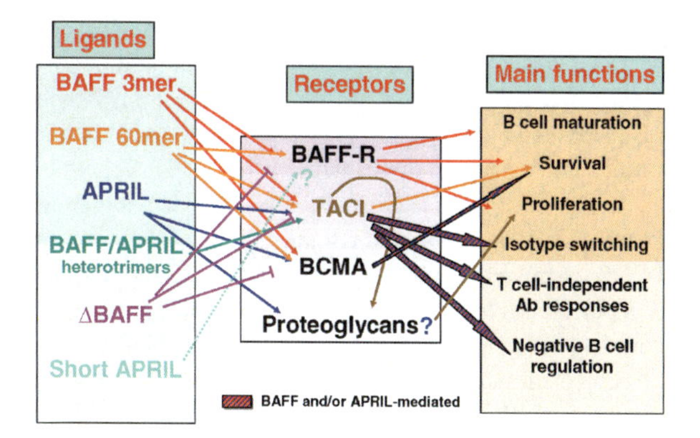

Fig. 7.1 Ligand/receptor interactions and function. Each ligand is depicted in a different colour to outline the specific binding to receptors and also functions unique to a particular ligand–receptor interaction. The *arrows* with *blue* and *red* stripes refer to functions that can be triggered by either BAFF or APRIL

encoding 30-amino acid residues between the furin-processing site and the THD. Mouse and human APRIL both contain a furin-processing site in exon 2 [18].

BAFF is an active ligand as a homotrimer [25], which contains two magnesium ions interacting with the side chains of Q234, N235, and N243 [26]. This is the main form of BAFF found in the circulation [27]. However, a 60-mer form of BAFF has been obtained at physiological pH conditions and this form is also able to bind to its receptors [28, 29]. BAFF 60-mers arise by the assembly of 20 BAFF trimers into a virus-like cluster. BAFF possesses a "flap" region, which is absent in APRIL, and which is necessary for the assembly of the 60-mer form [18, 28]. The "flap" region is also referred to as the DE loop, because it is formed by five amino acids inserted between β-strands D and E [3, 18, 30]. Interactions between BAFF trimers are extensive, as they are comprised of hydrogen bonds, salt bridges, and hydrophobic contacts [28]. For this reason, formation of the 60-mer form is irreversible at physiological pH; however, dissociation to trimers occurs at acidic pH. This pH dependency suggests that BAFF 60-mer assembly is not due to random aggregation [29]. Despite this evidence, claims were made that BAFF 60-mer assembly was merely an artefact due to fusing N-terminal tags of recombinant BAFF protein preparations used in these assays [25]. However, a construct of BAFF with no amino-terminal tag can form 60-mers in solution [31], and this suggests that BAFF 60-mer assembly is a naturally occurring event. The function of BAFF 60-mer may involve the recruitment of numerous receptor trimers to increase the local concentration of receptors and enhance signalling [28]. The BAFF 60-mer is the more active form of BAFF, and is therefore likely to be physiologically relevant. It appears to be the only form of BAFF triggering signals through the receptor transmembrane activator and calcium modulator and cyclophilin ligand interactor (TACI, TNFRSF13b) [27].

BAFF and APRIL are capable of forming heterotrimers when co-expressed [32]. These are biologically active and can be found circulating in the serum of both healthy humans and patients with systemic autoimmune rheumatic diseases [32]. It is possible that other members of the TNF family are similarly capable of forming heterotrimers with BAFF and/or APRIL, but this has not been investigated [32].

ΔBAFF is a splice variant of BAFF that was identified in humans and mice [33]. The transcript of ΔBAFF lacks the exon encoding the first β-sheet of the THD, which is exon 3 in humans and exon 4 in mice [33]. ΔBAFF is not released from the cell membrane, but can form multimers with BAFF such that the release of BAFF from the cell is inhibited. In this way, ΔBAFF acts as a negative regulator of BAFF [34]. ΔBAFF is less abundant than the full-length isoform of BAFF in lymphatic and CNS tissues [35]. Recently, two new transcript variants of BAFF have been identified in peripheral blood mononuclear cells [36], but functional studies are required to determine their structural and biological significance.

The APRIL locus is situated adjacent to the locus of TWEAK, another member of the TNF superfamily [19]. A splice variation event at the APRIL/TWEAK locus leads to the production of an APRIL/TWEAK hybrid ligand, named TWE–PRIL [19]. This ligand is a fusion of the APRIL extracellular domain with the TWEAK transmembrane portion, and is therefore suspected to be biologically active similarly to APRIL [37].

7.2.1.2 Expression

BAFF and APRIL are produced by a range of innate cells, including monocytes [38], macrophages, dendritic cells (DCs) [39, 40], neutrophils [41, 42], mast cells [43], and astrocytes [35] (Table 7.1). Innate cells act to defend the host from infection in a non-specific manner. They do not confer long-lasting protective immunity, but have a more immediate action than cells of the adaptive immune system. BAFF may play an important role in the activation of innate cells, as BAFF strongly induces monocyte survival, pro-inflammatory cytokine secretion, and co-stimulatory molecule expression [44]. Conversely, innate cells usually produce BAFF in response to pro-inflammatory factors and co-stimulatory molecules. For instance, BAFF is produced by macrophages or dendritic cells upon stimulation with LPS or IFN-γ (reviewed in [30]). Additionally, BAFF is up-regulated by IFN-γ, IL-10, and CD40L [45]. Pathogen-associated molecular patterns (PAMPs) such as LPS or peptidoglycans up-regulate the expression of BAFF and APRIL in macrophages and DCs via toll-like receptors (TLRs) [46, 47]. TLRs are means by which cells of the innate immune system recognise PAMPs and initiate T-cell-independent immune responses. They are especially relevant to marginal zone (MZ) B cells, which in the spleen are the first line of defence to blood-borne pathogens, and highly responsive to TLR activation [48].

BAFF is produced by astrocytes, which play an active role in cerebral innate immunity [35, 49]. Astrocytes in vitro produce high amounts of BAFF when stimulated with inflammatory cytokines [49]. BAFF, together with other soluble glial-derived cytokines, may be responsible for stimulating the adaptive immunity in an inflamed central nervous system (CNS) [49]. Astrocytes also produce APRIL, and APRIL expression is increased in patients with multiple sclerosis (MS) [50, 51], which is a progressive, debilitating CNS disorder. Also, BAFF is up-regulated in MS lesions [35]. MS patients have normal levels of BAFF/APRIL receptors [51], but they have elevated levels of APRIL in their plasma and cerebrospinal fluid (CSF), and in the case of a subset of MS patients, more BAFF mRNA was detected in their monocytes and T cells [50]. Whether BAFF and/or APRIL play a protective or accelerating role in MS remains unclear. Experiments using mouse experimental autoimmune encephalitis (EAE), which mimics MS, showed that treatment with BAFF/APRIL antagonist was able to alleviate the EAE symptoms [52]. This suggests that BAFF may play a pro-inflammatory role in EAE.

In addition to innate cells, BAFF mRNA has been detected in resting and stimulated T cells [40, 43, 53], stromal cells [54], and follicular dendritic cells (FDCs) [55]. Confirmation of BAFF expression at the protein level in these cell types has been difficult, presumably due to the low levels of expression [56], although activated T cells do produce APRIL [58, 59] and TWE-PRIL [37], APRIL is poorly expressed on bone marrow stromal cells, but APRIL is critical for plasmablast survival in the bone marrow [57].

Airway epithelial cells produce BAFF [60]. Stimulation of BEAS-2B primary human bronchial epithelial cells (PBECs) with TLR3 ligand resulted in up-regulation of BAFF and APRIL [60]. In mouse models of ovalbumin-induced airway inflammation, BAFF expression is up-regulated in alveolar-associated cells surrounding the bronchi [61], suggesting that BAFF may play a role in the pathogenesis of asthma. In fact, treatment with a BAFF/APRIL inhibitor reduced airway inflammation [61]. It is possible that the co-stimulatory effect of BAFF on T-cell activation [62] plays a critical role in this model and blocking it with a BAFF/APRIL inhibitor prevented exacerbated T-cell responses. However, mice over-expressing BAFF are protected in the same model of airway inflammation [63]. As these mice have elevated numbers of regulatory T cells (T_{reg}) [64], it is possible that T_{regs} contributed to the protection seen in the airway inflammation model. Collectively, these experiments reveal a complex role for BAFF/APRIL in asthma that still requires further explanations.

Studies of human RA synovium specimen revealed that both T and B cells produce BAFF in the joint [65]. Fibroblast-like synoviocytes (FLSs) in this study expressed BAFF but not on the cell surface. Another study showed that healthy FLSs stimulated with IFNg or TNF produce a strong BAFF mRNA signal [65]. FLSs isolated from the joint of RA patients spontaneously produce cytoplasmic BAFF protein but do not display any BAFF on the cell surface [66].

BAFF and APRIL expression is elevated in a subset of tumour cells. BAFF levels are high in the serum of patients with classical Hodgkin's lymphoma [67]. BAFF is also expressed by non–Hodgkin's lymphoma (NHL) cells [67–69]. APRIL, but not BAFF is increased in chronic lymphocytic leukaemia (CLL) serum [70].

The expression of BAFF and APRIL is regulated by various factors. For instance, lymphotoxin-beta receptor (LTβ-R) ligation induces BAFF expression in splenocytes [71]; suppressor of cytokine signalling-1 (SOCS-1) is a negative regulator of BAFF expression [72]. Mice deficient for SOCS-1 show aberrant expression of BAFF by DCs [72].

7.2.2 BAFF and APRIL Receptors

BAFF and APRIL share two receptors: B-cell maturation antigen (BCMA, TNFRSF17) and transmembrane activator and calcium modulator and cyclophilin ligand interactor (TACI, TNFRSF13b) (reviewed in [18, 20]). In addition, BAFF interacts selectively with a third receptor, BAFF receptor (BAFF-R, BR3, TNFRSF13c) (reviewed in [18, 20, 73]). BAFF-R, TACI, and BCMA have strong, moderate, and weak affinities for BAFF, respectively [74]. TACI binds BAFF and APRIL equally well, while BCMA has a much stronger affinity for APRIL than for BAFF [18, 74] (Fig. 7.1).

The interaction between BAFF/APRIL and their receptors is mediated by conserved cysteine-rich domains (CRDs) in the extracellular regions of the receptors. TACI contains two CRDs, whereas BCMA and BAFF-R contain only a single or partial CRD, respectively [75]. The complete CRDs are each ~40 amino acid residues in length. An alternative form of TACI exists where the N-terminal CRD is missing due to alternate splicing, and this form is fully capable of ligand-induced cell signalling. This suggests that the second CRD of TACI alone mediates complete affinity for both BAFF and APRIL [75]. Four residues on BCMA (Y13, I22, Q25, and R27) have been implicated in ligand specificity [76].

BAFF-R is the dominant receptor of BAFF expressed on naive B cells [43, 55]. In addition to having superior BAFF-binding affinity compared to TACI and BCMA, BAFF-R is activated by all forms of BAFF, and is the only BAFF receptor that responds strongly to BAFF 3-mer. BAFF-R is not expressed in immature B cells, but BAFF-R expression is induced as these cells progress through B-cell maturation [77]. BAFF-R is subsequently expressed in all mature B cells. T cells do not express TACI or BCMA, but express BAFF-R upon activation [43]. BAFF-R is also constitutively expressed on T_{reg} [78]. BAFF-R appears to be the receptor mediating BAFF-induced proliferation of anti-CD3-activated T cells [43].

APRIL, but not BAFF can interact with heparin sulphate proteoglycans (HSPGs) [79], which are structurally unrelated to TNF receptors, but are emerging as important players in the BAFF/APRIL ligand/receptor system. The basic residues with which APRIL binds glycosaminoglycans are independent to those used for binding to the receptors [79, 80]. This feature might be necessary for multimerisation of APRIL proteins on the extracelluar matrix, on the surface of HSPG-positive cells [18]. Transmembrane syndecans and the glycosylphosphatidylinositol (GPI)-anchored glycipans are the major cell membrane HSPGs. Their expression is not limited to haematopoietic cells; syndecan-1 (CD138) is an HSPG that is expressed on plasma cells [79], and is highly expressed on all myeloma cells. APRIL binds preferentially to HSPGs at the surface of syndecan-1^+ ($CD138^+$) cells to induce the expression of the pro-survival factor Bcl-X_L [57].

More than binding to APRIL, some HSPG such as syndecan-2 are capable of directly activating TACI-mediated signalling [81]. Heparin sulphate post-translational modifications of syndecan-2 are required for TACI to bind, and the spacing between these modifications may contribute to binding specificity. Similarly, syndecan-1 and syndecan-4 are also able to activate TACI [81]. In contrast to BAFF-R, TACI requires oligomeric BAFF or APRIL to be activated [27], so it is possible that HSPGs enhance APRIL multimerisation, which in turn enhances signalling of APRIL through TACI. Collaboration between APRIL and HSPGs appears to play an important role in TACI-mediated regulation of IgA production [82]. In conclusion, TACI is able to signal if activated by membrane-bound BAFF, cross-linked BAFF, cross-linked APRIL, or BAFF 60-mer but not trimeric BAFF [27]. The unresolved issue is the nature of the signal triggered by these different forms of ligands and how the availability of various ligand forms is controlled. BAFF 60-mer appear to promote survival signals via TACI [27], yet TACI can also trigger inhibitory signals [83], and the ligand responsible for the latter effect remains to be fully identified.

TACI is mainly expressed on B cells, especially on activated B cells, but early data suggested that TACI was also expressed on a subset of T cells [84] and two separate studies have confirmed a strong TACI signal not only in the spleen, but also in the thymus by Northern blot analysis [84, 85]. However, development of mouse and human TACI-specific monoclonal antibodies failed to confirm TACI expression on the surface of both mouse and human T cells [43, 86, 87]. This does not exclude

Table 7.1 Expression of BAFF, APRIL, and their receptors

Protein	Cell type	References
BAFF	-Neutrophils -Macrophages -Monocytes -Dendritic cells -Activated T cells -CD34+ cells from cord blood malignant B cells -EBV-infected B cells -Cytotrophoblast cells in the placenta -Epithelial cells -Astrocytes -Osteoclasts -Activated B cells -Fibroblast-like synoviocytes	[40, 42, 68, 203–208, 164, 180, 35, 66]
APRIL	-Monocytes -Macrophages -Dendritic cells -T cells -Plasmablasts -Glioma cells -Cancer cells -EBV-transformed B cells -Osteocytes	[37, 57, 58, 66, 184, 203, 209–213]
TWE-PRIL	-Primary T cells -Monocytic cell lines	[37]
BCMA	-Plasma cells -Plasmablasts -Tonsillar germinal centre B cells. -Fibroblast-like synoviocytes	[214, 43, 177, 215, 93, 43]
BAFF-R	-Peripheral B cells -Activated T cells -Regulatory T cells	[93, 43, 143, 216, 217]
TACI	-High on MZ and B1 B cells -Monocytes -$CD27^+$ human tonsillar memory B cells	[93, 43, 218, 44]

a possible expression of TACI on a small subset of T cells and/or under specific circumstances. Interestingly, the number of $CD4^+$ T cells is elevated in the Peyer's patches of $TACI^{-/-}$ mice [88] and the small intestine was another tissue where TACI expression was reproducibly high by Northern blot [84, 85]. If tissue preparations have included Peyer's patches, which contain B1 B cells, these express high levels of TACI [89] and may have contributed to the high TACI signal. TACI also appears to be expressed by human macrophages and to mediate their survival [44].

TLR activation affects the expression of BAFF/APRIL receptors [47, 64, 89, 90]. BAFF-R is increased by TLR4 ligand, LPS, but not the TLR9 ligand, CpG-containing oligodeoxynecleotides [91]. In comparison, TLR 4 and TLR 9 stimulation of B cells leads to an altered BAFF-binding capacity by strongly up-regulating

TACI and also, to a lesser extent, BAFF-R [91]. In conclusion, there is a very tight connection between TLR biology and the regulation of BAFF/APRIL receptors suggesting a role for BAFF/APRIL at the interface of the innate and adaptive immune system.

Looking at the human immune system, BAFF-R is widely expressed on human B-cell subsets such as naïve, memory B cells, and plasma cells (PCs [43], but not on PCs from the bone marrow (BM) and spleen [92, 93]. BCMA is expressed not only on PCs from tonsils, spleen, and BM [92, 93], but also on tonsillar memory B cells and in germinal centre (GC) B cells, the latter being TACI-negative and BAFF-R^{low} [43, 92]. TACI is expressed in CD27^{+} memory B cells, tonsillar, and BM PCs, in a subpopulation of activated CD27^{-}, non-GC cells [43, 64, 93] and in a small subset of naïve B cells in the blood and tonsils [93], consistent with the idea that, similar to the mouse system, TACI is an inducible receptor.

Many B-cell malignancies are associated with tumour cells expressing altered expression of BAFF, APRIL, and their receptors. These malignancies include: B-cell non-Hodgkin's lymphoma, Hodgkin's lymphoma, chronic lymphocytic leukaemia, multiple myeloma, and Waldenström's macroglobulinemia [68, 69, 94, 95].

7.3 Physiological Role of BAFF and APRIL

The first indication of a function for BAFF came from in vitro assays showing the specific survival of maturing transitional type 2 (T2) splenic B cells in cultures supplemented with BAFF, suggesting a role for this protein as a survival factor during B cell maturation in the spleen [96]. This model was confirmed when B cell maturation in BAFF-deficient animals was found to be impaired beyond the immature transitional type I (T1) stage that lies immediately prior to the T2 stage (Table 7.2) [97, 98]. This was a critical finding revealing that, in addition to a functional B cell receptor (BCR), immature B cells need BAFF-mediated survival signals to fully mature. Current evidence shows that all known BAFF receptors are expressed on B cells at differing levels depending on their maturation and/or activation state [30, 54, 99, 100]. For instance, TACI level is low in T1 B cells but high in T2 and marginal zone (MZ) B cells [43, 101]. Furthermore, BCR ligation up-regulates BAFF-R expression on B cells [100, 102] which promotes increased sensitivity to BAFF-mediated survival signals as B cells mature.

The role of APRIL is more elusive as conflicting results emerged from two independently generated APRIL$^{-/-}$ mouse models, one showing no obvious phenotype [101] while the other showed impaired isotype switching to IgA, bigger germinal centres (GCs) and increased numbers of effector T cells [103]. However, a recent study looking at both lines of APRIL$^{-/-}$ mice confirmed impaired production of IgA in the absence of APRIL (Table 7.2) [104].

BAFF-R is the key receptor triggering BAFF-mediated survival as mice deficient in this receptor display a phenotype similar to that of BAFF-null mice (Table 7.2) [3, 105, 106]. BCMA-deficient mice are born with no major immune defect apart from impaired survival of some PCs in the bone marrow (Table 7.2) [106]. TACI, in contrast, emerged as a negative regulator of B cell activation and expansion, as numbers of B cells are increased in TACI-deficient mice, B cells are hyper-responsive and animals eventually develop SLE-like disorders and lymphoid cancers (Table 7.2) [83, 106, 107, 108]. Moreover, T-independent type II antibody responses are impaired in these mice [108]. Whether TACI plays the same role in humans is now a matter of debate since TACI mutations in humans have been associated with immunodeficiency such as common variable immunodeficiency (CVID) [109]. However, CVID is occasionally associated with autoimmune manifestations [110].

Finally, BAFF expression plays an important role enforcing B cell self-tolerance [73]. In particular, the physiological expression of BAFF at limiting levels combined with the impaired ability of certain autoreactive B cells to respond to BAFF survival signals results in the elimination of such specificities from the normal B cell repertoire before they become mature long-lived cells [111, 112, 113]. The corollary of this is that up-regulation of BAFF expression in vivo can result in the rescue of self-reactive B cells from elimination [112]. Recent work showed that this effect explains only in part the greatly increased levels of autoantibody production and associated autoimmune manifestations observed in transgenic mice overexpressing BAFF [114, 115]. The link between BAFF and B cell-mediated autoimmune disease is explored in more detail below.

7.4 The BAFF/APRIL System and B-Cell Tolerance

7.4.1 Elements of B-Cell Tolerance

7.4.1.1 B-Cell Development and Maturation

B-cell development starts in the bone marrow (BM) from a common lymphoid progenitor and progresses through sequential developmental stages leading to the formation of immature B cells expressing a functional B cell receptor (BCR) [116]. Immature B cells then enter the blood and migrate to the spleen where B cell maturation takes place [117]. Immature B cells enter the spleen as transitional type 1 (T1) which can differentiate into transitional type 2 (T2) B cells [118]. Within the latter group, two subsets have been identified, T2 follicular (T2Fo) that give rise to follicular B cells and T2 MZ (T2MZ) precursors of MZ B cells [119, 120]. A third intermediate subset has been designated as T3, but these cells do not give rise to mature B cells and correspond to a small pool of anergic B cells sequestered from the normal maturation pathway [121, 122]. A fourth late transitional subpopulation of $CD21^{int}$ T2 B cells has been characterised, which is composed of B cells that have entered the cell cycle and which preferentially respond to BAFF-induced homeostatic proliferation. The $CD21^{int}$ T2 B cells are precursors of both follicular and MZ B cells [123].

For the most part, mechanisms responsible for whether a B cell should differentiate into a follicular versus a MZ B cell are not known. As the MZ B cell compartment contains more self-reactive B cells [124], the nature of the signal triggered through the BCR of these cells as opposed to follicular B cells may have decided their particular differentiation and localisation in the splenic MZ [125, 126]. It appears that certain BCR specificities are selected within the cycling CD21^{int} T2 compartment leading to preferential enrichment of these cells within the MZ B-cell compartment [123]. In conclusion, both the BCR and expression of BAFF are required for B-cell maturation [3, 127]. Yet, how these two signals interact to promote B-cell maturation and differentiation is unknown.

In the peritoneal cavity and the mesenteric lymph nodes reside B1 B cells composed of two subsets, B1a and B1b [128]. These B cells undergo renewal and may have originated from the fetal liver [128]. However, this possibility remains controversial as the bone marrow may also be a source of B1 B cells and the absence of the spleen prevents the maintenance of B1a B cells [129]. Regardless of their origin, B1 B cells do not appear to require BAFF to develop and survive [98, 130], although, it is possible that BAFF may play a stimulatory role during their activation as the number of B1 B cells increases in mice overexpressing BAFF as they age [131].

7.4.1.2 B-Cell Tolerance

Emerging self-reactive B cells are eliminated at various immune checkpoints throughout B-cell development, maturation, and activation [116, 132]. Developing B cells are tested for auto-reactivity in the BM after expression of the BCR on the cell surface [116]. Upon binding to self-antigen, B cells are either deleted, undergo receptor editing or become anergic [116, 132]. This safety feature is not absolute and a number of newly formed self-reactive B cells escape control and enter the periphery to be tested by the next immune checkpoint during B-cell maturation [116, 132]. In humans, numbers of self-reactive B cells drop from 40 to 20% in the mature B-cell compartment [133] except in patients with SLE or RA, suggesting that selection during B-cell maturation is affected in these patients [134, 135]. Not all self-reactive B cells are deleted; in fact some self-reactive B cells can be positively selected, often when the self-antigen is expressed at low levels [136–139]. These cells are known to enrich the MZ B-cell compartment [48, 123]. In any case, the nature and strength of the signal via the BCR will decide whether a self-reactive B cell is deleted, anergised or positively selected [125]. However, control mechanisms preventing positively selected self-reactive B cells to drive overt autoimmune reactions upon activation are not known.

The MZ B-cell compartment, which contains autoreactive B cells is localised in the splenic marginal zone, away from T-cell help [48, 140]. In fact MZ B cells respond better to T-independent antigens and it is possible that self-reactive B cells unlike memory B cells in this compartment are unable to differentiate into memory B cells [48]. Another interesting possibility is the immuno-regulatory nature of MZ precusor B cells, the T2MZ B cells, which can confer protection upon injection in an experimental mouse RA model, via the production of IL-10 likely to suppress T-cell function [141]. These cells have been labelled as B "regulatory" cells as an analogy to CD4 T regulatory cells (T_{regs}), which are also developing from mildly self-reactive precursor T cells [142].

7.4.2 *High BAFF Levels and Mechanisms of B-Cell Tolerance*

Early work on BAFF showed that addition of this factor in cultures of splenic B cells led to the preferential survival of immature transitional type 2 (T2) B cells and also marginal zone (MZ) B cells [96]. The data was suggestive of a critical role for BAFF during B-cell maturation in the spleen. This notion was later confirmed by the observation that B-cell maturation beyond the immature T1 B-cell stage was impaired in BAFF-deficient mice [98, 130]. However, lack of BAFF did not prevent the development of peritoneal B1 B cells. BAFF-mediated B cell survival is mediated via BAFF-R [97, 130, 143]. B-cell maturation in the spleen is a critical immune checkpoint during which self-reactive B cells are eliminated [116]. Activation of the B cell receptor (BCR) of immature cells rather than promoting proliferation, as it is the case for mature B cells, triggers cell death [116]. This feature is thought to delete maturing immature self-reactive B cells upon binding to self-antigen, and plays a major role in B-cell tolerance. For many years signalling through the BCR has been viewed as the central mechanism creaming away self-reactive B cells. Similarly, the BCR has been shown to be essential not just for B cell development but also B-cell maturation, survival, and maintenance [127], and as such, was thought to be the central molecule controlling B-cell development and tolerance. However, the discovery of BAFF has changed this notion, as production of this factor is equally important to the expression of a BCR on B cells for B-cell maturation and survival [3]. In addition, excessive BAFF production is likely to interfere with death signals triggered upon engagement of the BCR of self-reactive immature B cells and as such BAFF may corrupt the normal negative selection of potentially harmful autoreactive B cells [3]. Therefore, B-cell maturation is the sum of both BCR and BAFF signals and B-cell tolerance is maintained as long as these signals are well balanced. In addition, the nature and degree of BCR signalling as well as the presence of BAFF will decide whether an immature B cell will differentiate into a follicular or MZ B cell. The exact level of cooperation between the BCR and BAFF-R signalling pathways remains unclear but a couple of studies have outlined possible BCR/BAFF cooperative signal pathways [144, 145].

BAFF Tg mice develop autoimmunity linked to a loss of B-cell tolerance (Table 7.2) [3, 114]. B-cell tolerance is the fruit of carefully selected B cells at different immune checkpoints during B-cell development, maturation, and activation [132]. The question is where and when can excess BAFF corrupt B-cell tolerance. Animal models of BCR transgenic mice such as the hen egg lysozyme (HEL) model have been instrumental in dissecting the role of BAFF on B-cell tolerance [132]. A recent study used mice transgenic of the HEL-specific BCR crossed onto mice expressing a soluble form of HEL (sHEL-Tg x BCR-HEL-Tg) to address the role of excess BAFF on B-cell tolerance [111]. In the presence of physiological levels of BAFF, self-reactive B cells develop normally in the bone marrow but are rapidly eliminated in the periphery apart from a small population of anergic HEL-specific B cells [132]. The work by Lesley et al. showed that self-reactive HEL-specific B cells were more dependent on BAFF for their survival than non-self-reactive B cells [111]. Perhaps an issue with this study was to use non-self-reactive HEL-specific

B cells as control to compare with self-reactive HEL B cells. In sHEL-Tg x BCR-HEL-Tg mice, self-reactive/anergic B cells have a more immature phenotype than non-self-reactive B cells, and immature B cells as explained above are more dependent on BAFF for survival. Therefore, it is unclear whether self-reactive B cells in this model are more dependent on BAFF because they are self-reactive or because they are more immature [111]. The best way to answer this question would be to adoptively transfer self-reactive and non-self-reactive B cells at the same stage of maturation.

Another study by Thien et al. provided additional answers supporting the point made in the previous paragraph [112]. In this case, knockin HEL–BCR mice were used in which B cells, unlike HEL–BCR-Tg B cells, can switch from IgM to other Ig isotypes, and for this reason the mice were called switched HEL (SW_{HEL}) [112]. When SW_{HEL} mice were crossed onto membrane HEL-Tg (mHEL-Tg), this led to developmental arrest of B cells in the bone marrow and over-expression of BAFF in this system did not rescue B-cell development and did not corrupt B-cell tolerance [112]. This result was rather expected as developing B cells in the bone marrow do not express BAFF receptors and were not expected to survive negative selection [3]. When SW_{HEL} mice were crossed onto sHEL/BAFF double Tg mice, BAFF over-expression was able to suppress anergy and led to the maturation and expansion of self-reactive B cells [112]. One caveat with this approach was that self-reactive B cells in the cross did not compete with normal B cells, and negative selection of peripheral self-reactive B cells in this case is known to occurr at a later stage of B-cell maturation when B cells express BAFF-R and, therefore, can be rescued by BAFF [112]. Models using mixed bone marrow chimera allowing 10% of SW_{HEL} B cells to develop amongst 90% of normal B cells have shown that self-reactive B cells are deleted at an earlier B-cell maturation stage when they express little BAFF-R [112]. In addition these cells must compete with normal B cells for survival niches and entry into the B-cell follicle [146]. When this set up was tested in conditions of excessive BAFF production, negative selection and anergy of the self-reactive B cells was unaffected [112], suggesting that self-reactive B cells were eliminated prior to acquiring enough BAFF-R to be rescued by elevated BAFF levels.

This result was intriguing because it suggested that BAFF was unable to affect B-cell tolerance not only in the BM, but also in the periphery in condition mimicking physiological situations. Yet, one last aspect needed to be tested: the affinity variation of the BCR to the self-antigen. Indeed, in a model where SW_{HEL} B cells express both HEL-specific heavy and light chain, binding of the resulting BCR to HEL is very strong and leads to a very efficient deletion of self-reactive B cells [112]. As mentioned earlier, weak interaction of a self-reactive B cell to the self-antigen can lead to the positive selection of this cell rather than its elimination [136–139]. To test this possibility, SW_{HEL} mice, which do not express the transgene for the HEL-specific light chain were generated [112], allowing pairing of the HEL-specific heavy chain with available endogenous light chains. This led to the development of HEL-specific B cells with a wide range of affinity for HEL [112]. When BAFF over-expression was tested in this system, it showed that high-affinity

B cells were deleted normally compared to controls, yet excessive BAFF favoured the expansion of low-affinity self-reactive B cells in particular the cells that enriched the MZ compartment [112].

This work demonstrated that the effect of excessive BAFF production on B-cell tolerance is not as devastating as one would have thought considering that BAFF Tg mice develop severe and ultimately fatal autoimmune disorders [114]. Macrophages, neutrophils, and dendritic cells can normally produce significant amounts of BAFF upon activation by infectious agents and cytokines, yet infection and inflammation do not always lead to autoimmunity, confirming that BAFF production does not have an immediate devastating effect on B-cell tolerance but persistence of high BAFF production and additional events may collectively trigger autoimmune disorders. The only clue at this stage was the BAFF induced the expansion of low-affinity self-reactive B cells mostly MZ B cells [112] and whether these alone or in cooperation with other immune cells drive disease in BAFF Tg mice will now be detailed in the following sections.

Overexpression of APRIL in mice had no significant impact on B-cell development, survival, and function and as such is not thought to play a critical role in B-cell tolerance [58]. However, aging APRIL transgenic mice develop B1 B-cell lymphomas [147], yet whether APRIL also alters B1 B-cell tolerance perhaps require further investigation as like the MZ B-cell compartment the B1 compartment also contains self-reactive B cells [148].

7.5 The BAFF/APRIL System in Autoimmunity

Looking at the role of excess BAFF in altering key mechanisms controlling B-cell tolerance did not reveal a major breakdown in B-cell tolerance but in contrast the accumulation of fairly weak self-reactive B cells which participate in driving the severe autoimmune disorders developing in BAFF Tg mice, while possible, is unlikely to be the only mechanism. B cells are important pathogenic players in autoimmune conditions such as SLE but so are T cells, which can provide help to B cells and conversely B cells can be very active APC to T cells. This cooperation between T and B cells is critical in all known mouse models of SLE [12]. Therefore, it is important to also address the role of BAFF and APRIL in autoimmunity from the point of view of their function on T cells.

7.5.1 BAFF/APRIL and T Cells

7.5.1.1 BAFF and T-Cell Activation

As described above excess BAFF production does not prevent central and peripheral elimination of high-affinity self-reactive B cells but rather promotes the accumulation of low-affinity self-reactive B cells [112]. Yet, it is difficult to conceive that activated low-affinity self-reactive B cells could alone promote the severe autoimmune disease seen in BAFF Tg mice. As mentioned earlier, T cells as well as B cells play a central role in the pathogenesis of SLE [12]. Activated T cells express BAFF-R and BAFF promotes T-cell activation, differentiation into effector T cells, Bcl_2 expression, and T helper 1 (Th1) cytokine production [43, 63]. Moreover, BAFF Tg mice have a greater proportion of effector T cells compared to normal mice [63, 114]. However, expansion of the effector T-cell compartment was dependent on the presence of B cells as B-cell-deficient BAFF Tg mice had a normal effector T-cell compartment [63]. Therefore, while BAFF can directly signal into T cells via BAFF-R, B cells present in BAFF Tg mice play an additional role promoting the expansion of the effector T-cell compartment. CD4 T cells provide help to B cells but B cells can be very good APC to T cells, in particular MZ B cells which can efficiently present antigen to naïve T cells [149]. Therefore, a model is emerging whereby, excess BAFF promotes the expansion of low-affinity self-reactive B cells; many of these cells are MZ B cells capable of activating naïve T cells, leading to up-regulation of BAFF-R expression on activated T cells and subsequent binding of BAFF to BAFF-R which presumably fuels the activation process. Therefore, at this stage it seemed likely that disease in BAFF Tg mice like in many other mouse models of SLE is the sum of altered B and T cells functions, which together drive autoimmunity in these animals.

APRIL over-expression also stimulates CD4 and CD8 T-cell activation and survival, yet this function does not lead to autoimmune disorders [58]. In addition increased proportion of effector/memory CD4 T cells has been observed in APRIL$^{-/-}$ adding further confusion on the exact role of TACI in T cells [19]. APRIL binds to TACI and BCMA, whose expression could not be detected on the surface of human and mouse T cells [43], despite an early report suggesting that TACI was expressed on a subset of T cells [84]. This does not exclude a possible expression of TACI on a small subset of T cells and/or under specific circumstances. APRIL can also bind to proteoglycans expressed on T cells [79, 80], yet whether this translates into a signal inside T cells is not clear. Therefore, current data suggest that over-expression of APRIL can alter T-cell function but this effect might be indirect.

7.5.1.2 BAFF and Regulatory T Cells

As mentioned above a model involving both T and B cells driving disease in BAFF Tg mice seemed likely, until we made an unusual observation looking at T_{regs} in BAFF Tg mice [64]. The T_{regs} numbers were augmented in the thymus, blood, spleen, and lymph nodes [64], a finding that was inconsistent with a major role for T cells in driving disease in BAFF Tg mice as T_{regs} purified from BAFF Tg mice were shown to actively suppress T-cell activation ex vivo [64]. Therefore, it is likely that increased numbers of T_{regs} in BAFF Tg mice play a censoring role on activated T cells. A look at pDC activation also did not reveal any particular change in BAFF Tg mice compared to control animals [64] in contrast to other mouse models of SLE [10]. The involvement of T_{regs} and the apparent lack of pDC role in BAFF Tg mice added confusion to a picture where the role of T cells in driving disease in BAFF Tg mice seemed less and less certain.

7.5.2 B-Cell-Specific Lupus in BAFF Tg Mice

Findings that numbers of T_{regs} were elevated in BAFF Tg mice raised the doubt on the exact contribution of T cells as drivers of the disease in these mice. To directly address the role of T cells in BAFF Tg mice, BAFF Tg mice lacking T cells were developed [64]. To our surprise T-cell-deficient BAFF Tg mice developed an autoimmune disease undistinguishable from that of T-cell-sufficient BAFF Tg mice [64]. This result revealed that while BAFF may have an effect on T cells, this effect was dispensable for the development of autoimmunity in BAFF Tg mice. This interesting observation obliged us to look at T-cell-independent mechanisms activating B cells and how these may have affected BAFF Tg mice. T-cell-independent stimuli for B cells are generally ligands for Toll-like receptors (TLRs) such as LPS for TLR4, or CpG-ODN for TLR9 [150]. We and others showed that TLR7 and TLR9 activation up-regulated the expression of TACI on B cells [47, 64, 90, 91], a BAFF and APRIL receptor, which is in fact critical for T-cell-independent B-cell immune responses [108, 151]. In addition, stimulation of B cells with BAFF augmented TLR9 and TLR7 expression in B cells suggesting cooperation between BAFF and TLR7/9 in T-cell-independent activation possibly via the expression and function of TACI [64]. Stimulation of TLR9 on self-reactive B cells in the presence of BAFF led to greater levels of autoantibodies produced [64]. Moreover, expression of MyD88 in B cells, an adaptor critical for TLR signalling, is required for disease in BAFF Tg mice [64]. Therefore, the following model has been suggested (see Fig. 7.2): 1. Excess levels of BAFF expand self-reactive MZ and B1 B cells [112]. 2. BAFF signals promote TLR activation following internalisation of autoreactive B-cell receptors bound to either DNA or to immune complexes containing nucleic acids [47, 64, 90, 91, 152]. 3. BAFF enhances TLR signals, which led to the production of pro-inflammatory IgG2c and IgG2b antibodies. Autoantibodies deposit in the kidney and promote inflammation through complement activation. The weak autoreactive nature of the B cells is possibly important but their phenotype (MZ B cells) is perhaps more important as this subset of B cells is a particular strong responder to TLR activation [153] and express high levels of TACI [64, 43], a receptor key for T-cell-independent immune responses [108, 151]. While the data suggest that TACI may play a role in TLR-MyD88-mediated T-cell-independent autoimmune disease, its exact role in autoimmunity remains confusing as TACI-deficient mice also develop autoimmune disorders and lymphomas [83]. It has to be noted that autoimmune manifestations in $TACI^{-/-}$ mice are quite variable depending on where the line is kept, suggesting again that the microbial environment, hence TLR biology may play a role in the outcome of TACI deficiency (R.J. Bram, personal communication).

As previously mentioned, BAFF-R, as the survival receptor for B cells, probably plays the initial role as the promoter of self-reactive B-cell expansion. Yet, a recent intriguing study looking at A/WySnJ mice, which are BAFF-R mutant mice, showed that these animals develop SLE-like symptoms as they age [154]. This phenotype has not been reported for $BAFF\text{-}R^{-/-}$ mice and questions the mechanism driving disease in the A/WySnJ mice. We have reported that A/WySnJ T cells are

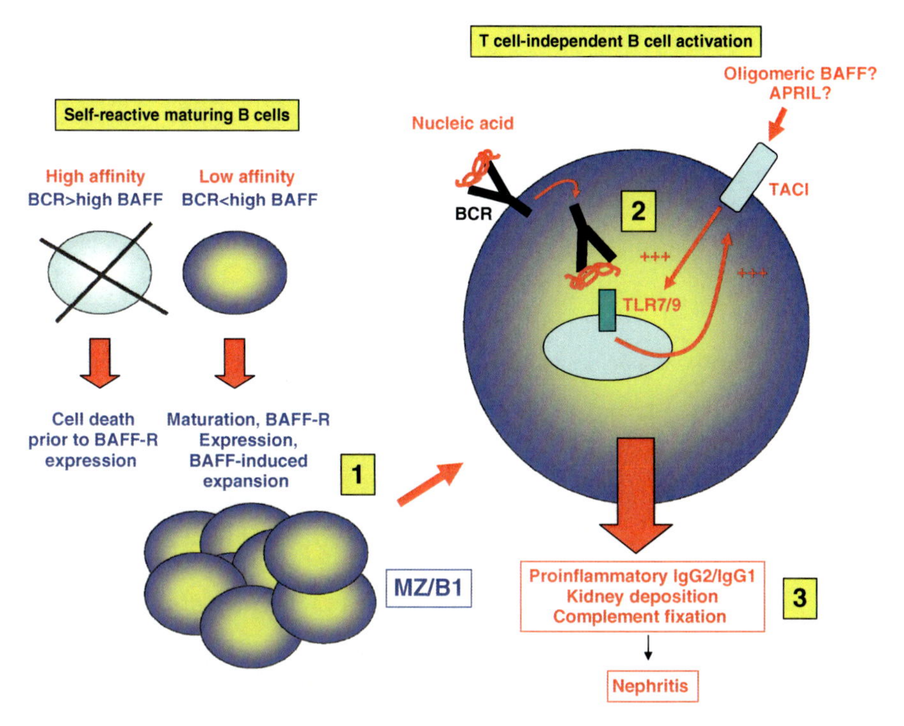

Fig. 7.2 The effect of excess BAFF on B-cell tolerance, T-cell-independent activation of autoreactive B cells and production of pro-inflammatory autoantibodies. 1. Excess levels of BAFF expand self-reactive MZ and B1 B cells [112]. 2. BAFF signals promote TLR activation following internalisation of autoreactive B-cell receptors bound to either DNA or to immune complexes containing nucleic acids [47, 64, 90, 91, 152]. 3. BAFF enhances TLR signals, which lead to production of pro-inflammatory IgG2c and IgG2b antibodies. Autoantibodies deposit in the kidney and promote inflammation through complement fixation

more responsive to activation compared to control A/J T cells, therefore a defect in T-cell activation may underlie the disease mechanism in these animals [43]. Yet, as BAFF-R is expressed on T_{regs} [78] and inactivation of these cells leads to autoimmunity [142], it will be useful to observe ageing BAFF-R-deficient animals more closely.

7.6 The BAFF/APRIL System and Autoimmunity in Humans

As discussed previously, BAFF Tg mice first develop a disease mimicking systemic lupus erythematosus (SLE) then sialadenitis looking like Sjögren' syndrome (SS) as they age [114, 155], and have an increased risk of lymphoma, in particular when they lack TNF [131]. Interestingly, these three diseases are those where the data are the strongest to support the role of BAFF in human pathogenesis.

7.6.1 BAFF and SLE

7.6.1.1 Serum Levels of BAFF and SLE

Since the first study in 2001 showing an increase in serum BAFF in a human disease by Zhang et al. [156], all the surveys which assessed the levels of BAFF in the serum of patients with SLE found increased levels compared with controls [105]. Persistent or transient, over-expression of BAFF is present in around 50% of human patients with SLE [157].

7.6.1.2 A Possible Association Between BAFF and Activity of SLE

Some of the cross-sectional studies revealed a correlation between serum BAFF and anti-DNA levels, but not with lupus activity [158]. Some longitudinal studies found that serum BAFF levels could be predictive of the activity of the disease three months later (measured by the SLE disease activity index (SLEDAI)). One very interesting study looking at 245 SLE patients from four different US medical centres, which have been followed for an average of 15 months, showed that serum BAFF levels were more predictive of a lupus flare than the level of anti-DNA antibodies [158].

Blood PBMC BAFF mRNA levels appear to associate better with disease activity than serum BAFF protein levels. Stohl et al. have found that although BAFF mRNA levels correlated significantly with BAFF protein levels, BAFF mRNA levels were more closely associated with serum immunoglobulin levels and SLE disease activity index scores than BAFF protein levels [158].

Analysis of cells from PBMC showed that BAFF is expressed and secreted physiologically by monocytes [21]. In some pathologic conditions, T cells can also express BAFF [45]. Recently, Morimoto et al. found that $CD4^+$ and $CD8^+$ T cells from patients with active SLE expressed intra-cellular BAFF, whereas T cells from normal subject did not [45].

7.6.2 BAFF and Sjögren's Syndrome (SS)

7.6.2.1 Serum BAFF in SS

One year after the first implication of BAFF in human lupus, Groom et al. demonstrated that BAFF transgenic mice developed sialadenitis as they age, and that higher serum BAFF levels were detected in patients with SS [155]. This finding was confirmed in 2003 with the correlation between serum BAFF levels and total IgG as well as rheumatoid factor levels in SS patients [159]. Moreover, in this study, patients with anti-SSA/SSB autoantibodies had higher levels of serum BAFF than patients without autoantibodies. Thereafter, like in SLE, all the studies which assessed serum BAFF levels in SS confirmed this observation and most studies confirmed the correlation between high BAFF levels and autoantibody levels (reviewed in [1]).

Interestingly, Batten et al. observed the increased occurrence of lymphomas in BAFF Tg mice lacking TNF, with similarities to B lymphomas developing as a complication is a subset of SS patients [155, 1]. Moreover, the study by Groom et al. showed that B cells infiltrated the salivary gland of BAFF Tg mice and these have a MZ-like phenotype [155]. In humans, MZ/memory B cells express CD27 and disappear from the blood of SS patients while accumulating in salivary and lacrimal glands [160]. Others have also detected transitional type II B cells in these tissues [161], a subset also present outside of the spleen in BAFF Tg mice [96].

7.6.2.2 Non-lymphoid and Lymphoid Cells Express BAFF in SS

The first report by from Groom et al. showed the presence of BAFF within the salivary lymphoid infiltrate characteristic of this disease [155]. Later, Lavie et al. demonstrated that both the T cells in the infiltrates and the salivary epithelial cells expressed BAFF [162]. This was confirmed by subsequent studies [163]. In the latter study, Daridon et al. detected BAFF expression not only in the salivary epithelial cells, but also in B cells infiltrating salivary glands [163]. Thus, salivary B cells, which are the target of BAFF and express the different BAFF receptors, express the ligand, leading to an autocrine pathway for BAFF secretion and activation of B cells. A recent study confirmed BAFF expression by activated B cells [164] and this aspect is likely to change the views on ways BAFF locally regulates B-cell function/survival. Like in SLE, Lavie et al. found that in SS patients contrary to normal controls, T cells from PBMC expressed and secreted BAFF [38]. However, the expression of BAFF in T cells is much lower than that of monocytes, the professional secretor of BAFF [21].

7.6.2.3 Elevated BAFF Levels May Explain the Lack of Efficacy of Anti-TNF Therapies in SS

Two randomised controlled trials, one with Infliximab [165] and one with Etanercept [166], demonstrated the absence of efficacy of TNF blockers in SS. A recent study offered a potential explanation for the poor efficacy of anti-TNF treatments. On Etanercept and not on placebo, primary (p) SS patients (pSS) experienced increased production of type 1 interferon and BAFF, which could explain the absence of improvement with anti-TNF therapy [167]. In addition, work by Batten et al. using TNF-deficient BAFF Tg mice showed no protection of BAFF Tg mice but rather increased incidence of B lymphomas in these animals [131]. This data suggest that rather than protecting SS patients, anti-TNF treatments may exacerbate BAFF-mediated pathogenic mechanisms.

7.6.3 BAFF and Other Autoimmune Diseases

In rheumatoid arthritis (RA), studies looking at BAFF serum levels have been controversial with some studies showing elevated BAFF levels [168] but others showing normal levels [156, 169]. However, in most studies, BAFF levels are increased in the synovial fluid, which suggests a local production of the cytokine with this disease [169].

BAFF serum levels are also elevated in a number of other systemic autoimmune diseases. In hepatitis C virus (HCV)-associated cryoglobulinemia, BAFF levels are greater in patients with vasculitis or associated lymphoma compared to patients with asymptomatic cryoglobulinaemia [170, 171]. Serum BAFF levels are also elevated in patients with Wegener granulomatosis [35], systemic sclerosis [172], idiopathic thrombocytopenic purpura (ITP) [173], and multiple sclerosis [50].

7.6.4 BAFF Receptors Expression and Autoimmune Diseases

Very few studies focused on BAFF receptors expression in human autoimmune diseases. Carter et al. described a decrease of free BAFF-R in lupus, linked to BAFF occupancy, but the quantitative protein level of the receptor was normal [92]. Conversely, Sellam et al. recently observed decreased BAFF-R expression on B cells of patients with SLE and SS [174]. This reduction of BAFF-R expression correlated with the activity of the disease assessed by the SLEDAI in SLE and by the presence of systemic manifestations in SS. No change in BAFF mRNA was observed suggesting that decrease of BAFF-R protein was the result of post-transcriptional regulation [174]. There was a negative correlation between BAFF-R expression on B cells and serum BAFF level, suggesting that a high serum BAFF levels negatively regulate BAFF-R expression on B cells either by internalisation of the receptor or by shedding from the membrane [174].

7.6.5 APRIL and Human Autoimmunity

7.6.5.1 The APRIL–TACI Axis

The role of APRIL in autoimmunity is less clear than that of BAFF. Indeed, APRIL transgenic mice develop neither B-cell abnormalities nor serological and/or clinical signs of autoimmunity [58].

Interestingly, APRIL could have a dual effect via TACI:

1) A B-cell stimulatory effect mainly responsible for immunoglobulin class switch. Indeed, TACI signalling is involved in immunoglobulin class switch and TACI mutations are found in 10% of patients with common variable immunodeficiency (CVID) [109].
2) A negative signal on B cells explaining hyperactivity of B cells in TACI$^{-/-}$ mice [151, 108]. Accordingly, APRIL could serve as a homeostatic down modulator of B-cell hyperactivity induced by BAFF.

7.6.5.2 APRIL and SLE or RA

A protective effect of APRIL in SLE has been suggested [175]. Stohl et al. demonstrated an inverse correlation between circulating APRIL levels and serological and clinical activity in patients with SLE [175]. Conversely, Koyama et al. found increased serum levels of APRIL in SLE patients, correlating with musculoskeletal manifestations [176].

One possible pathological role for APRIL may be a role in stimulating B cells. A recent study described increased APRIL serum levels in patients with RA [177]. Moreover in this study, APRIL was able to stimulate fibroblast-like synoviocytes [177], which express the BCMA receptor [177]. Another study with RA patients showed the expression of both BAFF and APRIL in the rheumatoid synovitis, but only APRIL mRNA levels in the synovium were associated with the formation of ectopic germinal centres in this tissue [178].

Although APRIL transgenic mice do not present features of autoimmunity, they can develop B1 B-cell-derived cancers as they age [147]. Therefore, the possibility that a local overproduction of APRIL could affect B-cell function and plays a role in some autoimmune diseases, especially in RA, remains possible.

7.6.6 BAFF Secretion by Resident Cells in Target Organs of Autoimmunity: A New Concept

As indicated above, several groups have demonstrated that salivary epithelial cells may express and secrete BAFF, both in patients with SS and healthy subjects [179, 180, 163, 162]. Interestingly, BAFF expression is largely increased by stimulation with type 1 or type 2 interferon (IFN) [179]. Patients with SS seem to be more sensible to the effect of type 1 interferon-induced BAFF expression and secretion by salivary epithelial cells [179]. The same findings have been described in RA and multiple sclerosis [66, 35]. In RA, fibroblastic-like synoviocytes express BAFF in response to type-2 IFN stimulation [66]. This expression appeared increased after co-stimulation with TNF, although this remains controversial. In multiple sclerosis, astrocytes in culture express BAFF, and this expression is also increased in response to type 2 IFN [35]. Finally, not only transformed B cells, but also activated B cells appear to produce BAFF [164], how this may affect B-cell function locally in inflamed tissues remains to be investigated.

7.6.7 BAFF: A Possible Bridge Between Innate and Adaptive Immunity in Autoimmune Diseases

In recent years, it has been clearly demonstrated that SLE and SS share a lot of common features, especially an IFN signature present in PBMC and targeted organs in the two diseases (salivary glands in SS and kidneys in SLE) [181–183]. Parallel to this IFN signature is the elevated BAFF levels detected in target organs of SS, the salivary glands and the conjunctival epithelial cells [182]. Litinskiy et al. demonstrated that type 1 IFN can induce BAFF expression and secretion by monocytes and follicular cells [184]. As indicated above, in vitro, type 1 IFN induces BAFF expression and secretion by salivary epithelial cells [179]. Moreover, recent work demonstrated that stimulation of salivary epithelial cells, target cells of autoimmunity, by poly(I:C) or infection of the cells using a retrovirus (a double-stranded RNA virus) induced a robust expression and secretion of BAFF through pathways dependent or not on TLR activation and IFN [180]. Thus, resident cells of target organs of autoimmunity are not just passive bystanders, but also may play an active role by secreting BAFF after innate immune stimulation, resulting in the activation/survival of autoreactive B lymphocytes.

This mechanism is interesting as it may explain the frequent observation of a possible relationship between viral infections and autoimmune diseases [185], which might be common to other organ-specific autoimmune diseases such as RA and multiple sclerosis, since synoviocytes and astrocytes also secrete BAFF after IFN stimulation [66, 35].

7.7 Targeting BAFF and/or APRIL: Progress and Challenges

As mentioned above, BAFF is clearly implicated in the pathogenesis of SLE and SS, a fact that supported further development of BAFF antagonists such as Belimumab in the clinic [186]. The role of APRIL is less clear and often controversial, but could be more important in the local stimulation of B cells within the synovium of RA patients. Thus, neutralising BAFF and APRIL may provide additional benefits for the treatment of human autoimmunity. To date three different drugs have been designed (Fig. 7.3):

- Belimumab is a monoclonal anti-BAFF antibody which targets only BAFF [187–189].
- Atacicept is a TACI-Fc molecule which targets both BAFF and APRIL [190–193].
- BR3-Fc targets only BAFF [194].

To date, two large phase-II studies (400–500 patients each) have been presented with belimumab. In RA, the results are rather disappointing with around 30% of ACR 20 response in all belimumab groups versus 15% in the placebo group [186, 189, 195, 196]. This may be explained by the fact that, as indicated above, B-cell activation in RA may not be driven only by BAFF [196]. In SLE, the results are more encouraging [189, 195]. Although the primary end point (decrease of SLEDAI of more than three points) was not achieved in the whole study including 449 patients, the analyses restricted to the 70% of patients with antinuclear antibodies or anti-DNA antibodies showed a significant effect of belimumab for decreasing

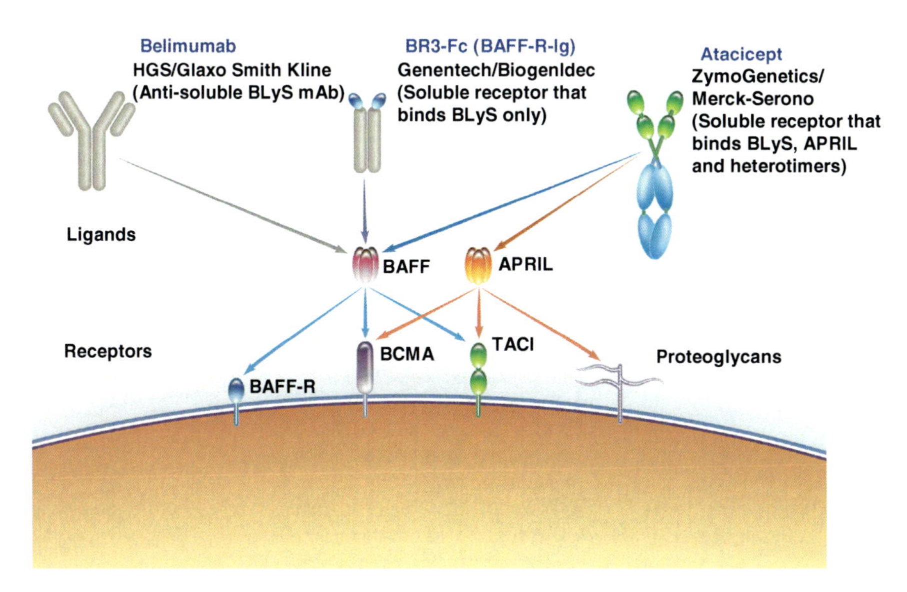

Fig. 7.3 BAFF-specific therapies currently tested in the clinic

activity of the disease measured by SLEDAI and anti-DNA antibody level [195]. Phase-III studies are in progress in SLE to confirm these preliminary results and phase-II studies in SS should begin. Phase-II studies with Atacicept and BR3-Fc are in progress with RA patients. As a blocker of both APRIL and BAFF Atacicept may offer additional benefits, and current data from phase-Ib clinical trials in SLE and RA show positive results consistent with the expected mechanism of action [192, 193].

Another possible use for anti-BAFF therapy is just after treatment with Rituximab (anti-CD20). Treatment with the B-cell-depleting agent Rituximab leads to increased serum BAFF levels as a general effect of the therapy, regardless of the autoimmune problem treated, whether RA, SS, or SLE [197–199]. Increased serum BAFF levels have first been attributed to reduced numbers of B cells normally present to bind BAFF. However, two independent studies [197–199] showed that, beyond this mechanical increase of BAFF after disappearance of B cells, there was an active homeostatic feedback characterised by increased BAFF mRNA expression in monocytes in response to the treatment. This suggests that B cells do not just "mop" BAFF present in the serum but variation in their numbers plays a role in regulating BAFF expression. Increased BAFF levels following Rituximab treatment is a potentially concerning aspect which could favour the stimulation of new autoimmune B cells as they repopulate the immune system of the patients, in particular during B-cell maturation where excessive survival may lead to the emergence of self-reactive B cells. Using BAFF-targeted therapies after Rituximab treatment to avoid this possible adverse effect may improve safety and efficacy of Rituximab.

Table 7.2 Understanding the BAFF/APRIL system through genetic manipulations and pharmacology

Models	Phenotype/notes	References
$BAFF^{-/-}$	Lack of T2 and MZ B cells and major loss of mature B cells in secondary lymphoid organs Reduced baseline serum levels of Ig and reduced Ig responses to T-cell-dependent and T-cell-independent antigens	[97, 98, 143] [97, 98]
BAFF-Tg	B-cell hyperplasia B cells with an MZ-like phenotype infiltrate the salivary glands of BAFF-Tg mice Elevated serum titres of multiple autoantibodies: anti-ss/dsDNA, rheumatoid factor, ANA, anti-chromatin autoantibodies Expanded effector T-cell compartment Ageing leads to the development of a secondary pathology reminiscent of Sjögren's Syndrome: severe sialadenitis, decreased saliva production, and destruction of submaxillary glands Increased T_{reg} numbers T-cell-independent disease	[114, 115] [97, 114, 115] [131] [155] [64]
Human soluble recombinant BAFF intraperitoneal injection	Enhanced immune response: increased B-cell proliferation, stronger NK cell activities, dose-dependent increase of $CD4^+$ T cell% (no effect on $CD8^+$ T cell%) B-cell hyperplasia Disrupted splenic B- and T-cell zones Hypergammaglobulinemia	[23]
$APRIL^{-/-}$	Impaired IgA class switching, now confirmed in both lines of $APRIL^{-/-}$ mice	[103, 104] [101]
APRIL-Tg	Increased T-cell survival in vivo and in vitro via up-regulation of pro-survival oncogene Bcl-2 B1-cell neoplasia	[58] [147]
$TACI^{-/-}$	Elevated B-cell numbers Twofold increase in circulating and splenic B cells due to higher proliferation rate Defective T-independent type-II response Fatal autoimmune glomerulonephritis, proteinuria, and elevated levels of circulating antibodies No class switch recombination Increased numbers of $CD4^+$ T cells on the Peyer's patches	[83, 151] [108] [108] [83] [219] [151]

Table 7.2 (continued)

Models	Phenotype/notes	References
TACI-Ig treatment	Impaired survival of plasma cells in vitro and in vivo Inhibited production of collagen-specific Abs and disease progression in a mouse model of rheumatoid arthritis (RA)	[220] [97]
TACI-Ig Tg	Fewer T2 and mature B cells Reduced circulating Ig	[97] [97]
$TACI^{-/-} \times BCMA^{-/-}$	Similar to $TACI^{-/-}$	[221]
$BCMA^{-/-}$	Reduced capacity to sustain long-lived bone marrow PCs	[220]
BCMA transfection or upregulation by IL-4/IL-6	Enhanced Ag presentation by mature B cells in a murine B-cell line	[222]
$BAFF\text{-}R^{-/-}$	Major loss of mature and transitional B cells Cannot carry out T-cell-dependent Ab formation Impaired longevity of the GC reaction Impaired class-switch recombination to produce IgG, IgA, and IgE in response to BAFF Impaired proliferation and cytokine secretion by T cells	[143, 151, 214, 221] [221] [223, 224] [219] [78]
A/WySnJ (BAFF-R mutant)	Same B-cell defect as $BAFF\text{-}R^{-/-}$ mice Develop SLE as they age	[154]

7.8 Conclusion

Almost a decade after the discovery of BAFF and its receptors, a number of new concepts have been established, although new questions have also emerged. BAFF is a critical survival factor for B cells which, together with signals from the BCR, is essential for B-cell maturation and survival [3]. This effect is mediated via BAFF-R and the signalling events driving this function have now been extensively mapped [200, 201]. Dissecting the role of BAFF in autoimmunity has revealed a number of surprises, in particular the role of excess BAFF in driving T-cell-independent autoimmune disorders, at least in mice [64]. Thus BAFF overproduction in BAFF Tg mice may be a unique experimental model for spontaneous lupus, distinct from the T-dependent models. These findings raise the possibility of a subset of patients suffering from similar B-cell-dominant forms of autoimmune disease. Recent studies have also revealed an interesting relationship between TACI expression, function, and TLR expression and activation [64]. As TACI emerges as the possible additional key player in this system, a number of puzzling elements about this receptor must be addressed, in particular its role in promoting T-cell-independent B-cell responses while repressing/controlling B-cell activation at the same time. Multimeric/crosslinked ligands but not soluble trimers activate TACI [27], yet the importance and regulation of various ligand forms in vivo remains unknown.

BAFF is a key cytokine, expression of which is dysregulated in number of autoimmune diseases, especially in SLE and SS. Emerging studies suggest that blocking both BAFF and/or APRIL may offer an added benefit, especially in SLE. However, not all B cells require BAFF for their survival, in particular long-lived memory B cells [202]. Depending on the patient type, these resistant B cells may play a more important role in disease, hence limiting the efficacy of anti-BAFF/APRIL treatment. Whether blocking APRIL in addition to BAFF will offer additional benefit is a possibility, but this still requires additional studies to understand the biological basis. TACI, which is more highly expressed on B cells, may play a role in autoimmunity following activation via TLR-dependent mechanisms. As such targeting TACI-expressing B cells may be a promising therapeutic option. One limitation of therapies targeting CD20$^+$ B cells is their relative inability to eliminate antibody-producing cells, which express low levels of CD20. In contrast, BCMA is expressed on plasma cells, albeit at low levels. Therefore, combining BCMA- and CD20-specific depleting treatments may be an avenue to improve B-cell depletion.

Hopefully the BAFF/APRIL system will provide novel strategies for treating autoimmune diseases and cancers. Regardless, the discovery of BAFF, APRIL, and their receptors has been been a major advance for immunology and has led to an unravelling of the meaning of life for B cells.

Acknowledgments We thank Charles Mackay for critical reading of the manuscript. Fabienne Mackay is supported by the National Health and Medical Research Council of Australia and the New South Wales Lupus association.
The authors declare having no conflict of interest.

References

1. Mackay F, Groom JR, Tangye SG. An important role for B-cell activation factor and B cells in the pathogenesis of Sjogren's syndrome. Curr Opin Rheumatol 2007;19(5):406–13.
2. Mackay F, Silveira PA, Brink R. B cells and the BAFF/APRIL axis: fast-forward on autoimmunity and signaling. Curr Opin Immunol 2007;19(3):327–36.
3. Mackay F, Schneider P, Rennert P, Browning JL. BAFF and APRIL: a tutorial on B cell survival. Ann Rev Immunol 2003;21:231–64.
4. Mills JA. Systemic Lupus Erythematosus. N Engl J Med 1994;330:1871–9.
5. Kang YM, Zhang X, Wagner UG, et al. CD8 T cells are required for the formation of ectopic germinal centers in rheumatoid arthritis. J Exp Med 2002;195:1325–36.
6. Davidson A, Aranow C. Pathogenesis and treatment of systemic lupus eythematosus nephritis. Curr Opin Rheumatol 2006;18:268–475.
7. Eisenberg D, Rahman A. Systemic lupus erythematosus-2005 *annus mirabillis*. Nat Clin Pract Rheum 2005;2:145–52.
8. Carroll MC. A protective role for innate immunity in systemic lupus erythematosus. Nat Rev Immunol 2004;4:825–31.
9. McGaha TL, Sorrentino B, Ravetch JV. Restoration of tolerance in lupus by targeted inhibitory receptor expression. Science 2005;307:590–3.

10. Marshak-Rothstein A. Toll-like receptors in systemic autoimmune disease. Nat Rev Immunol 2006;6:823–35.
11. Rahman AH, Eisenberg RA. The role of toll-like receptors in systemic lupus erythematosus. Springer Semin Immunopathol 2006;28:131–43.
12. Hoffman RW. T cells in the pathogenesis of systemic lupus erythematosus. Clin Immunol 2004;113(1):4–13.
13. Singh RR. SLE: translating lessons from model systems to human disease. Trends Immunol 2005;26:572–9.
14. Davidson A, Wang X, Mihara M, et al. Co-stimulatory blockade in the treatment of murine systemic lupus erythematosus (SLE). Ann NY Acad Sci 2003;987:188–98.
15. Kalled SL, Cutler AH, Datta SK, Thomas DW. Anti-CD40 ligand antibody treatment of SNF1 mice with established nephritis: preservation of kidney function. J Immunol 1998;160(5):2158–65.
16. Vinuesa CG, Cook MC, Angelucci C, et al. A RING-type ubiquitin ligase family member required to repress follicular helper T cells and autoimmunity. Nature 2005;435(7041):452–8.
17. Thatayatikom A, White AJ. Rituximab: a promising therapy in systemic lupus erythematosus. Autoimmun Rev 2006;5:18–24.
18. Bossen C, Schneider P. BAFF, APRIL and their receptors: structure, function and signaling. Semin Immunol 2006;18:263–75.
19. Dillon SR, Gross JA, Ansell SM, Novak AJ. An APRIL to remember: novel TNF ligands as therapeuthic targets. Nat Rev Drug Discov 2006;5(3):235–46.
20. Salzer U, Birmelin J, Bacchelli C, et al. Sequence analysis of TNFRSF13b, encoding TACI, in patients with systemic lupus erythematosus. J Clin Immunol 2007;27(4): 372–7.
21. Nardelli B, Belvedere O, Roschke V, et al. Synthesis and release of B-lymphocyte stimulator from myeloid cells. Blood 2001;97(1):198–204.
22. Lopez-Fraga M, Fernandez R, Albar JP, Hahne M. Biologically active APRIL is secreted following intracellular processing in the Golgi apparatus by furin convertase. EMBO 2001;2:945–51.
23. Moore PA, Belvedere O, Orr A, et al. BlyS: member of the tumor necrosis factor family and B lymphocyte stimulator. Science 1999;285:260–3.
24. Schneider P, Mackay F, Steiner V, et al. BAFF, a novel ligand of the tumor necrosis factor (TNF) family, stimulates B-cell growth. J Exp Med 1999;189:1747–56.
25. Zhukovsky EA, Lee J-L, Villegas M, Chan C, Chu S, Mroske C. Is TALL-1 a trimer or a virus-like cluster? Nature 2004;427:413–4.
26. Oren DA, Li Y, Volovik Y, et al. Structural basis of BLyS receptor recognition. Nat Struct Biol 2002;9:288–92.
27. Bossen C, Cachero TG, Tardivel A, et al. TACI, unlike BAFF-R, is solely activated by oligoimeric BAFF and APRIL to support survival of activated B cells and plasmablast. Blood 2008;111(3):1004–12.
28. Liu Y, Xu L, Opalka N, Kappler J, Shu H-B, Zhang G. Crystal structure of sTALL-1 reveals a virus-like assembly of TNF ligands. Cell 2002;108:383–94.
29. Hong X, Kappler J, Liu YJ, Xu L, Shu H-B, Zhang G. Is TALL-1 a trimer or a virus-like cluster? Nature 2004;427:414.
30. Mackay F, Ambrose C. The TNF family members BAFF and APRIL: the growing complexity. Cytokine Growth Factor Rev 2003;14:311–24.
31. Cachero TG, Schwartz IM, Qian F, et al. Formation of virus-like clusters is an intrinsic property of the tumor necrosis factor family member BAFF (B Cell Activating Factor). Biochemistry 2006;45(7):2006–13.
32. Roschke V, Sosnovtseva S, Ward CD, et al. BLyS and APRIL form biologically active heteromers that are expressed in patients with systemic immune-based rheumatic diseases. J Immunol 2002;169:4314–21.

33. Gavin AL, AÏt-Azzouzene D, Ware CF, Nemazee D. DBAFF, an alternate splice isoform that regulates receptor binding and biopresentation of the B cell survival cytokine, BAFF. J Biol Chem 2003;278:38220–8.
34. Gavin AL, Duong B, Skog P, et al. deltaBAFF, a splice isoform of BAFF, opposes full-length BAFF activity in vivo in transgenic mouse models. J Immunol 2005;175(1):319–28.
35. Krumbholz M, Theil D, Derfuss T, et al. BAFF is produced by astrocytes and up-regulated in multiple sclerosis lesions and primary central nervous system lymphoma. J Exp Med 2005;201(2):195–200.
36. Smirnova AS, Andrade-Oliveira V, Gerbase-DeLima M. Identification of new splice variants of the genes BAFF and BCMA. Mol Immunol 2008;45(4):1179–83.
37. Pradet-Balade B, Medema JP, Lopez-Fraga M, et al. An endogenous hybrid mRNA encodes TWE-PRIL, a functional cell surface TWEAK-APRIL fusion protein. EMBO J 2002;21:5711–20.
38. Lavie F, Miceli-Richard C, Ittah M, Sellam J, Gottenberg JE, Mariette X. B-cell activating factor of the tumour necrosis factor family expression in blood monocytes and T cells from patients with primary Sjogren's syndrome. Scand J Immunol 2008;67(2):185–92.
39. Nardelli B, Belvedere O, Roschke V, et al. Synthesis and release of B-lymphocyte stimulator from myeloid cells. Blood 2000;97:198–204.
40. Huard B, Arlettaz L, Ambrose C, et al. BAFF production by antigen-presenting cells provides T cell co-stimulation. Int Immunol 2004;16:467–75.
41. Scapini P, Carletto A, Nardelli B, et al. Proinflammatory mediators elicit secretion of the intracellular B-lymphocyte stimulator pool (BLyS) that is stored in activated neutrophils: implications for inflammatory diseases. Blood 2005;105(2):830–7.
42. Scapini P, Nardelli B, Nadali G, et al. G-CSF-stimulated neutrophils are a prominent source of functional BLyS. J Exp Med 2003;197:297–302.
43. Ng LG, Sutherland A, Newton R, et al. BAFF-R is the principal BAFF receptor facilitating BAFF co-stimulation of B and T cells. J Immunol 2004;173:807–17.
44. Chang SK, Arendt BK, Darce JR, Wu X, Jelinek DF. A role for BLyS in the activation of innate immune cells. Blood 2006;108(8):2687–94.
45. Morimoto S, Nakano S, Watanabe T, et al. Expression of B-cell activating factor of the tumour necrosis factor family (BAFF) in T cells in active systemic lupus erythematosus: the role of BAFF in T cell-dependent B cell pathogenic autoantibody production. Rheumatology (Oxford, England) 2007;46(7):1083–6.
46. Hardenberg G, Planelles L, Schwarte CM, et al. Specific TLR ligands regulate APRIL secretion by dendritic cells in a PKR-dependent manner. Eur J Immunol 2007;37(10): 2900–11.
47. Katsenelson N, Kanswal S, Puig M, Mostowski H, Verthelyi D, Akkoyunlu M. Synthetic CpG oligodeoxynucleotides augment BAFF- and APRIL-mediated immunoglobulin secretion. Eur J Immunol 2007;37(7):1785–95.
48. Martin F, Kearney JF. Marginal-zone B cells. Nat Rev Immunol 2002;2:323–35.
49. Farina C, Aloisi F, Meinl E. Astrocytes are active players in cerebral innate immunity. Trends Immunol 2007;28(3):138–45.
50. Thangarajh M, Gomes A, Masterman T, Hillert J, Hjelmstrom P. Expression of B-cell-activating factor of the TNF family (BAFF) and its receptors in multiple sclerosis. J Neuroimmunol 2004;152(1–2):183–90.
51. Thangarajh M, Masterman T, Hillert J, Moerk S, Jonsson R. A proliferation-inducing ligand (APRIL) is expressed by astrocytes and is increased in multiple sclerosis. Scand J Immunol 2007;65(1):92–8.
52. Huntington ND, Tomioka R, Clavarino C, et al. A BAFF antagonist suppresses experimental autoimmune encephalomyelitis by targeting cell-mediated and humoral immune responses. Int Immunol 2006;18(10):1473–85.
53. Huard B, Schneider P, Mauri D, Tschopp J, French LE. T cell costimulation by the TNF ligand BAFF. J Immunol 2001;167:6225–31.

54. Gorelik L, Gilbride K, Dobles M, Kalled SL, Zandman D, Scott ML. Normal B cell homeostasis requires B cell activation factor production by radiation-resistant cells. J Exp Med 2003;198:937–45.
55. Zhang X, Park CS, Yoon SO, et al. BAFF supports human B cell differentiation in the lymphoid follicles through distinct receptors. Int Immunol 2005;17(6):779–88.
56. Tangye SG, Bryant VL, Cuss AK, Good KL. BAFF, APRIL and human B cell disorders. Semin Immunol 2006;18(5):305–17.
57. Belnoue E, Pihlgren M, McGaha TL, et al. APRIL is critical for plasmablast survival in the bone marrow and poorly expressed by early-life bone marrow stromal cells. Blood 2008;111(5):2755–64.
58. Stein JV, Lopez-Fraga M, Elustondo FA, et al. APRIL modulates B and T cell immunity. J Clin Invest 2002;109:1587–98.
59. Hahne M, Kataoka T, Schroter M, et al. APRIL, a new ligand of the tumor necrosis factor family, stimulates tumor growth. J Exp Med 1998;188:1185–90.
60. Kato A, Truong-Tran AQ, Scott AL, Matsumoto K, Schleimer RP. Airway epithelial cells produce B cell-activating factor of TNF Family by an IFN-beta-dependent mechanism. J Immunol 2006;177(10):7164–72.
61. Moon EY, Ryu SK. TACI:Fc scavenging B cell activating factor (BAFF) alleviates ovalbumin-induced bronchial asthma in mice. Exp Mol Med 2007;39(3):343–52.
62. Ng LG, Mackay CR, Mackay F. The BAFF/APRIL system: life beyond B lymphocytes. Mol Immunol 2005;42(7):763–72.
63. Sutherland AP, Ng LG, Fletcher CA, et al. BAFF augments certain Th1-associated inflammatory responses. J Immunol 2005;174(9):5537–44.
64. Groom JR, Fletcher CA, Walters SN, et al. BAFF and MyD88 signals promote a lupuslike disease independent of T cells. J Exp Med 2007;204(8):1959–71.
65. Nakajima K, Itoh K, Nagatani K, et al. Expression of BAFF and BAFF-R in the synovial tissue of patients with rheumatoid arthritis. Scandinavian J Rheumatol 2007;36(5):365–72.
66. Ohata J, Zvaifler NJ, Nishio M, et al. Fibroblast-like synoviocytes of mesenchymal origin express functional B cell-activating factor of the TNF family in response to proinflammatory cytokines. J Immunol 2005;174(2):864–70.
67. Tecchio C, Nadali G, Scapini P, et al. High serum levels of B-lymphocyte stimulator are associated with clinical-pathological features and outcome in classical Hodgkin lymphoma. Br J Haematol 2007;137(6):553–9.
68. Novak AJ, Grote DM, Stenson M, et al. Expression of BLyS and its receptors in B-cell non-Hodgkin lymphoma: correlation with disease activity and patient outcome. Blood 2004;104(8):2247–53.
69. He B, Chadburn A, Jou E, Schattner EJ, Knowles DM, Cerutti A. Lymphoma B cells evade apoptosis through the TNF family members BAFF/BLyS and APRIL. J Immunol 2004;172(5):3268–79.
70. Planelles L, Castillo-Gutierrez S, Medema JP, Morales-Luque A, Merle-Beral H, Hahne M. APRIL but not BLyS serum levels are increased in chronic lymphocytic leukemia: prognostic relevance of APRIL for survival. Haematologica 2007;92(9):1284–5.
71. Dejardin E, Droin NM, Delhase M, et al. The lymphotoxin-beta receptor induces different patterns of gene expression via two NF-kappaB pathways. Immunity 2002;17(4):525–35.
72. Hanada T, Yoshida H, Kato S, et al. Suppressor of cytokine signaling-1 is essential for suppressing dendritic cell activation and systemic autoimmunity. Immunity 2003;19(3): 437–50.
73. Brink R. Regulation of B cell self-tolerance by BAFF. Semin Immunol 2006;18(5):276–83.
74. Day ES, Cachero TG, Qian F, et al. Selectivity of BAFF/BLyS and APRIL for binding to the TNF family receptors BAFFR/BR3 and BCMA. Biochemistry 2005;44(6):1919–31.
75. Hymowitz SG, Patel DR, Wallweber HJ, et al. Structures of APRIL-receptor complexes: like BCMA, TACI employs only a single cysteine-rich domain for high affinity ligand binding. J Biol Chem 2005;280(8):7218–27.

76. Patel DR, Wallweber HJ, Yin J, et al. Engineering an APRIL-specific B cell maturation antigen. J Biol Chem 2004;279(16):16727–35.
77. Cancro MP, Kearney JF. B cell positive selection: road map to the primary repertoire? J Immunol 2004;173(1):15–9.
78. Ye Q, Wang L, Wells AD, et al. BAFF binding to T cell-expressed BAFF-R costimulates T cell proliferation and alloresponses. Eur J Immunol 2004;34(10):2750–9.
79. Ingold K, Zumsteg A, Tardivel A, et al. Identification of proteoglycans as the APRIL-specific binding partners. J Exp Med 2005;201(9):1375–83.
80. Hendriks J, Planelles L, de Jong-Odding J, et al. Heparan sulfate proteoglycan binding promotes APRIL-induced tumor cell proliferation. Cell Death Differ 2005;12(6):637–48.
81. Bischof D, Elsawa SF, Mantchev G, et al. Selective activation of TACI by syndecan-2. Blood 2006;107(8):3235–42.
82. Sakurai D, Hase H, Kanno Y, Kojima H, Okumura K, Kobata T. TACI regulates IgA production by APRIL in collaboration with HSPG. Blood 2007;109(7):2961–7.
83. Seshasayee D, Valdez P, Yan M, Dixit VM, Tumas D, Grewal IS. Loss of TACI causes fatal lymphoproliferation and autoimmunity, establishing TACI as an inhibitory BLyS receptor. Immunity 2003;18:279–88.
84. von Bulow GU, Bram RJ. NF-AT activation induced by a CAML-interacting member of the tumor necrosis factor receptor superfamily. Science 1997;278:138–41.
85. Xia X-Z, Treanor J, Senaldi G, et al. TACI is a TRAF-interacting receptor for TALL-1, a tumor necrosis factor family member involved in B cell regulation. J Exp Med 2000;192:137–43.
86. Mackay F, Cancro MP. Travelling with the BAFF/BLyS family: are we there yet? Semin Immunol 2006;18(5):261–2.
87. Mackay F, Leung H. The role of the BAFF/APRIL system on T cell function. Semin Immunol 2006;18(5):284–9.
88. Yan M, Brady JR, Chan B, et al. Identification of a novel receptor for B lymphocyte stimulator that is mutated in a mouse strain with severe B cell deficiency. Curr Biol 2001;11:1547–52.
89. Ng LG, Ng CH, Woehl B, et al. BAFF costimulation of Toll-like receptor-activated B-1 cells. Eur J Immunol 2006;36(7):1837–46.
90. Treml LS, Carlesso G, Hoek KL, et al. TLR stimulation modifies BLyS receptor expression in follicular and marginal zone B cells. J Immunol 2007;178(12):7531–9.
91. Acosta-Rodriguez EV, Craxton A, Hendricks DW, et al. BAFF and LPS cooperate to induce B cells to become susceptible to CD95/Fas-mediated cell death. Eur J Immunol 2007;37(4):990–1000.
92. Carter RH, Zhao H, Liu X, et al. Expression and occupancy of BAFF-R on B cells in systemic lupus erythematosus. Arthritis Rheum 2005;52(12):3943–54.
93. Darce JR, Arendt BK, Wu X, Jelinek DF. Regulated expression of BAFF-binding receptors during human B cell differentiation. J Immunol 2007;179(11):7276–86.
94. Shivakumar L, Ansell S. Targeting B-lymphocyte stimulator/B-cell activating factor and a proliferation-inducing ligand in hematologic malignancies. Clin Lymphoma Myeloma 2006;7(2):106–8.
95. Chiu A, Xu W, He B, et al. Hodgkin lymphoma cells express TACI and BCMA receptors and generate survival and proliferation signals in response to BAFF and APRIL. Blood 2007;109(2):729–39.
96. Batten M, Groom J, Cachero TG, et al. BAFF mediates survival of peripheral immature B lymphocytes. J Exp Med 2000;192:1453–65.
97. Gross JA, Dillon SR, Mudri S, et al. TACI-Ig neutralizes molecules critical for B cell development and autoimmune disease: impaired B cell maturation in mice lacking BLyS. Immunity 2001;15:289–302.
98. Schiemann B, Gommerman JL, Vora K, et al. An essential role for BAFF in the normal development of B cells through a BCMA-independent pathway. Science 2001;293:2111–4.

99. Pelletier M, Thompson JS, Qian F, et al. Comparison of soluble decoy IgG fusion proteins of BAFF-R and BCMA as antagonists for BAFF. J Biol Chem 2003;278:33127–33.
100. Smith SH, Cancro MP. B cell receptor signals regulate BLyS receptor levels in mature B cells and their immediate progenitors. J Immunol 2003;170:5820–3.
101. Varfolomeev E, Kischkel F, Martin F, et al. APRIL-deficient mice have normal immune system development. Mol Cell Biol 2004;24:997–1006.
102. Walmsley MJ, Ooi SKT, Reynolds LF, et al. Critical roles for Rac1 and Rac2 GTPases in B cell development and signaling. Science 2003;302:459–62.
103. Castigli E, Scott S, Dedeoglu F, et al. Impaired IgA class switching in APRIL-deficient mice. Proc Natl Acad Sci USA 2004;101(11):3903–8.
104. Hardenberg G, van Bostelen L, Hahne M, Medema JP. Thymus-independent class switch recombination is affected by APRIL. Immunol Cell Biol 2008;86(6):530–4.
105. Mackay F, Sierro F, Grey ST, Gordon TP. The BAFF/APRIL system: an important player in systemic rheumatic diseases. Curr Dir Autoimmun 2005;8:243–65.
106. Kalled SL, Ambrose C, Hsu YM. The biochemistry and biology of BAFF, APRIL and their receptors. Curr Dir Autoimmun 2005;8:206–42.
107. von Bulow GU, Russell H, Copeland NG, Gilbert DJ, Jenkins NA, Bram RJ. Molecular cloning and functional characterization of murine transmembrane activator and CAML interactor (TACI) with chromosomal localization in human and mouse. Mamm Genome 2000;11:628–32.
108. von Bulow G-U, van Deursen JM, Bram RJ. Regulation of the T-independent humoral response by TACI. Immunity 2001;14:573–82.
109. Castigli E, Wilson SA, Garibyan L, et al. TACI is mutant in common variable immunodeficiency and IgA deficiency. Nat Genet 2005;37(8):829–34.
110. Goldacker S, Warnatz K. Tackling the heterogeneity of CVID. Curr Opin Allergy Clin Immunol 2005;5(6):504–9.
111. Lesley R, Xu Y, Kalled SL, et al. Reduced competitiveness of autoantigen-engaged B cells due to increased dependence on BAFF. Immunity 2004;20(4):441–53.
112. Thien M, Phan TG, Gardam S, et al. Excess BAFF rescues self-reactive B cells from peripheral deletion and allows them to enter forbidden follicular and marginal zone niches. Immunity 2004;20(6):785–98.
113. Ait-Azzouzene D, Gavin AL, Skog P, Duong B, Nemazee D. Effect of cell:cell competition and BAFF expression on peripheral B cell tolerance and B-1 cell survival in transgenic mice expressing a low level of Igkappa-reactive macroself antigen. Eur J Immunol 2006;36(4):985–96.
114. Mackay F, Woodcock SA, Lawton P, et al. Mice transgenic for BAFF develop lymphocytic disorders along with autoimmune manifestations. J Exp Med 1999;190:1697–710.
115. Khare SD, Sarosi I, Xia X-Z, et al. Severe B cell hyperplasia and autoimmune disease in TALL-1 transgenic mice. Proc Natl Acad Sci USA 2000;97:3370–5.
116. Hardy RR, Hayakawa K. B cell development pathways. Ann Rev Immunol 2001;19:595–621.
117. Allman D, Srivastava B, Lindsley RC. Alternative route to maturity: branch points and pathways for generating follicular and marginal zone B cells. Immunol Rev 2004;197: 147–60.
118. Carsetti R, Rosado MM, Wardmann H. Peripheral development of B cells in mouse and man. Immunol Rev 2004;197:179–91.
119. Su TT, Rawlings DJ. Transitional B lymphocyte subsets operate as distinct checkpoints in murine splenic B cell development. J Immunol 2002;168(5):2101–10.
120. Su TT, Guo B, Wei B, Braun J, Rawlings DJ. Signaling in transitional type 2 B cells is critical for peripheral B-cell development. Immunol Rev 2004;197:161–78.
121. Allman D, Lindsley RC, DeMuth W, Rudd K, Shinton SA, Hardy RR. Resolution of three nonproliferative immature splenic B cell subsets reveals multiple selection points during peripheral B cell maturation. J Immunol 2001;167(12):6834–40.

122. Teague BN, Pan Y, Mudd PA, et al. Cutting edge: transitional T3 B cells do not give rise to mature B cells, have undergone selection, and are reduced in murine lupus. J Immunol 2007;178(12):7511–5.
123. Meyer-Bahlburg A, Andrews SF, Yu KO, Porcelli SA, Rawlings DJ. Characterization of a late transitional B cell population highly sensitive to BAFF-mediated homeostatic proliferation. J Exp Med 2008;205(1):155–68.
124. Chen X, Martin F, Forbush KA, Perlmutter RM, Kearney JK. Evidence for selection of a population of multi-reactive B cells into the splenic marginal zone. Int Immunol 1997;9:27–41.
125. Martin F, Kearney JF. Positive selection from newly formed to marginal zone B cells depends on the rate of clonal production, CD19 and *btk*. Immunity 2000;12:39–49.
126. Cariappa A, Tang M, Parng C, et al. The follicular versus marginal zone B lymphocyte cell fate decision is regulated by aiolos, Btk, and CD21. Immunity 2001;14:603–15.
127. Lam KP, kuhn R, Rajewsky K. In vivo ablation of surface immunoglobulin on mature B cells by inducible gene targeting results in rapid death. Cell 1997;90:1073–83.
128. Berland R, Fernandez L, Kari E, et al. Toll-like receptor 7-dependent loss of B cell tolerance in pathogenic autoantibody knockin mice. Immunity 2006;25(3):429–40.
129. Wardemann H, Boehm T, Dear N, Carsetti R. B-1a B cells that link the innate and adaptive immune responses are lacking in the absence of the spleen. J Exp Med 2002;195:771–80.
130. Gross JA, Johnston J, Mudri S, et al. TACI and BCMA are receptors for a TNF homologue implicated in B-cell autoimmune disease. Nature 2000;404:995–9.
131. Batten M, Fletcher C, Ng L, et al. TNF deficiency fails to protect BAFF transgenic mice against autoimmunity and reveals a predisposition to B cell lymphomas. J Immunol 2004;172:812–22.
132. Goodnow CC, Cyster JG, Hartley SB, et al. Self-tolerance check points in B cell development. Adv Immunol 1995;59:279–369.
133. Wardemann H, Yurasov S, Schaefer A, Young JW, Meffre E, Nussenzweig MC. Predominant autoantibody production by early human B cell precursors. Science 2003;301(5638):1374–7.
134. Samuels J, Ng YS, Coupillaud C, Paget D, Meffre E. Impaired early B cell tolerance in patients with rheumatoid arthritis. J Exp Med 2005;201(10):1659–67.
135. Yurasov S, Wardemann H, Hammersen J, et al. Defective B cell tolerance checkpoints in systemic lupus erythematosus. J Exp Med 2005;201(5):703–11.
136. Levine MH, haberman AM, Sant'Angelo DB, et al. A B-cell receptor-specicfic selection step governs immature to mature B cell differentiation. Proc Natl Acad Sci USA 2000;97:2743–8.
137. Gaudin E, Hao Y, Rosado MM, Chaby R, Girard R, Freitas AA. Positive selection of B cells expressing low densities of self-reactive BCRs. J Exp Med 2004;199(6):843–53.
138. Cyster JG, Healy JI, Kishihara K, Mak TW, Thomas ML, Goodnow CC. Regulation of B-lymphocyte negative and positive selection by tyrosine phosphatase CD45. Nature 1996;381(6580):325–8.
139. Hayakawa K, Asano M, Shinton SA, et al. Positive selection of natural autoreactive B cells. Science 1999;285(5424):113–6.
140. Balazs M, Martin F, Zhou T, Kearney JF. Blood dendritic cells interact with splenic marginal zone B cells to initiate T-independent Immune responses. Immunity 2002;17:341–52.
141. Evans JG, Chavez-Rueda KA, Eddaoudi A, et al. Novel suppressive function of transitional 2 B cells in experimental arthritis. J Immunol 2007;178(12):7868–78.
142. Kim JM, Rudensky A. The role of the transcription factor Foxp3 in the development of regulatory T cells. Immunol Rev 2006;212:86–98.
143. Thompson JS, Bixler SA, Qian F, et al. BAFF-R, a novel TNF receptor that specifically interacts with BAFF. Science 2001;293:2108–11.
144. Patke A, Mecklenbrauker I, Erdjument-Bromage H, Tempst P, Tarakhovsky A. BAFF controls B cell metabolic fitness through a PKC beta- and Akt-dependent mechanism. J Exp Med 2006;203(11):2551–62.
145. Craxton A, Draves KE, Gruppi A, Clark EA. BAFF regulates B cell survival by downregulating the BH3-only family member Bim via the ERK pathway. J Exp Med 2005;202(10):1363–74.

146. Cyster JG, Hartley SB, Goodnow CC. Competition for follicular niches excludes self-reactive cells from the recirculating B-cell repertoire. Nature 1994;371:389–95.
147. Planelles L, Carvalho-Pinto CE, Hardenberg G, et al. APRIL promotes B-1 cell-associated neoplasm. Cancer cell 2004;6(4):399–408.
148. Berland R, Wortis HH. Origins and functions of B-1 Cells with notes on the role of CD5. Ann Rev Immunol 2002;20:253–300.
149. Attanavanich K, Kearney JF. Marginal zone, but not follicular B cells, are potent activators of naive CD4 T cells. J Immunol 2004;172:803–11.
150. Meyer-Bahlburg A, Rawlings DJ. B cell autonomous TLR signaling and autoimmunity. Autoimmun Rev 2008;7(4):313–6.
151. Yan M, Wang H, Chan B, et al. Activation and accumulation of B cells in TACI-deficient mice. Nature Immunol 2001;2:638–43.
152. Tian J, Avalos AM, Mao SY, et al. Toll-like receptor 9-dependent activation by DNA-containing immune complexes is mediated by HMGB1 and RAGE. Nat Immunol 2007;8(5):487–96.
153. Pillai S, Cariappa A, Moran ST. Marginal zone B cells. Annu Rev Immunol 2005;23:161–96.
154. Mayne CG, Amanna IJ, Nashold FE, Hayes CE. Systemic autoimmunity in BAFF-R-mutant A/WySnJ strain mice. Eur J Immunol 2008;38(2):587–98.
155. Groom J, Kalled SL, Cutler AH, et al. Association of BAFF/BLyS overexpression and altered B cell differentiation with Sjögren's syndrome. J Clin Invest 2002;109:59–68.
156. Zhang J, Roschke V, Baker KP, et al. A role for B lymphocyte stimulator in Systemic Lupus Erythematosus. J Immunol 2001;166:6–10.
157. Stohl W. Targeting B lymphocyte stimulator in systemic lupus and other autoimmune rheumatic disorders. Expert Opin Ther Tar 2004;8:177–89.
158. Stohl W, Metyas S, Tan SM, et al. B lymphocyte stimulator overexpression in patients with systemic lupus erythematosus: longitudinal observations. Arthritis Rheum 2003;48(12):3475–86.
159. Mariette X, Roux S, Zhang J, et al. The level of BLyS (BAFF) correlates with the titre of autoantibodies in human Sjogren's syndrome. Ann Rheum Dis 2003;62(2):168–71.
160. Hansen A, Odendahl M, Reiter K, et al. Diminished peripheral blood memory B cells and accumulation of memory B cells in the salivary glands of patients with Sjögren's syndrome. Arthritis Rheum 2002;46:2160–71.
161. Daridon C, Pers JO, Devauchelle V, et al. Identification of transitional type II B cells in the salivary glands of patients with Sjogren's syndrome. Arthritis Rheum 2006;54(7):2280–8.
162. Lavie F, Miceli-Richard C, Quillard J, Roux S, Leclerc P, Mariette X. Expression of BAFF (BLyS) in T cells infiltrating labial salivary glands from patients with Sjogren's syndrome. J Pathol 2004;202(4):496–502.
163. Daridon C, Devauchelle V, Hutin P, et al. Aberrant expression of BAFF by B lymphocytes infiltrating the salivary glands of patients with primary Sjogren's syndrome. Arthritis Rheum 2007;56(4):1134–44.
164. Chu VT, Enghard P, Riemekasten G, Berek C. In vitro and in vivo activation induces BAFF and APRIL expression in B cells. J Immunol 2007;179(9):5947–57.
165. Mariette X, Ravaud P, Steinfeld S, et al. Inefficacy of infliximab in primary Sjogren's syndrome: results of the randomized, controlled Trial of Remicade in Primary Sjogren's Syndrome (TRIPSS). Arthritis Rheum 2004;50(4):1270–6.
166. Sankar V, Brennan MT, Kok MR, et al. Etanercept in Sjogren's syndrome: a twelve-week randomized, double-blind, placebo-controlled pilot clinical trial. Arthritis Rheum 2004;50(7):2240–5.
167. Mavragani CP, Niewold TB, Moutsopoulos NM, Pillemer SR, Wahl SM, Crow MK. Augmented interferon-alpha pathway activation in patients with Sjogren's syndrome treated with etanercept. Arthritis Rheum 2007;56(12):3995–4004.
168. Cheema GS, Roschke V, Hilbert DM, Stohl W. Elevated serum B lymphocyte stimulator levels in patients with systemic immune-based rheumatic diseases. Arthritis Rheum 2001;44:1313–9.

169. Tan S, Xu D, Roschke V, et al. Local production of B lymphocyte stimulator protein and APRIL in arthritic joints of patients with inflammatory arthritis. Arthritis and Rheum 2003;48:982–92.
170. Fabris M, Quartuccio L, Sacco S, et al. B-Lymphocyte stimulator (BLyS) up-regulation in mixed cryoglobulinaemia syndrome and hepatitis-C virus infection. Rheumatology (Oxford, England) 2007;46(1):37–43.
171. Sene D, Limal N, Ghillani-Dalbin P, Saadoun D, Piette JC, Cacoub P. Hepatitis C virus-associated B-cell proliferation–the role of serum B lymphocyte stimulator (BLyS/BAFF). Rheumatology (Oxford, England) 2007;46(1):65–9.
172. Matsushita T, Hasegawa M, Yanaba K, Kodera M, Takehara K, Sato S. Elevated serum BAFF levels in patients with systemic sclerosis: enhanced BAFF signaling in systemic sclerosis B lymphocytes. Arthritis Rheum 2006;54(1):192–201.
173. Yu HM, Liu YF, Hou M. BAFF – an essential survival factor for B cells: Links to genesis of ITP and may be of therapeutic target. Med hypotheses 2008;70(1):40–2.
174. Sellam J, Miceli-Richard C, Gottenberg JE, et al. Decreased B cell activating factor receptor expression on peripheral lymphocytes associated with increased disease activity in primary Sjogren's syndrome and systemic lupus erythematosus. Ann Rheum Dis 2007;66(6):790–7.
175. Stohl W, Metyas S, Tan SM, et al. Inverse association between circulating APRIL levels and serological and clinical disease activity in patients with systemic lupus erythematosus. Ann Rheum Dis 2004;63(9):1096–103.
176. Koyama T, Tsukamoto H, Masumoto K, et al. A novel polymorphism of the human APRIL gene is associated with systemic lupus erythematosus. Rheumatology (Oxford, England) 2003;42(8):980–5.
177. Nagatani K, Itoh K, Nakajima K, et al. Rheumatoid arthritis fibroblast-like synoviocytes express BCMA and are stimulated by APRIL. Arthritis Rheum 2007;56(11):3554–63.
178. Seyler TM, Park YW, Takemura S, et al. BLyS and APRIL in rheumatoid arthritis. J Clin Invest 2005;115:3083–92.
179. Ittah M, Miceli-Richard C, Eric Gottenberg J, et al. B cell-activating factor of the tumor necrosis factor family (BAFF) is expressed under stimulation by interferon in salivary gland epithelial cells in primary Sjogren's syndrome. Arthritis Res Ther 2006;8(2):R51.
180. Ittah M, Miceli-Richard C, Gottenberg JE, et al. Viruses induce high expression of BAFF by salivary gland epithelial cells through TLR- and type-I IFN-dependent and -independent pathways. Eur J Immunol 2008;38(4):1058–64.
181. Bave U, Nordmark G, Lovgren T, et al. Activation of the type I interferon system in primary Sjogren's syndrome: a possible etiopathogenic mechanism. Arthritis Rheum 2005;52(4):1185–95.
182. Gottenberg JE, Cagnard N, Lucchesi C, et al. Activation of IFN pathways and plasmacytoid dendritic cell recruitment in target organs of primary Sjogren's syndrome. Proc Natl Acad Sci USA 2006;103(8):2770–5.
183. Hjelmervik TO, Petersen K, Jonassen I, Jonsson R, Bolstad AI. Gene expression profiling of minor salivary glands clearly distinguishes primary Sjogren's syndrome patients from healthy control subjects. Arthritis Rheum 2005;52(5):1534–44.
184. Litinskiy M, Nardelli B, Hilbert BM, et al. DCs induce CD40-independent immunoglobulin class switching through BLyS and APRIL. Nat Immunol 2002;3:822–9.
185. Rose NR. The role of infection in the pathogenesis of autoimmune disease. Semin Immunol 1998;10:5–13.
186. Ding C, Jones G. Belimumab human genome sciences/Cambridge antibody technology/GlaxoSmithKline. Curr Opin Investig Drugs 2006;7(5):464–72.
187. Baker KP, Edwards BM, Main SH, et al. Generation and characterization of LymphoStat-B, a human monoclonal antibody that antagonizes the bioactivities of B Lymphocyte Stimulator. Arthritis Rheum 2003;48:3253–65.
188. Halpern WG, Lappin P, Zanardi T, et al. Chronic administration of belimumab, a BLyS antagonist, decreases tissue and peripheral blood B-lymphocyte populations in

cynomolgus monkeys: pharmacokinetic, pharmacodynamic, and toxicologic effects. Toxicol Sci 2006;91(2):586–99.
189. Bhat P, Radhakrishnan J. B lymphocytes and lupus nephritis: new insights into pathogenesis and targeted therapies. Kidney Int 2008;73(3):261–8.
190. Nestorov I, Munafo A, Papasouliotis O, Visich J. Pharmacokinetics and biological activity of atacicept in patients with rheumatoid arthritis. J Clin Pharmacol 2008;48(4):406–17.
191. Munafo A, Priestley A, Nestorov I, Visich J, Rogge M. Safety, pharmacokinetics and pharmacodynamics of atacicept in healthy volunteers. Eur J Clin Pharmacol 2007;63(7):647–56.
192. Tak PP, Thurlings RM, Rossier C, et al. Atacicept in patients with rheumatoid arthritis: results of a multicenter, phase Ib, double-blind, placebo-controlled, dose-escalating, single- and repeated-dose study. Arthritis Rheum 2008;58(1):61–72.
193. Dall'Era M, Chakravarty E, Wallace D, et al. Reduced B lymphocyte and immunoglobulin levels after atacicept treatment in patients with systemic lupus erythematosus: results of a multicenter, phase Ib, double-blind, placebo-controlled, dose-escalating trial. Arthritis Rheum 2007;56(12):4142–50.
194. Lin WY, Gong Q, Seshasayee D, et al. Anti-BR3 antibodies: a new class of B-cell immunotherapy combining cellular depletion and survival blockade. Blood 2007;110(12):3959–67.
195. Sabahi R, Anolik JH. B-cell-targeted therapy for systemic lupus erythematosus. Drugs 2006;66(15):1933–48.
196. Looney RJ. B cell-targeted therapy for rheumatoid arthritis: an update on the evidence. Drugs 2006;66(5):625–39.
197. Lavie F, Miceli-Richard C, Ittah M, Sellam J, Gottenberg JE, Mariette X. Increase of B cell-activating factor of the TNF family (BAFF) after rituximab treatment: insights into a new regulating system of BAFF production. Ann Rheum Dis 2007;66(5):700–3.
198. Pers JO, Devauchelle V, Daridon C, et al. BAFF-modulated repopulation of B lymphocytes in the blood and salivary glands of rituximab-treated patients with Sjogren's syndrome. Arthritis Rheum 2007;56(5):1464–77.
199. Toubi E, Kessel A, Slobodin G, et al. Changes in macrophage function after rituximab treatment in patients with rheumatoid arthritis. Ann Rheum Dis 2007;66(6):818–20.
200. Gardam S, Sierro F, Basten A, Mackay F, Brink R. TRAF2 and TRAF3 signal adapters act cooperatively to control the maturation and survival signals delivered to B cells by the BAFF receptor. Immunity 2008;28(3):391–401.
201. Sasaki Y, Derudder E, Hobeika E, et al. Canonical NF-kappaB activity, dispensable for B cell development, replaces BAFF-receptor signals and promotes B cell proliferation upon activation. Immunity 2006;24(6):729–39.
202. Benson MJ, Dillon SR, Castigli E, et al. Cutting Edge: The Dependence of Plasma Cells and Independence of Memory B Cells on BAFF and APRIL. J Immunol 2008;180(6):3655–9.
203. Craxton A, Magaletti D, Ryan EJ, Clack EA. Macrophage- and dendritic cell-dependent regulation of human B-cell proliferation requires the TNF family ligand BAFF. Blood 2003;101:4464–71.
204. Hase H, Kanno Y, Kojima M, et al. BAFF/BLyS can potentiate B-cell selection with the B-cell coreceptor complex. Blood 2004;103(6):2257–65.
205. He B, Raab-Traub N, Casali P, Cerutti A. EBV-encoded latent membrane protein 1 cooperates with BAFF/BLyS and APRIL to induce T cell-independent Ig heavy chain class switching. J Immunol 2003;171(10):5215–24.
206. Phillips TA, Ni J, Hunt JS. Cell-specific expression of B lymphocyte (APRIL, BLyS)- and Th2 (CD30L/CD153)-promoting tumor necrosis factor superfamily ligands in human placentas. J Leukoc Biol 2003;74(1):81–7.
207. Shu H-B, Hu W-H, Johnson H. TALL-1 is a novel member of the TNF family that is down-regulated by mitogens. J Leuk Biol 1999;65:680–3.
208. Tao W, Hangoc G, Hawes JW, Si Y, Cooper S, Broxmeyer HE. Profiling of differentially expressed apoptosis-related genes by cDNA arrays in human cord blood CD34+ cells treated with etoposide. Exp Hematol 2003;31(3):251–60.

209. Moreaux J, Legouffe E, Jourdan E, et al. BAFF and APRIL protect myeloma cells from apoptosis induced by IL-6 deprivation and dexamethasone. Blood 2004;103(8):3148–57.
210. Roth W, Wagenknecht B, Klummp A, et al. APRIL, a new member of the tumor necrosis factor family, modulates death-induced apoptosis. Cell Death Differ 2001;8:403–10.
211. Moreaux J, Legouffe E, Jourdan E, et al. BAFF and APRIL protect myeloma cells from apoptosis induced by interleukin 6 deprivation and dexamethasone. Blood 2004;103(8):3148–57.
212. Moreaux J, Cremer FW, Reme T, et al. The level of TACI gene expression in myeloma cells is associated with a signature of microenvironment dependence versus a plasmablastic signature. Blood 2005;106(3):1021–30.
213. Abe M, Kido S, Hiasa M, et al. BAFF and APRIL as osteoclast-derived survival factors for myeloma cells: a rationale for TACI-Fc treatment in patients with multiple myeloma. Leukemia 2006;20(7):1313–5.
214. Avery DT, Kalled SL, Ellyard JI, et al. BAFF selectively enhances the survival of plasmablasts generated from human memory cells. J Clin Invest 2003;112(2):286–97.
215. Tarte K, De Vos J, Thykjaer T, et al. Generation of polyclonal plasmablasts from peripheral blood B cells: a normal counterpart of malignant plasmablasts. Blood 2002;100(4):1113–22.
216. Gorelik L, Cutler AH, Thill G, et al. BAFF regulates CD21/35 and CD23 expression independently of its B cell survival function. J Immunol 2004;172:762–6.
217. Hsu BL, Harless SM, Lindsley RC, Hilbert DM, Cancro MP. BLyS enables survival of transitional and mature B cells through distinct mediators. J Immunol 2002;168:5993–6.
218. Moreaux J, Hose D, Jourdan M, et al. TACI expression is associated with a mature bone marrow plasma cell signature and C-MAF overexpression in human myeloma cell lines. Haematologica 2007;92(6):803–11.
219. Castigli E, Wilson SA, Scott S, et al. TACI and BAFF-R mediate isotype switching in B cells. J Exp Med 2005;201(1):35–9.
220. O'Connor BP, Raman VS, Erickson LD, et al. BCMA is essential for the survival of long-lived bone marrow plasma cells. J Exp Med 2004;199:91–7.
221. Shulga-Morskaya S, Dobles M, Walsh ME, et al. B cell-activating factor belonging to the TNF family acts through separate receptors to support B cell survival and T cell-independent antibody formation. J Immunol 2004;173(4):2331–41.
222. Yang M, Hase H, Legarda-Addison D, Varughese L, Seed B, Ting AT. B cell maturation antigen, the receptor for a proliferation-inducing ligand and B cell-activating factor of the TNF family, induces antigen presentation in B cells. J Immunol 2005;175(5):2814–24.
223. Rahman ZSM, Rao SP, Kalled SL, Manser T. Normal induction but attenuated progression of germinal center responses in BAFF and BAFF-R signaling-deficient mice. J Exp Med 2003;198:1157–69.
224. Vora KA, Wang LC, Rao SP, et al. Germinal centers formed in the absence of B cell-activating factor belonging to the TNF family exhibit impaired maturation and function. J Immunol 2003;171:547–51.

Chapter 8
Systemic Immune-Based Rheumatic Diseases: Blissless States of BLySfulness

William Stohl

Abstract B lymphocyte stimulator (BLyS) is a vital B-cell survival and differentiation factor. Overexpression of BLyS in mice can lead to clinical and serological features of systemic lupus erythematosus (SLE) and Sjögren's syndrome (SS). Treatment with BLyS antagonists of mice with established SLE ameliorates disease progression and enhances survival. Moreover, similar treatment of mice with inflammatory arthritis ameliorates the ongoing inflammation and subsequent joint destruction. In humans, BLyS overexpression is common in patients with several systemic immune-based rheumatic diseases (SIRDs), including SLE, rheumatoid arthritis (RA), SS, scleroderma, dermatomyositis, ANCA-associated vasculitis, and mixed cryoglobulinemia. Phase-I and phase-II clinical trials with BLyS antagonists in SLE and RA have documented in vivo biological activities along with favorable safety profiles for these agents and have pointed to beneficial effects in subsets of patients. These features collectively point to BLyS as an attractive therapeutic target in human SIRDs.

Keywords APRIL · B cells · B lymphocyte stimulator (BLyS) · rheumatoid arthritis · Sjögren's syndrome · systemic immune-based rheumatic disease · systemic lupus erythematosus

8.1 Introduction

It is no secret that current treatment options for patients with systemic immune-based rheumatic diseases (SIRDs) are inadequate and fraught with serious toxicities. Although many SIRD patients respond clinically to corticosteroids, cytotoxic

W. Stohl (✉)
Division of Rheumatology, Department of Medicine, University of Southern California Keck School of Medicine, Los Angeles, CA 90033, USA

M.P. Cancro (ed.), *BLyS Ligands and Receptors,* Contemporary Immunology,
DOI 10.1007/978-1-60327-013-7_8,

drugs, and/or tumor necrosis factor α (TNF) antagonists, the responses are almost always subtotal, and complications from the medications are far-too-commonly more pernicious than is the underlying disease itself. The emergence of BLyS (B-lymphocyte stimulator) as a likely pathogenic contributor to several SIRDs coupled to the exquisite specificity of BLyS binding to B cells suggests that neutralization of BLyS might selectively affect (pathogenic) B-cell function without perturbing the function of other cell types (lymphoid and non-lymphoid). Accordingly, the prospect of therapeutically targeting BLyS has brought cautious optimism to both clinicians and their patients that a safe and efficacious treatment modality is on the horizon.

8.2 BLyS and Its Receptors

8.2.1 General Properties of BLyS

BLyS (also known as BAFF, TALL-1, THANK, TNFSF13B, and zTNF4) is a 285-amino acid member of the TNF ligand superfamily and is expressed as a type-II transmembrane protein [1–6]. Cleavage of BLyS by a furin protease from the cell surface results in release of a soluble, biologically active 17-kDa molecule [3, 7]. Release of BLyS from human peripheral blood monocytes can be triggered through the cross-linking of FcγRI, but not FcγIIa, with IgG or C-reactive protein [8]. In neutrophils, processing of full-length BLyS to the soluble form takes place intracellularly [9, 10], although BLyS may be expressed at the surface of these cells as well [11].

Since BLyS may be either membrane bound or soluble (Table 8.1), the biological activities of each must be individually analyzed. Although membrane-bound BLyS has potent in vitro activity in an assay based on its multimeric binding to a non-physiological cell target [12], it remains unresolved whether membrane-bound BLyS has biological activity in vivo and, if so, what that biological activity is. This is not merely a point of academic interest but potentially has profound clinical ramifications. Different BLyS antagonists may differentially neutralize soluble versus membrane-bound BLyS, so depending upon the discrete in vivo functions of soluble versus membrane-bound BLyS, strikingly disparate biologic outcomes may ensue

Table 8.1 Different forms assumed by BLyS

Membrane-bound BLyS
Soluble BLyS
BLyS homotrimers
BLyS/APRIL heterotrimers
BLyS virus-like 60-mers
BLyS multimers (not necessarily 60-mers)
Receptor-bound BLyS
ΔBLyS

following treatment with different antagonists that nominally are directed against the same target (BLyS).

The uncertain role for membrane-bound BLyS notwithstanding, it is clear that soluble BLyS is biologically active and circulates in trimeric form [3, 13]. Whether higher-order multimers of BLyS also exist under physiologic conditions is a subject of debate. Based on the elongated "DE loop" manifest by BLyS and the resulting structural "flap" that ensues, some laboratories have induced BLyS to assemble into virus-like clusters of 60 monomers under appropriate in vitro conditions [14–16]. However, other laboratories have not been able to duplicate this feat [17, 18]. It has been suggested that multimeric self-assembly of BLyS is an artifactual consequence of the manner in which BLyS is "tagged" for in vitro experimentation [19], although this notion itself has been challenged [20]. Multimeric BLyS (not necessarily comprised of 60-mers) has been detected in the plasma of mice that constitutively overexpress BLyS [12], but comparable observations in humans are still lacking.

Even if they truly do exist, the in vivo biologic activity and roles for these multimers remain entirely speculative at present. This point may have vital clinical ramifications in that large BLyS multimers may be far more resistant to neutralization by BLyS antagonists than are smaller BLyS trimers. Accordingly, a given BLyS antagonist may promote widely disparate clinical responses among patients with different SIRDs or even among patients with the "same" SIRD consequent to differences in the distribution of soluble BLyS between trimeric and multimeric fractions.

Additional complexity arises from the fact that at least two isoforms of BLyS are expressed. The full-length BLyS mRNA isoform codes for the biologically active full-length protein, whereas the alternatively spliced ΔBLyS mRNA isoform codes for a protein with a small peptide deletion [21]. ΔBLyS does not bind to surface-expressed BLyS receptors and, therefore, has no agonistic activity. Moreover, ΔBLyS forms biologically inactive heterotrimers with full-length BLyS, thereby functioning as a dominant-negative antagonist of BLyS activity. Indeed, selective overexpression of the ΔBLyS isoform can result in functional BLyS neutralization in vivo [22]. Accordingly, differential up- or downregulation of ΔBLyS expression relative to expression of full-length BLyS may have enormous consequences for development and/or maintenance of disease. How ΔBLyS expression is normally regulated and whether ΔBLyS expression is dysregulated in any of the SIRDs are largely unknown. From a therapeutic perspective, the individual full-length BLyS and ΔBLyS homeostatic responses of the host to BLyS neutralization may considerably affect the nature and degree of the net clinical response.

8.2.2 Systemic and Local BLyS Expression

BLyS is systemically expressed by myeloid-lineage cells (monocytes, dendritic cells, macrophages, neutrophils) [1–3, 5, 7, 9], bone marrow-derived radiation-resistant stromal cells [23], and, to some degree, T cells [3, 24]. Although B cells express varying levels of BLyS mRNA, surface expression of BLyS on B cells is very limited [25]. In myeloid-lineage cells, BLyS expression can be upregulated by multiple cytokines, including interferon (IFN)γ, IFNα, and transforming growth factor (TGF) β [7, 9, 26, 27]. Inasmuch as IFN, especially IFNα, has been ascribed a pathogenic role in several SIRDs [28–31], one may entertain the possibility that at least part of that pathogenicity is effected through BLyS.

Of note, the intracellular signaling pathways leading to BLyS upregulation can widely differ among different BLyS inducers. For example, TGFβ drives BLyS promoter activity via a Smad3/4-dependent pathway, whereas IFNγ drives BLyS promoter activity via a PKA- and CREB-dependent pathway [27]. Although it might be pathophysiologically more "correct" to therapeutically target BLyS *production* in disease states characterized by BLyS overexpression, this would likely be operationally difficult, inasmuch as multiple intracellular signaling pathways would have to be concurrently affected. A far more tractable therapeutic approach would be to target the protein end product itself, an approach that is currently being implemented (see Section 8.5 below).

Of note, BLyS is not expressed by peripheral blood B cells, but its expression can be detected in several tonsillar B-cell subsets [32]. Indeed, the ability of certain B cells to produce BLyS may represent an autocrine pathway of survival in neoplastic B cells [33–36]. The role, if any, played by B-cell-derived BLyS in the survival and/or function of non-neoplastic autoreactive B cells remains speculative. Importantly, therapeutic B-cell depletion in human SIRD patients leads to *increased*, rather than *decreased*, circulating levels of BLyS [37–40], suggesting that the contribution of B-cell-derived BLyS to the total BLyS pool is, at most, very small.

In addition to systemic BLyS expression, BLyS is expressed locally in a tissue-specific manner. For example, BLyS is expressed by astrocytes in the central nervous system (CNS) [41]. Such expression may be crucial to survival of virus-specific antibody-secreting cells that control viral persistence in the CNS [42]. Expression of BLyS has also been documented in human placental mesenchymal cells, trophoblasts, and decidua, with expression being higher in the tissues from normal pregnant women (collected at the time of elective therapeutic abortions) than in tissues from recurrent spontaneous abortion patients at the same gestational age [43–45]. Although the importance of reduced BLyS expression to spontaneous abortion is nebulous at present, one must be sensitive to the possibility that BLyS antagonists may adversely affect pregnancy outcomes.

BLyS is also locally expressed by fibroblast-like synoviocytes (FLSs) in synovial tissue [46, 47]. This observation may be especially germane to patients with rheumatoid arthritis (RA) in that pro-inflammatory cytokines routinely present in inflamed joint spaces, such as IFN-γ and TNF-α, may upregulate such BLyS expression. This augmented BLyS expression by FLS can blunt the B-cell-depleting effects of rituximab in the joint and, thereby, limit the clinical effectiveness of rituximab therapy [48]. Of note, reactive oxygen species, typically generated in the course of inflammatory responses, can augment BLyS expression [49]. Thus, in the context of inflammatory diseases in which B cells contribute to the pathologic outcome, anti-inflammatory therapies may not only primarily attenuate ongoing inflammation but may also secondarily attenuate BLyS-driven pathology.

8.2.3 BLyS Receptors

Three BLyS receptors (BCMA, TACI, and BAFFR [also called BR3]) have been identified, and their expression is largely (but not exclusively) limited to B cells [50–53]. BLyS binds strongly to B cells, weakly to T cells and monocytes, but not to NK cells [1, 54, 55]. With the exception of bone marrow plasma cells, human B-lineage cells from naive through plasma cells uniformly surface express BAFFR [56], although this observation is not universally accepted [57]. In any case, the vast majority, if not all, of the BLyS that binds to human peripheral blood B cells does so via surface BAFFR and/or TACI, with little, if any, BLyS binding via BCMA [34]. In contrast, human tonsil and bone marrow memory B cells and plasma cells surface express BCMA (along with TACI) [56]. There may be a reciprocal relationship between BAFFR and BCMA expression, inasmuch as levels of BCMA mRNA in terminally differentiated plasma cells are much greater than those in mature B cells [58, 59], and in vitro-generated plasmablasts upregulate surface BCMA and downregulate surface BAFFR and TACI [60]. Inhibition of terminal differentiation in vitro by addition of anti-Ig not only reduces Ig production but also blocks the switch in expression of BLyS receptors from a BAFFR-predominant profile to a BCMA-predominant profile with little effect on TACI expression [56].

Expression of BLyS receptors can be modified not just through B-cell antigen receptor (BCR) engagement but through engagement of Toll-like receptors (TLRs) as well. In murine late transitional and mature spleen B cells, BCR engagement leads to upregulation of BAFFR mRNA without discernible effects on TACI or BCMA mRNA [61]. In contrast, CpG-containing oligodeoxynucleotides (ODNs) (TLR9 ligands) and lipopolysaccharide (LPS) (TLR4 ligand) preferentially increase TACI expression among spleen follicular and marginal zone (MZ) B cells [62]. Thus, the expression profiles of BLyS receptors among discrete B-cell populations are likely highly dynamic. Subtle variations in these expression profiles may have profound ramifications for the generation and/or regulation of autoimmune responses. Moreover, differences in expression profiles of individual BLyS receptors may greatly affect the clinical success in therapeutic targeting of BLyS.

8.2.4 Consequences of BLyS Binding to Surface BLyS Receptors

8.2.4.1 Consequences of BLyS Binding to B Cells

At the molecular level, binding of BLyS to its receptors on the B cell triggers a complex intracellular scheme. Several TNF receptor-associated factors (TRAFs), including TRAF1, TRAF2, TRAF3, TRAF5, and TRAF6, interact with one or more of the three BLyS receptors [54, 63–65]. Phospholipase C-2, NF-κB1 (canonical NF-κB pathway), and NF-κB2 (non-canonical NF-κB pathway) are all activated [66–69], HSH2 expression is upregulated [70], nuclear accumulation of PKC is prevented [71], surface expression of CD19 is enhanced [72, 73], and B-cell survival is increased [60, 69, 74–78] (Table 8.2). Upregulation of cyclooxygenase-2,

Table 8.2 Functions ascribed to BLyS

Functions ascribed to BLyS
Promotes B-cell survival
Promotes B-cell "metabolic fitness"
Promotes B-cell differentiation
Promotes Ig class switching
Promotes B-cell chemotaxis (in concert with certain chemokines)

production of prostaglandin E_2, and increased expression of Mcl-1 may be especially important in BLyS-driven viability and continued cell cycling of recently stimulated B cells [79, 80]. Enhanced survival may also, at least in part, be secondary to BLyS-induced upregulation of Pim family members [80] and downregulation of Bim [81]. In fact, BLyS has little effect on survival of Bim-deficient B cells [82]. Moreover, BLyS-driven B cell survival may, in part, be related to upregulation of Bcl-2 and/or Bcl-x_L [75], inasmuch as B cells with enforced overexpression of Bcl-x_L are protected from the premature death that ensues in the absence of BLyS signaling [83]. The NF-κB1 pathway may be especially vital in that its constitutive activation in B cells obviates the need for BLyS/BAFFR interactions in normal B-cell development [84]. Nevertheless, long-term survival of B cells promoted by BLyS appears to be mediated primarily via the NF-κB2 pathway [85] which, in turn, may be heavily dependent upon the ability of BLyS/BAFFR interactions to reverse its constitutive repression by TRAF2 and TRAF3 [86]. Indeed, in the absence of TRAF3, BLyS/BAFFR interactions are no longer needed for B-cell survival. NF-κB2 activity is greatly upregulated in TRAF3-deficient B cells, and such B cells are insensitive to in vivo treatment with a BLyS antagonist [87].

BLyS not only promotes B-cell survival but also promotes B cell "metabolic fitness" through a PKC- and Akt-dependent pathway [80, 88]. B cells cultured with BLyS display increased cell size, cellular protein content, and mitochondrial membrane potential. Indeed, BLyS directs gene transcription to proteins required for cell-cycle progression and glycolytic metabolism, the latter likely being especially important at inflammatory sites where oxygen tension is low. Whether these responses in autoreactive B cells differ from those in non-autoreactive B cells is unknown, so the ramifications for autoimmunity remain uncertain.

Somewhat paradoxically, it is possible that BLyS may promote B-cell death rather than B-cell survival under certain conditions. Although survival of anti-IgM- or anti-CD40-stimulated murine spleen B cells is enhanced by supplementing the cultures with BLyS, survival of the same B cells stimulated with LPS is impaired by supplementation with BLyS [89]. It must be emphasized that this observation is an in vitro one in a murine system, so whether a similar outcome would occur in vivo in humans remains entirely speculative. Since the critical death pathway is a Fas (CD95)-dependent one [89], defects in Fas-driven cell death could convert any BLyS-driven B-cell death into BLyS-driven B-cell survival. In the context of autoreactive B cells, this could potentially augment an ongoing autoimmune response.

In addition to its role as a B-cell survival and "metabolic fitness" factor, BLyS also serves as a B-cell differentiation factor, promoting the differentiation of immature B cells to mature B cells, including MZ B cells [90, 91]. Moreover, BLyS promotes Ig class switching and Ig production by B cells via engagement of TACI and BAFFR [26, 92, 93] and synergizes with IL-21 in driving plasma cell differentiation [94] and with TLR3 ligands in promoting B-cell expansion [95]. Accordingly, protracted neutralization of BLyS could, in principle, adversely affect global humoral immunity. In vivo experience to date with BLyS antagonists in human SLE and RA indicates that this is not the case in the short term [96, 97]. Whether this concern actually becomes clinically meaningful in the long term remains to be empirically established.

Based on in vitro observations, BLyS may enhance chemotaxis under certain conditions. The addition of BLyS to the chemokines CCL21, CXCL12, or CXCL13 increased chemotaxis of either naive or memory human B cells to the respective chemokines alone. This enhancement could be blocked with antibodies to BAFFR [98], pointing to BLyS/BAFFR interactions being vital. Of note, the BLyS-mediated enhancement of chemotaxis was greatest for memory B cells in response to CXCL13. Since CXCL13 plays a vital role in B-cell entry into lymphoid follicles and in germinal center organization, the synergy between BLyS and CXCL13 raises the specter of pathogenic B cells being recruited and sequestered in follicle-like structures, thereby contributing to the maintenance of the underlying SIRD.

The effects of BLyS are not limited to "conventional" B2 B cells but extend to "non-conventional" B1 B cells as well. Peritoneal B1 ($CD5^+$) B cells treated in culture with BLyS demonstrate upregulation of surface CD21, activation of the NF-κB2 pathway, and enhanced survival. Moreover, BLyS enhances expression of activation markers by, proliferation of, and cytokine production by TLR ligand-stimulated B1 B cells [99].

8.2.4.2 Consequences of BLyS Binding to Dendritic Cells (DCs)

In addition to robust binding of BLyS to B cells, BLyS also binds modestly to DC, likely via engagement of TACI. Although the in vivo consequences of such binding have not been directly studied, in vitro studies point to profound ramifications for the immune system. DC treated with BLyS undergoes several morphologic changes, including increased cell size, greater number of dendrites, and abundant clustering. Surface expression of the costimulatory molecules CD80 and CD86 is increased, production of pro-inflammatory cytokines (IL-6, IL-1β, TNFα) is triggered, and the BLyS-treated DC can drive $CD4^+$ T cells to produce Th1 cytokines (IFNγ) but not Th2 cytokines (IL-4, IL-5, IL-10) [100]. This may lie at the heart of the Th1-like polarization of the immune response in BLyS-Tg mice [101]. Whether a similar skewing of the immune response occurs in humans who overexpress BLyS remains to be determined. If so, then the systemic effects of BLyS antagonism may extend far beyond B cells and humoral immunity.

8.2.5 A Proliferation-Inducing Ligand (APRIL), the Confounding Relative of BLyS

Any discussion of BLyS must incorporate a discussion of its "kissing cousin", APRIL (also known as TALL-2, TRDL-1, and TNFSF13A), a 250-amino acid protein member of the TNF ligand superfamily. In contrast to BLyS, APRIL is not expressed as a surface protein but is processed intracellularly and released in its soluble, biologically active form [102]. (A hybrid molecule [TWE—PRIL] between the APRIL C-terminal and the cytoplasmic and transmembrane domains of TWEAK [another TNF ligand superfamily member] is surface-expressed and has APRIL-like activity in vitro [103]. The relevance of TWE–PRIL to in vivo biology is entirely speculative at present.)

Although APRIL does not bind to BAFFR [52], its three-dimensional structure is sufficiently similar to that of BLyS to permit APRIL to bind to the other two BLyS receptors (BCMA and TACI) [104–110]. This ability of APRIL to bind to BCMA and TACI may not only be important for systemic immune (and autoimmune) responses, but it likely has severe consequences for local pathology in RA patients. Since FLS from these patients express BCMA, their stimulation by APRIL in the synovial fluids (SF) of these patients can drive production of several pro-inflammatory cytokines [111, 112] and likely aggravate the ongoing disease process. Indeed, the association between SLE and a polymorphism in the *April* gene in a Japanese cohort [113] and possibly in Hispanic and African-American populations [114] supports the relevance of APRIL to at least some SIRDs.

In addition to circulating as individual homotrimers, APRIL and BLyS spontaneously form APRIL/BLyS heterotrimers in vivo which are detectable in the circulation [115]. However, APRIL has a much shorter "DE loop" than does BLyS which precludes development of the "flap" that is essential to formation of higher-order multimers [116]. Of note, murine APRIL is alternatively spliced, resulting in deletion of one amino acid (alanine at position 112). Although this shorter protein can weakly bind to BAFFR [117], its impact on in vivo biology remains entirely unknown at present.

Full-length APRIL can costimulate B cells in vitro and in vivo [26, 106, 107], although it does so with considerably less potency than that of BLyS [118]. APRIL also contributes to Ig class switching, especially for IgA [93, 119, 120]. In addition to its binding to BCMA and TACI, APRIL binds to heparan sulfate proteoglycans (HSPGs), whereas BLyS does not [121, 122]. The ability of APRIL, but not BLyS, to bind HSPG (e.g., syndecans) on plasmablasts may explain why APRIL is so much better at supporting plasmablast survival in vitro than is BLyS [123].

This in vitro pro-survival capacity of APRIL notwithstanding, neither complete deficiency of APRIL nor its constitutive overexpression has dramatic effects in vivo. APRIL-deficient mice bearing a non-autoimmune-prone genetic background have been described as being phenotypically normal [124] or as having selective deficiencies in circulating IgA levels and IgA responses to mucosal challenges

[125]. Similarly, APRIL-transgenic (Tg) mice, which constitutively overexpress APRIL, manifest only subtle immunologic abnormalities [126]. Most noteworthy, no serologic or clinical autoimmune features have been appreciated in such APRIL-overexpressing mice.

Nevertheless, APRIL may play an important role in modifying the net biologic effects of BLyS. Although APRIL/BLyS heterotrimers have BLyS-like biologic activity in vitro [115], their in vivo biologic activity remains uncertain. In principle, the biologic activity of APRIL/BLyS heterotrimers may be greater than, less than, or equal to that of BLyS homotrimers. Accordingly, therapeutic neutralization of APRIL concomitant with neutralization of BLyS might be beneficial, harmful, or neutral in the context of autoimmunity. This can potentially have profound ramifications for the types of BLyS antagonists chosen for use in the clinic. Indeed, differences in various immunologic parameters were appreciated among SLE-prone (NZB × NZW)F1 mice depending on whether they were treated with TACI-Ig (which neutralizes both BLyS and APRIL) or with BAFFR-Ig (which neutralizes only BLyS) [127]. Since TACI binds not only to BLyS and APRIL but to certain syndecans as well [128], some of the immunologic differences between TACI-Ig-treated mice and BAFFR-Ig-treated mice may relate to syndecan-mediated effects. Specific in vivo neutralization of APRIL without concomitant neutralization of BLyS would permit better dissection of the role that APRIL plays in autoimmunity. A reagent that specifically binds and neutralizes murine APRIL but neither binds nor neutralizes murine BLyS has, in fact, been generated [129], so additional insights into this area may be shortly forthcoming.

Although BLyS-specific antagonists have not been compared head-to-head in humans against agents which antagonize both BLyS and APRIL, differences between the two types of antagonists may well emerge. Two studies are noteworthy. In the first, a longitudinal study of SLE patients, serum APRIL levels modestly, but significantly, inversely correlated with disease activity and modestly, but significantly, inversely correlated with serum anti-dsDNA titers in anti-dsDNA-positive patients. Overall, serum APRIL levels, despite being much greater than serum BLyS levels, tended to usually (but not uniformly) remain in the normal range, whereas serum BLyS levels frequently climbed above the normal range, and changes in serum levels of BLyS and APRIL over time were usually discordant [130].

A second study of SLE, cross-sectional in nature, documented increased, rather than normal, circulating APRIL levels in SLE patients [131]. However, since changes in circulating APRIL levels with time were not assessed in the second study, the findings may not necessarily be in conflict with those of the first. Although other interpretations are plausible, one intriguing interpretation is that APRIL is a downregulator of disease activity, at least in human SLE. The mechanism through which APRIL (perhaps) effects such downregulation may be by complexing to BLyS as heterotrimers. Empiric experience will ultimately determine the relative clinical efficacies of antagonists specific for BLyS and those able to antagonize both BLyS and APRIL.

8.3 BLyS/BLyS Receptor Deficiency In Vivo

8.3.1 BLyS-Deficient Mice

One might anticipate that the absence of a factor which promotes B-cell survival and B-cell differentiation would lead to profound alterations in B-cell number and/or function. This, in fact, is the case. Non-autoimmune-prone mice that are genetically BLyS deficient display considerable global reductions in B cells beyond the transitional type-1 (T1) stage. This includes the vast majority of mature B2 B cells (although numbers of peritoneal B1 B cells remain nearly intact) and leads to marked reductions in baseline serum Ig levels and Ig responses to T-cell-dependent (TD) and T-cell-independent (TI) antigens [132, 133]. Of note, a similar phenotype develops in mice that are transgenic for TACI-Ig by virtue of the constitutively expressed TACI-Ig continuously neutralizing endogenously produced BLyS [134].

8.3.2 BLyS Receptor Deficiency

Remarkably, the phenotypes of mice genetically deficient in individual BLyS receptors are highly disparate. In two of the cases, the phenotypes do not resemble that of BLyS-deficient mice at all, strongly pointing to non-overlapping functions for the individual BLyS receptors.

8.3.2.1 BCMA-Deficient Mice

The most straightforward phenotype among BLyS receptor-deficient mice belongs to BCMA-deficient mice (bearing a non-autoimmune-prone genetic background) in that they are almost "normal". They harbor normal numbers of lymphocytes and lymphocyte subsets; the in vitro function of these cells is normal; and these mice manifest no obvious immunodeficiency in vivo [133, 135]. Nevertheless, closer inspection of BCMA-deficient mice has demonstrated that, after immunization, they do not harbor as many antigen-specific long-lived Ig-secreting cells in their bone marrow, as do BCMA-intact mice [136]. Indeed, it has recently been shown that bone marrow plasma cell survival is markedly impaired by neutralization of both BLyS and APRIL but not by neutralization of either of these cytokines alone. This is consistent with the upregulated expression of BCMA on plasma cells. In contrast, survival and function of memory B cells is (largely) independent of either BLyS or APRIL [137], consistent with the very low (absent) expression of BCMA by these cells.

It must be stressed that the near-normal phenotype of non-autoimmune-prone BCMA-deficient mice does not exclude a potential pathogenic role for BCMA in the context of autoimmunity or autoimmune disease. Engagement of BCMA can trigger B cells to upregulate surface expression of costimulatory molecules which, in turn, enhances the ability of the B cells to serve as antigen-presenting cells (APCs)

[138]. The APC function of B cells may be as important, if not more important, than their autoantibody-producing capacities in promoting autoimmune disease, as has compellingly been shown in SLE-prone MRL-*lpr/lpr* mice [139–141]. Thus, it is entirely plausible that BLyS overexpression in an autoimmune-prone host will drive increased signaling via BCMA which, in turn, will lead to increased B-cell expression of costimulatory molecules and APC function. Autoimmune-prone BCMA-deficient mice are presently being created to assess the effects of BCMA deficiency in the context of autoimmunity.

8.3.2.2 TACI-Deficient Mice

The phenotype of mice genetically deficient in TACI dramatically differs from those of either BLyS-deficient or BCMA-deficient mice. Strikingly, TACI-deficient mice harbor increased numbers of B cells rather than the decreased or normal numbers of B cells observed in BLyS-deficient or BCMA-deficient mice, respectively [142, 143]. As TACI-deficient mice age, they develop elevated circulating titers of autoantibodies, Ig deposition in their kidneys with concomitant glomerulonephritis (GN), and premature death [144]. In vitro treatment of normal murine or human B cells with anti-TACI mAb blocks B-cell responses to agonists [144, 145], strongly suggesting that TACI transmits a negative signal to B cells.

Nevertheless, the physiologic function of TACI is not limited to simply transmitting a negative signal to B cells. Under certain in vitro conditions, engagement of TACI leads to increased plasma cell differentiation and increased IgG1 and IgE production [146]. Moreover, whereas TACI-intact B cells treated with CpG-containing ODN secrete Ig in response to BLyS, identically treated and stimulated TACI-deficient B cells do not [147]. Indeed, mutations of the human *Taci* gene are associated with common varied immunodeficiency and IgA deficiency [148, 149], strongly suggesting that the intact (wild-type) TACI molecule plays a vital role in normal Ig class switching and Ig production.

Of note, TACI-deficient mice, despite their B-cell expansion and serological and clinical autoimmunity, manifest impaired Ig responses to TI antigens [142, 143]. Immunization of TACI-deficient quasi-monoclonal mice with the relevant TI-2 antigen leads to greater B-cell expansion but less IgG production than those following identical immunization of TACI-intact quasi-monoclonal mice [150]. Furthermore, B cells from newborn mice, on which surface TACI expression is markedly reduced in comparison to that on B cells from adult mice, do not produce Ig in response to BLyS (or APRIL). However, if these newborn B cells are exposed in vitro or in vivo to CpG-containing ODN, surface TACI expression is upregulated, and the cells acquire responsiveness to BLyS (or APRIL). Along these lines, addition of CpG-containing ODN to a TI-2 antigen at the time of immunization of newborn TACI-intact, but not TACI-deficient, mice markedly augments the in vivo antibody response in comparison to that achieved through immunization with the TI-2 antigen alone [151].

The effects of TACI deficiency are not limited to humoral immunity. TACI-deficient B cells are incapable of cooperating with (BLyS-intact) dendritic cells to prime a $CD8^+$ CTL response [152]. Thus, TACI has pleiotropic effects on the immune system, so subtle differences in the manner by which TACI is engaged may trigger different intracellular signaling pathways and result in divergent phenotypes. Autoimmune-prone TACI-deficient mice are presently being generated to assess the effects of TACI deficiency in the context of autoimmunity.

8.3.2.3 BAFFR-Deficient Mice

In contrast to BCMA- or TACI-deficient mice, A/WySnJ mice (which bear a mutated *baffr* gene) and BAFFR-deficient mice display deficiencies in mature B-cell number and antibody responses qualitatively similar to those of BLyS-deficient mice [52, 53, 153, 154]. When injected with exogenous BLyS, A/WySnJ, and BAFFR-deficient mice do not undergo splenic B lymphocytosis, whereas similarly treated A/J and BLyS-deficient control mice do [77, 154]. Moreover, in bone marrow chimeric mice harboring B cells that bear the mutated *baffr* gene and B cells that bear the wild-type *baffr* gene, the B cells bearing the mutated *baffr* gene display decreased in vivo survival [77]. Taken together, these observations strongly point to BLyS/BAFFR interactions as essential for the pro-survival effects of BLyS on peripheral B cells.

Although one would apriori predict that decreased B-cell survival and a decreased B cell load protects a host from development of humoral autoimmunity and concomitant immunopathology, A/WySnJ mice, as they age, unexpectedly develop elevated serum titers of IgG anti-dsDNA antibodies and clinically overt GN [155]. Since the mutant BAFFR is surface-expressed and binds BLyS [52], circulating BLyS levels in these mice are not significantly different than those in congenic A/J mice that express wild-type BAFFR [155]. Of note, retroviral-mediated gene transfer of the wild-type *baffr* gene to A/WySnJ mice fully restores B-cell development. Moreover, congenic mice homozygous for the wild-type *baffr* gene not only manifest normal B-cell development, but do not develop the autoimmune features, strongly suggesting that it is the mutated *baffr* gene in A/WySnJ mice that leads to the autoimmunity. It is possible that BLyS engagement of the mutant BAFFR results in *altered* intracellular signaling rather than in *absent* signaling. This altered signaling, perhaps in concert with signaling through TACI and/or BCMA, may favor survival and/or expansion of autoreactive B cells. Studies with BAFFR-deficient mice (as opposed to BAFFR-mutant mice) may help clarify these enigmatic observations.

8.4 BLyS Overexpression In Vivo

8.4.1 BLyS Overexpression In Vivo in Mice

If underexpression (deficiency) of BLyS leads to considerable reductions in B cells and circulating Ig, then it should not be surprising that overexpression of BLyS, either through exogenous administration of BLyS or endogenous overproduction, has the converse effects. Administration of exogenous BLyS to mice at the time of immunization with antigen enhances in vivo antigen-specific antibody production [75]. Moreover, repeated administration of BLyS to mice without specific antigenic immunization results in B-cell expansion and polyclonal hypergammaglobulinemia [1]. Increased levels of BLyS may relax the selection stringency as B cells mature through the transitional stages, thereby dampening the elimination of autoreactive B cells [156]. Thus, not only is some of the increased Ig production in BLyS-overexpressing mice likely directed to ambient environmental (foreign) antigens, but also some of the increased Ig production is likely directed to self-antigens. Indeed, constitutive overexpression of BLyS in BLyS-Tg mice bearing a non-autoimmune-prone genetic background often leads to SLE-like features (elevated circulating titers of multiple autoantibodies and renal pathology), with clinical disease (increased proteinuria) being manifest by 8 months (but not 5 months) of age [6, 157, 158]. Although BLyS overexpression may rescue autoreactive B cells only when the proportion of competing (non-autoreactive) B cells is small [159], the development of autoimmunity in BLyS-Tg mice (bearing a full Ig repertoire) may reflect BLyS-driven *expansion* of autoreactive B cells that have escaped tolerance by a BLyS-independent pathway rather than through BLyS-driven *selection* of otherwise-tolerant autoreactive B-cell populations.

In any case, BLyS overexpression also has striking effects on hosts that otherwise already have an underlying diathesis to autoimmunity. Congenic C57BL/6 (B6) mice bearing the *Sle1* or *Nba2* locus (derived from NZW and NZB mice, respectively) lose tolerance to nucleosomal autoantigens [160, 161]. These mice harbor an incomplete genetic predisposition to development of SLE in that they spontaneously develop elevated circulating titers of IgG anti-chromatin autoantibodies but rarely develop renal disease [161–163]. Introduction of a BLyS Tg into these mice dramatically accelerates development of target-organ (kidney) pathology [164]. Of note, acceleration of clinical disease (increased proteinuria and/or death) is not observed, suggesting that even in an autoimmune-prone host, factors beyond just BLyS overexpression are required for development of clinical disease.

Even without "artificial" Tg-driven BLyS overexpression, BLyS overexpression is a feature of murine SLE, in that circulating BLyS levels are elevated in (NZB $\times$ NZW)F1 (BWF1) and MRL-*lpr/lpr* mice at the time of disease onset [6]. It may be that BLyS facilitates phenotypic expression of a latent SLE diathesis, with greater cumulative BLyS exposure leading to greater penetrance of the autoimmune phenotype.

Importantly, SLE is not the only phenotype associated with BLyS overexpression. Although many BLyS-Tg mice succumb to SLE nephritis by ~12 months of age, not all do. Of those that do not, many develop a phenotype resembling Sjögren's syndrome, with enlarged salivary glands, periductal infiltrates, acinar destruction, and reduced production of saliva [165]. BLyS overexpression is also a feature of inflammatory arthritis, as evidenced by the upregulation of BLyS during the development of collagen-induced arthritis (CIA) [166].

8.4.2 BLyS Overexpression In Vivo in Human SIRDs

8.4.2.1 BLyS Overexpression in SLE

Overexpression of BLyS in association with autoimmune disease is not limited to mice but is also a feature of several human SIRDs (Table 8.3). An association between elevated circulating BLyS levels and human rheumatic disease was first noted in SLE, with circulating BLyS levels being increased in as many as 50% of SLE patients [167–169]. A large 2-year longitudinal study documented a significant, albeit modest, correlation between circulating BLyS protein levels and clinical disease activity [170]. Of note, a separate smaller study (containing both cross-sectional and longitudinal elements) demonstrated a substantially stronger correlation between disease activity and BLyS mRNA levels in blood leukocytes [171], and additional small cross-sectional studies confirmed these observations [172, 173]. Although normal T cells express little BLyS at most, both $CD4^+$ and $CD8^+$ T cells from SLE patients express substantial levels of intracellular BLyS [174], raising the possibility that some of the systemic BLyS overexpression in SLE may arise from (autoreactive) T cells (in addition to arising from myeloid-lineage cells). In any case, these studies collectively indicate that BLyS overexpression not only promotes development of disease (as gleaned from murine in vivo studies) but that it likely also actively contributes to the ongoing maintenance of disease in SLE patients.

Table 8.3 Human SIRDs associated with BLyS overexpression

SLE
RA
Sjogrën's syndrome
Scleroderma
Dermatomyositis
ANCA-associated vasculitis
Mixed cryoglobulinemia

8.4.2.2 BLyS Overexpression in RA

Elevated circulating BLyS levels are observed not just in SLE patients but in patients with other SIRDs as well. Early reports described elevated circulating BLyS levels in a substantial fraction of RA patients [167, 168]. However, it subsequently became apparent that rheumatoid factor (RF) in the samples could lead to spurious elevations in the BLyS determinations. Reassuringly, more recent reports using an assay that minimizes (eliminates) the potential confounding effects of RF have confirmed elevated circulating BLyS levels in RA, albeit not to the extent initially believed [37, 171]. Importantly, BLyS levels in SF from RA patients (and from patients with other inflammatory arthritides) are routinely greater than those in corresponding serum [112], consistent with local production of BLyS in the inflamed joints [175] and mediated, at least in part, via the ability of TNF to promote release of BLyS from infiltrating neutrophils [11]. Moreover, microarray analysis of synovial tissues from RA patients has documented that among all TNF ligand superfamily members, it is

BLyS that is upregulated to the greatest extent [176]. Local BLyS production may limit the extent and/or duration of B-cell depletion in clinically involved joints that ensues following rituximab therapy [48] and, thereby, mitigate the clinical efficacy of such B-cell-targeted therapy.

Of note, serum levels of BLyS are elevated not just in patients with established RA, but they are also elevated in patients with early RA (disease duration ≤ 1 year). Moreover, the decline over time in serum autoantibody levels and disease activity following initiation of treatment in early RA is paralleled by a decline in serum BLyS levels [177], consistent with a contributory role for BLyS in the disease process. Equally noteworthy is that among RA patients treated with TNF antagonists, circulating BLyS levels decline in those clinically responding well but not in those responding poorly [178], raising the possibility that at least part of the salutary clinical response to TNF antagonists may be mediated through downregulation of BLyS expression.

Although BLyS levels in SF from patients with any of a myriad of inflammatory arthritides are elevated relative to levels in SF from patients with noninflammatory arthritides [112], synovial tissue BLyS expression may be especially dysregulated in RA patients. A direct comparison of BLyS, APRIL, and TACI mRNA levels in synovial tissue samples exhibiting lymphoid neogenesis from patients with RA or spondyloarthritis documented much greater BLyS and TACI, but not APRIL, expression in the RA synovial tissue samples [179]. The BLyS/TACI axis may be especially important in contributing to synovial pathology in RA.

8.4.2.3 BLyS Overexpression in Sjögren's Syndrome

Circulating BLyS levels may be especially high in patients with Sjögren's syndrome [31, 73, 165, 180–182]. Elevated salivary BLyS levels may importantly contribute to the periodontal disease that often plagues these patients [183], and local overexpression of BLyS in the salivary glands, perhaps induced by dsRNA viruses that engage TLR3 on salivary gland epithelial cells, likely contributes to the pathologic changes [184–187]. Salivary gland biopsies have revealed that the predominant phenotypes of the B cells in the ectopic germinal centers are those of T2 and MZ B cells [188], precisely those B cells expected to expand in response to BLyS overstimulation. Consistent with this notion is the strong expression of BLyS localized to the T2 B-cell area. Moreover, BLyS is expressed by a large percentage of B cells that infiltrate the salivary glands [189], raising the possibility of an autocrine survival pathway for these B cells.

8.4.2.4 BLyS Overexpression in Other SIRDs

Elevated circulating BLyS levels are also found in patients with scleroderma, dermatomyositis, ANCA-associated vasculitis, or mixed cryoglobulinemia [190–196]. In scleroderma patients, skin thickness correlates with serum BLyS levels, and local BLyS expression is increased in the skin of patients with early diffuse cutaneous scleroderma. Moreover, declining serum BLyS levels over time correlate with regression of skin sclerosis, whereas increasing serum BLyS levels over time correlate with new onset or worsening of organ involvement [190]. Along these lines in patients with hepatitis virus C-associated mixed cryoglobulinemia, clinical remission in response to therapy was associated with a normalization of serum BLyS levels, whereas no such normalization occurred in patients who did not respond to therapy [196]. These intriguing correlations notwithstanding, the precise pathogenetic connection between circulating BLyS levels and clinical features in these disorders remains to be better and more fully elucidated.

8.5 Therapeutic Targeting of BLyS

8.5.1 Elimination/Neutralization of BLyS in Murine Disease

It must be stressed that elimination/neutralization of BLyS is not synonymous with absolute loss of B cells. Indeed, the reduction in B cells in BLyS-deficient mice, albeit substantial, is not total, and the magnitude of the reduction among mature B cells has previously been overstated due to concomitant effects of BLyS deficiency on surface expression of CD21 and CD23 (markers typically used to delineate discrete maturational subsets of B cells) [197]. This incomplete B-cell depletion points to some BLyS-independent means of B-cell survival. Although studies in hen egg lysozyme (HEL)/anti-HEL double-Tg mice had suggested that survival of autoreactive B cells is more dependent upon BLyS than is survival of non-autoreactive B cells [198, 199], considerable levels of IgG autoantibodies did develop with age in BLyS-deficient SLE-prone NZM 2328 mice despite the life-long total absence of BLyS [200]. Moreover, global T-cell dysregulation in otherwise non-autoimmune-prone B6 mice has been shown to promote rapid development of IgG autoantibodies in a BLyS-independent manner [201]. These latter two findings indicated that not only can at least some autoreactive B cells survive without BLyS, but at least some of these autoreactive B cells can differentiate into autoantibody-secreting cells via an as yet undefined BLyS-independent pathway.

The development of serological autoimmunity notwithstanding, BLyS deficiency in the otherwise SLE-prone NZM 2328 mice did have a beneficial clinical effect. Although glomerular Ig deposition and proliferative GN developed in many of the BLyS-deficient NZM 2328 mice, clinical autoimmunity (severe proteinuria, premature death) was markedly attenuated [200]. This suggests that in hosts with a strong diathesis to autoimmune disease, elimination of BLyS may have a disproportionately greater effect on clinical features of disease than on serological or pathological features of the disease. Based on observations in closely related BWF1 mice, a key step in development of clinical renal disease may be the infiltration of BLyS-expressing macrophages into the renal parenchyma [202]. Absent of such infiltration, clinical renal disease is nearly undetectable despite copious glomerular Ig deposition and hypercellularity.

It must be stressed that the dramatic effect that BLyS deficiency had on clinical disease in NZM 2328 mice might not necessarily be duplicated in all murine

SLE models (or human SLE patients). Observations in SLE-prone MRL-*lpr* mice highlight the need to examine the roles of BLyS and each individual BLyS receptor in the context of systemic autoimmunity. Although MRL-*lpr* mice bearing a mutant BAFFR derived from A/WySnJ mice did manifest the anticipated reduction in spleen B cells, the autoantibody production and renal immunopathology in these mice were identical to those in MRL-*lpr* mice bearing the wild-type BAFFR [203]. Importantly, clinical disease (severe proteinuria) developed by 3–4 months, in stark contrast to the paucity of clinical disease (including severe proteinuria) in BLyS-deficient NZM 2328 mice. Whether these disparate clinical findings between these two strains of mice are related to background genetic differences between NZM 2328 and MRL-*lpr* mice or whether they reflect the consequences of differential BLyS-triggered signaling through TACI and/or BCMA (absent in BLyS-deficient mice; present in BAFFR-mutant mice) remains to be experimentally determined.

In any case, the clinical response observed in NZM 2328 mice by genetic ablation of BLyS is paralleled by more standard therapeutic interventions. Both BWF1 and MRL-*lpr/lpr* mice manifest dramatic clinical responses (decreased disease progression and improved survival) to treatment with BLyS antagonists [6, 68, 127, 204]. Moreover, TACI-Ig has both preventive and therapeutic effects in the CIA model of RA [132, 205]. In addition, early and repeated administration of BAFFR-Ig to tight-skin mice (murine scleroderma model) attenuates development of dermatologic disease (although it has no effect on development of pulmonary emphysema or cardiac hypertrophy) [206]. Furthermore, BCMA-Ig and BAFFR-Ig therapeutically ameliorate established hyperthyroidism in a murine model of Graves' disease [207]. Thus, BLyS antagonism in the mouse is beneficial in multiple rheumatic and autoimmune disease states, suggesting that the clinical potential of BLyS antagonism in the human arena may extend to multiple rheumatic and autoimmune disorders as well.

Although not directed to a rheumatic or autoimmune disease, results from a murine model of multiple myeloma may have ramifications for bone integrity following treatment with a BLyS antagonist. Bone fragments containing a human multiple myeloma cell line were transplanted into SCID mouse recipients. Treatment of these mice with a neutralizing anti-BLyS antibody not only led to a reduction in tumor burden and increased survival, but anti-BLyS treatment also inhibited recruitment of osteoclasts to the bone fragments. The reduction in osteoclast numbers resulted in decreased numbers of lytic bony lesions [208]. One can speculate that an added bonus of BLyS-antagonist therapy in human SIRDs may be amelioration (stabilization) of the frequently attendant osteoporosis.

8.5.2 Therapeutic Neutralization of BLyS in Human Disease

No BLyS antagonist has yet been approved by the FDA for use in any rheumatic (or non-rheumatic) disease, so the accrued human experience with BLyS antagonists is limited to the context of clinical trials (Table 8.4). Based on the encouraging results in murine SLE and inflammatory arthritis models, it had been anticipated by some that therapeutic targeting of BLyS in human SIRDs would have dramatically salutary effects on disease activity. To date, this expectation has not been realized, but absent additional studies and evaluations, it would be premature and unwise to delete BLyS antagonists from the future armamentarium of the practicing rheumatologist.

Table 8.4 BLyS antagonists in clinical trials

Belimumab
Atacicept
BR3-Fc
AMG 623

8.5.2.1 Treatment with Belimumab (Anti-BLyS mAb)

The greatest experience to date with BLyS antagonists has accrued with belimumab, a fully human IgG1λ mAb that binds and neutralizes soluble BLyS [209]. Preclinical studies with belimumab in cynomolgus monkeys demonstrated no cage-side toxicities and reversible effects on B-cell numbers in peripheral blood and secondary lymphoid tissues [210]. Similarly, belimumab was shown to be safe in a randomized, double-blind, placebo-controlled phase-I trial in SLE, in which the prevalence of adverse events was no different in patients treated with belimumab at any tested dose (1, 4, 10, or 20 mg/kg) than in placebo-treated patients. Moreover, belimumab was demonstrated to be biologically active by virtue of reduction in peripheral blood B cells in belimumab-treated patients but not in placebo-treated patients [211]. No clinical efficacy was demonstrated, although the small number of patients ($n = 70$) and very brief treatment schedules (single infusion or two infusions 3 weeks apart) and follow-up period (12 weeks after final infusion) likely precluded demonstration of clinical benefit.

Clinical benefit by belimumab was demonstrated in a subsequent 24-week, randomized, double-blind, placebo-controlled phase-II trial in RA [212]. In this trial, patients with active RA ($n = 283$) were treated with one of three doses of belimumab (1, 4, or 10 mg/kg) or placebo at weeks 0, 2, 4, and every 4 weeks thereafter through 24 weeks. Belimumab or placebo was added to standard-of-care therapy, with a major caveat being that the subjects could not concurrently be receiving biologic therapy (e.g., TNF antagonists, rituximab). No dose-dependent clinical response was observed, suggesting that the lowest dose tested exerted the maximal effect. Among all belimumab-treated patients, 29% experienced a clinical response (defined as an ACR20 response), whereas only 16% of the placebo-treated patients had an ACR20 response. Among the belimumab-treated patients, peripheral blood B cells declined by ~20%, and serum RF levels declined by ~30% (compared to no change in placebo-treated patients for either) [213]. At face value, the low response rate among belimumab-treated patients was disappointing, but it must be viewed in relation to the very low response rate among the placebo-treated patients. No additional clinical trials of belimumab in RA are presently ongoing, so the therapeutic niche for belimumab in RA remains uncertain.

The therapeutic niche for belimumab is equally as unsettled in SLE, inasmuch as the concurrent 52-week, randomized, double-blind, placebo-controlled phase-II trial of belimumab in SLE failed to statistically show clinical efficacy [214]. In this trial, patients with active SLE ($n = 449$) were treated with belimumab or placebo as in the phase-II RA trial, except that the double-blind treatment lasted for 52 weeks rather than for only 24 weeks. The trial failed to meet its co-primary endpoints (disease activity at 24 weeks and time to first flare during the 52 weeks) when considering the entire SLE cohort. However, reduced disease activity was demonstrable at 52 weeks (but not at 24 weeks) in the $\sim$70% of patients who were "seropositive" (ANA titer $\geq$1:80 and/or positive for anti-dsDNA antibodies) at entry, raising the hope that belimumab may be clinically efficacious in a substantial subset of SLE patients. Indeed, two separate large phase-III trials ($n = 810$ in each) of belimumab in "seropositive" SLE are presently under way.

It is tempting to speculate that the slow onset of clinical efficacy by belimumab in the SLE phase-II trial may have been related to the persistently increased occupancy by BLyS of BAFFR on SLE B cells, even at times when circulating levels of soluble BLyS are not high [215]. For clinical efficacy, BLyS antagonists may not only have to neutralize circulating BLyS but may also have to overcome the extended tight binding of BLyS to its receptors. For this reason, BLyS antagonists may not be ideal agents during the "induction" phase of treatment but may be better suited for "maintenance" therapy. Future clinical trials will be needed to test this possibility.

8.5.2.2 Treatment with Atacicept (TACI–Ig Fusion Protein)

Another BLyS antagonist currently undergoing clinical trials in RA and SLE is atacicept, a fusion protein between one of the BLyS receptors (TACI) and the Fc portion of IgG. Atacicept, in contrast to belimumab, binds and neutralizes both BLyS and APRIL. Accordingly, its biologic and clinical effects could substantially differ from those of belimumab.

Preclinical studies in mice and cynomolgus monkeys documented reversible subtotal depletion of B cells and reduction in circulating Ig (especially IgM) levels, with the only apparent systemic toxicity being transient elevations in liver-derived transaminases (without any histologic changes in the livers) [216]. In humans, favorable safety and tolerability were demonstrated in a randomized, double-blind, placebo-controlled of atacicept in SLE [96]. SLE patients ($n = 49$) received a single dose of atacicept (0.3, 1, 3, or 9 mg/kg) or placebo or four weekly doses of atacicept (1 or 3 mg/kg) or placebo. Dose-dependent reductions in peripheral blood B cells and in circulating Ig levels were noted, but, as the case with the phase-I trial of belimumab in SLE, clinical efficacy could not be demonstrated due to the limited treatment and limited follow-up period. A phase-II/III trial of atacicept in SLE patients with nephritis has recently been initiated.

Favorable safety and tolerability profiles were also demonstrated for atacicept in a randomized, double-blind, placebo-controlled phase-Ib trial in RA [97, 217]. These patients ($n = 73$) received either single doses of atacicept (70, 210, or 630 mg) or multiple doses of atacicept at 2-week intervals (3 doses of 70 mg, 3 doses of

210 mg, or 7 doses of 420 mg). Schedule- and dose-dependent reductions in circulating Ig and autoantibody levels were observed as early as after the first dose of ataticept, which continued with repeated dosing. In the cohort of 19 patients that received 7 doses over a 3-month period, positive trends on some disease activity measures (e.g., ACR20, DAS28) were noted. Based on the encouraging results of the phase-Ib trial, a phase-II trial of atacicept in RA patients has been initiated.

8.5.2.3 Treatment with BR3-Fc Fusion Protein

Limited clinical information is available on a third BLyS antagonist, BR3-Fc (a fusion protein between BAFFR [BR3] and IgG Fc). BR3-Fc, which binds and neutralizes BLyS but not APRIL, was shown, in preclinical studies, to effect subtotal B-cell depletion in the blood and secondary lymphoid tissues of cynomolgus monkeys [218]. In humans, BR3-Fc was shown to be safe and tolerable in a single-dose phase-I trial in RA ($n = 56$) in which patients received BR3-Fc (0.1, 0.3, 1, 3, or mg/kg intravenously or 0.3 or 3 mg/kg subcutaneously) or matching placebo [219]. Peripheral blood B cells, predominantly naive B cells, decreased following an initial transient increase. No changes were appreciated in circulating Ig levels, and clinical responses were not reported.

A subsequent multi-dose phase-I trial in RA ($n = 35$) yielded similar results [220]. Patients received 9 doses of BR3-Fc (150 or 300 mg every week or 300 mg every 4 weeks) or matching placebo. Minimal changes were observed in circulating Ig levels, and peripheral blood B cells (predominantly naive B cells) were reduced by a median ~55%. Establishment of clinical efficacy will require additional trials.

8.5.2.4 Treatment with AMG 623

Even less information is available with regard to the fourth BLyS antagonist being tested in clinical trials, AMG 623, a fusion between the Fc portion of IgG and a peptide sequence selected for its ability to bind with high affinity to BLyS. In a double-blind, placebo-controlled phase-I trial, SLE patients ($n = {\sim}60$) received 4 weekly doses of AMG 623 (0.3, 1, or 3 mg/kg subcutaneously or 6 mg/kg intravenously) or matching placebo [221, 222]. A dose-independent decrease in naive and total peripheral blood B cells was accompanied by an actual increase in memory B cells. Clinical responses were not reported, so the relevance of the disparate changes among B-cell subsets to clinical parameters remains unknown. Perhaps the tentatively planned phase-II trial with AMG 623 in SLE will address this issue.

8.6 BLyS Antagonism as Part of Combination Therapy

Although current clinical trials are evaluating BLyS antagonists as single agents (superimposed upon "standard-of-care" therapy), the most rational utilization of BLyS antagonists may lie in their combination with other B-cell-depleting agents (e.g., rituximab). The clinical efficacy of rituximab-based therapy in RA notwithstanding, it is well established that achievement of complete clinical remission with this approach is not common [223–225]. Experience to date with rituximab in SLE also points to incomplete clinical efficacy [226–229], and a recent phase-II/III study in SLE patients without nephritis failed to demonstrate any significant clinical efficacy [229a].

An incomplete (inadequate) clinical response to rituximab therapy may be due to one or both of two reasons. First, depletion of B cells from extravascular sites following rituximab therapy is incomplete. The renal allografts explanted from two individuals who had been unsuccessfully treated with rituximab for chronic rejection were analyzed for tissue B cells and BLyS expression. In each case, nodular aggregates of B cells in the interstitium of the graft (as part of ectopic "tertiary" lymphoid organs) were observed, with B cells accounting for 1–4% of the lymphoid infiltrate despite the profound depletion of B cells ($\leq$5 cells/mm^3) from the peripheral blood. Cultures of small cortical fragments from the explanted allografts yielded (pathogenic) IgG with anti-donor specificity. Of note, BLyS expression was greatly increased in these allografts relative to that in normal kidneys (as assessed by RT-PCR and immunohistochemistry), raising the possibility that BLyS overexpression may have contributed to B-cell survival in the allografts [230].

Incomplete extravascular depletion of B cells following rituximab therapy extends to non-ectopic secondary lymphoid tissues as well. In mice treated with rituximab-like agents, B-cell depletion from secondary lymphoid tissues is subtotal even when depletion of B cells from peripheral blood is total [231]. In an autoimmune host, it is likely that some pathogenic autoreactive B cells would survive rituximab therapy. With time, these pathogenic autoreactive B cells would likely re-expand and become capable of effecting clinically relevant damage. However, when mice were treated with a rituximab-like agent in combination with a BLyS antagonist, B-cell depletion from secondary lymphoid tissues was near total. In an autoimmune host, this near-total B-cell depletion would presumably encompass pathogenic autoreactive B cells as well. Although the leap from mouse to man is a very great one, one can envision that the combination of rituximab along with a BLyS antagonist would increase the likelihood of eliminating pathogenic autoreactive B cells and, thereby, increase the likelihood of inducing a complete clinical remission. (The presumed enhancement in clinical efficacy would have to be balanced against a presumed increased risk of infection that would arise as residual numbers of B cells decline.)

Second, BLyS almost certainly directly counteracts the B-cell-depleting effects of rituximab. In mice, genetically devoid of B cells, circulating BLyS levels are markedly elevated without comparable elevation in BLyS mRNA levels [198]. This suggests that a physical reduction in B cells (which are the predominant cells that express BLyS receptors) eliminates a major sink and clearance pathway for circulating BLyS. If correct, then therapeutic depletion of B cells should lead to increases in circulating BLyS levels, and B-cell repletion should be associated with reductions in circulating BLyS levels.

Recent observations in rituximab-treated RA and SLE patients support this notion [37, 38, 40, 232]. Following rituximab-induced B-cell depletion, circulating BLyS levels rise and remain elevated during the entire duration of peripheral blood B-cell depletion. With return of B cells to the peripheral blood, circulating BLyS levels decline and circulating BLyS levels routinely return to pre-treatment baseline levels by the onset of clinical relapse. The period during which circulating BLyS levels remain elevated may play a key role in promoting re-emergence of pathogenic autoreactive B cells. Accordingly, neutralization of BLyS following rituximab-based therapy could delay the re-emergence of such pathogenic autoreactive B cells and prolong the period of clinical remission.

8.7 Concluding Remarks

Despite the great excitement generated by BLyS antagonism in murine models of SLE and RA, no unbridled exuberance has yet been generated from the experience to date with BLyS antagonism in human SLE, RA, or other SIRDs. Since mice are not simply small furry humans with tails, it should not surprise anyone that differences have emerged between responses to therapy in murine disease and responses to therapy in human disease. Although there may yet be several bumps along the road, it is the opinion of this author that BLyS antagonism will ultimately assume an important role in the management of patients with SLE, RA, and/or other SIRDs. BLySlessness may turn out to be quite blissful for SIRD patients.

References

1. Moore PA, Belvedere O, Orr A, et al. BLyS: member of the tumor necrosis factor family and B lymphocyte stimulator. Science 1999;285:260–3.
2. Shu H-B, Hu W-H, Johnson H. TALL-1 is a novel member of the TNF family that is down-regulated by mitogens. J Leukocyte Biol 1999;65:680–3.
3. Schneider P, MacKay F, Steiner V, et al. BAFF, a novel ligand of the tumor necrosis factor family, stimulates B cell growth. J Exp Med 1999;189:1747–56.
4. Mukhopadhyay A, Ni J, Zhai Y, Yu G-L, Aggarwal BB. Identification and characterization of a novel cytokine, THANK, a TNF homologue that activates apoptosis, nuclear factor-κB, and c-Jun NH2-terminal kinase. J Biol Chem 1999;274:15978–81.
5. Tribouley C, Wallroth M, Chan V, et al. Characterization of a new member of the TNF family expressed on antigen presenting cells. Biol Chem 1999;380:1443–7.
6. Gross JA, Johnston J, Mudri S, et al. TACI and BCMA are receptors for a TNF homologue implicated in B-cell autoimmune disease. Nature 2000;404:995–9.
7. Nardelli B, Belvedere O, Roschke V, et al. Synthesis and release of B-lymphocyte stimulator from myeloid cells. Blood 2001;97:198–204.
8. Li X, Su K, Ji C, et al. Immune opsonins modulate BLyS/BAFF release in a receptor-specific fashion. J Immunol 2008;181:1012–8.
9. Scapini P, Nardelli B, Nadali G, et al. G-CSF-stimulated neutrophils are a prominent source of functional BLyS. J Exp Med 2003;197:297–302.
10. Scapini P, Carletto A, Nardelli B, et al. Proinflammatory mediators elicit the secretion of the intracellular B-lymphocyte stimulator pool (BLyS) that is stored in activated neutrophils: implications for inflammatory diseases. Blood 2005;105:830–7.
11. Assi LK, Wong SH, Ludwig A, et al. Tumor necrosis factor α activates release of B lymphocyte stimulator by neutrophils infiltrating the rheumatoid joint. Arthritis Rheum 2007;56:1776–1786.

12. Bossen C, Cachero TG, Tardivel A, et al. TACI, unlike BAFF-R, is solely activated by oligomeric BAFF and APRIL to support survival of activated B cells and plasmablasts. Blood 2008;111:1004–12.
13. Kanakaraj P, Migone T-S, Nardelli B, et al. BLyS binds to B cells with high affinity and induces activation of the transcription factors NF-κB and ELF-1. Cytokine 2001;13: 25–31.
14. Liu Y, Xu L, Opalka N, Kappler M, Shu H-B, Zhang G. Crystal structure of sTALL-1 reveals a virus-like assembly of TNF family ligands. Cell 2002;108:383–94.
15. Kim HM, Yu KS, Lee ME, et al. Crystal structure of the BAFF-BAFF-R complex and its implications for receptor activation. Nat Struct Biol 2003;10:342–8.
16. Liu Y, Hong X, Kappler J, et al. Ligand-receptor binding revealed by the TNF family member TALL-1. Nature 2003;423:49–56.
17. Karpusas M, Cachero TG, Qian F, et al. Crystal structure of extracellular human BAFF, a TNF family member that stimulates B lymphocytes. J Mol Biol 2002;315:1145–54.
18. Oren DA, Li Y, Volovik Y, et al. Structural basis of BLyS receptor recognition. Nat Struct Biol 2002;9:288–92.
19. Zhukovsky EA, Lee J-O, Villegas M, Chan C, Chu S, Mroske C. Is TALL-1 a trimer of a virus-like cluster? Nature 2004;427:413–4.
20. Cachero TG, Schwartz IM, Qian F, et al. Formation of virus-like clusters is an intrinsic property of the tumor necrosis factor family member BAFF (B cell activating factor). Biochemistry 2006;45:2006–13.
21. Gavin AL, Aït-Azzouzene D, Ware CF, Nemazee D. ΔBAFF, an alternate splice isoform that regulates receptor binding and biopresentation of the B cell survival cytokine, BAFF. J Biol Chem 2003;278:38220–8.
22. Gavin AL, Duong B, Skog P, et al. ΔBAFF, a splice isoform of BAFF, opposes full length BAFF activity in vivo in transgenic mouse models. J Immunol 2005;175: 319–28.
23. Gorelik L, Gilbride K, Dobles M, Kalled SL, Zandman D, Scott ML. Normal B cell homeostasis requires B cell activation factor production by radiation-resistant cells. J Exp Med 2003;198:937–45.
24. Yoshimoto K, Takahashi Y, Ogasawara M, et al. Aberrant expression of BAFF in T cells of systemic lupus erythematosus, which is recapitualted by a human T cell line, Loucey. Int Immunol 2006;18:1189–96.
25. Chu VT, Enghard P, Riemekasten G, Berek C. In vitro and in vivo activation induces BAFF and APRIL expression in B cells. J Immunol 2007;179:5947–57.
26. Litinskiy MB, Nardelli B, Hilbert DM, et al. DCs induce CD40-independent immunoglobulin class switching through BLyS and APRIL. Nat Immunol 2002;3:822–9.
27. Kim H-A, Jeon S-H, Seo G-Y, Park J-B, Kim P-H. TGF-β1 and IFN-γ stimulate mouse macrophages to express BAFF via different signaling pathways. J Leukocyte Biol 2008;83:1431–9.
28. Baechler EC, Batliwalla FM, Karypis G, et al. Interferon-inducible gene expression signature in peripheral blood cells of patients with severe lupus. Proc Natl Acad Sci USA 2003;100:2610–5.
29. Bennett L, Palucka AK, Arce E, et al. Interferon and granulopoiesis signatures in systemic lupus erythematosus blood. J Exp Med 2003;197:711–23.
30. Kirou KA, Lee C, George S, et al. Coordinate overexpression of interferon-α-induced genes in systemic lupus erythematosus. Arthritis Rheum 2004;50:3958–67.
31. Mavragani CP, Niewold TB, Moutsopoulos NM, Pillemer SR, Wahl SM, Crow MK. Augmented interferon-α pathway activation in patients with Sjögren's syndrome treated with etanercept. Arthritis Rheum 2007;56:3995–4004.
32. He B, Chadburn A, Jou E, Schattner EJ, Knowles DM, Cerutti A. Lymphoma B cells evade apoptosis through the TNF family members BAFF/BLyS and APRIL. J Immunol 2004;172:3268–79.

33. Novak AJ, Bram RJ, Kay NE, Jelinek DF. Aberrant expression of B-lymphocyte stimulator by B chronic lymphocytic leukemia cells: a mechanism for survival. Blood 2002;100: 2973–9.
34. Novak AJ, Darce JR, Arendt BK, et al. Expression of BCMA, TACI, and BAFF-R in multiple myeloma: a mechanism for growth and survival. Blood 2004;103:689–94.
35. Novak AJ, Grote DM, Stenson M, et al. Expression of BLyS and its receptors in B-cell non-Hodgkin lymphoma: correlation with disease activity and patient outcome. Blood 2004;104:2247–53.
36. Elsawa SF, Novak AJ, Grote DM, et al. B-lymphocyte stimulator (BLyS) stimulates immunoglobulin production and malignant B-cell growth in Waldenström macroglobulinemia. Blood 2006;107:2882–8.
37. Cambridge G, Stohl W, Leandro MJ, Migone T-S, Hilbert DM, Edwards JCW. Circulating levels of B lymphocyte stimulator in patients with rheumatoid arthritis following rituximab treatment: relationships with B cell depletion, circulating antibodies, and clinical relapse. Arthritis Rheum 2006;54:723–32.
38. Seror R, Sordet C, Guillevin L, et al. Tolerance and efficacy of rituximab and changes in serum B cell biomarkers in patients with systemic complications of primary Sjögren's syndrome. Ann Rheum Dis 2007;66:351–7.
39. Lavie F, Miceli-Richard C, Ittah M, Sellam J, Gottenberg J-E, Mariette X. Increase of B cell-activating factor of the TNF family (BAFF) after rituximab treatment: insights into a new regulating system of BAFF production. Ann Rheum Dis 2007;66: 700–3.
40. Cambridge G, Isenberg DA, Edwards JCW, et al. B cell depletion therapy in systemic lupus erythaematosus: relationships among serum B lymphocyte stimulator levels, autoantibody profile and clinical response. Ann Rheum Dis 2008;67:1011–6.
41. Krumbholz M, Theil D, Derfuss T, et al. BAFF is produced by astrocytes and up-regulated in multiple sclerosis lesions and primary central nervous system lymphoma. J Exp Med 2005;201:195–200.
42. Tschen S-I, Stohlman SA, Ramakrishna C, Hinton DR, Atkinson RD, Bergmann CC. CNS viral infection diverts homing of antibody-secreting cells from lymphoid organs to the CNS. Eur J Immunol 2006;36:603–12.
43. Phillips TA, Ni J, Hunt JS. Cell-specific expression of B lymphocyte (APRIL, BLyS)- and Th2 (CD30L/CD153)-promoting tumor necrosis factor superfamily ligands in human placentas. J Leukocyte Biol 2003;74:81–7.
44. Guo W-J, Yang M-X, Zhang W-D, Liang L, Shao Q-Q, Kong B-H. Expression of BAFF in the trophoblast and decidua of normal early pregnant women and patients with recurrent spontaneous miscarriage. Chin Med J 2008;121:309–15.
45. Langat DL, Wheaton DA, Platt JS, Sifers T, Hunt JS. Signaling pathways for B cell-activating factor (BAFF) and a proliferation-inducing ligand (APRIL) in human placenta. Am J Pathol 2008;172:1303–11.
46. Ohata J, Zvaifler NJ, Nishio M, et al. Fibroblast-like synoviocytes of mesenchymal orgin express functional B cell-activating factor of the TNF family in response to proinflammatory cytokines. J Immunol 2005;174:864–70.
47. Alsaleh G, Messer L, Semaan N, et al. BAFF synthesis by rheumatoid synoviocytes is positively controlled by $\alpha 5\beta 1$ integrin stimulation and is negatively regulated by tumor necrosis factor α and toll-like receptor ligands. Arthritis Rheum 2007;56:3202–14.
48. Goodyear CS, Boyle DL, Silverman GJ. Secretion of BAFF by fibroblast-like synoviocytes from rheumatoid arthritis biopsies attenuates B-cell depletion by rituximab. Arthritis Rheum 2005;52:S290.
49. Moon E-Y, Lee J-H, Oh S-Y, et al. Reactive oxygen species augment B-cell-activating factor expression. Free Radic Biol Med 2006;40:2103–11.
50. Laabi Y, Gras M-P, Brouet J-C, Berger R, Larsen C-J, Tsapis A. The BCMA gene, preferentially expressed during B lymphoid maturation, is bidirectionally transcribed. Nucleic Acids Res 1994;22:1147–54.

51. von Bülow G-U, Bram RJ. NF-AT activation induced by a CAML-interacting member of the tumor necrosis factor receptor superfamily. Science 1997;278:138–41.
52. Thompson JS, Bixler SA, Qian F, et al. BAFF-R, a novel TNF receptor that specifically interacts with BAFF. Science 2001;293:2108–11.
53. Yan M, Brady JR, Chan B, et al. Identification of a novel receptor for B lymphocyte stimulator that is mutated in a mouse strain with severe B cell deficiency. Curr Biol 2001;11:1547–52.
54. Xia X-Z, Treanor J, Senaldi G, et al. TACI is a TRAF-interacting receptor for TALL-1, a tumor necrosis factor family member involved in B cell regulation. J Exp Med 2000;192:137–43.
55. Chang SK, Arendt BK, Darce JR, Wu X, Jelinek DF. A role for BLyS in the activation of innate immune cells. Blood 2006;108:2687–94.
56. Darce JR, Arendt BK, Wu X, Jelinek DF. Regulated expression of BAFF-binding receptors during human B cell differentiation. J Immunol 2007;179:7276–86.
57. Ellyard JI, Avery DT, Mackay CR, Tangye SG. Contribution of stromal cells to the migration, function and retention of plasma cells in human spleen: potential roles of CXCL12, IL-6 and CD54. Eur J Immunol 2005;35:699–708.
58. Underhill GH, George D, Bremer EG, Kansas GS. Gene expression profiling reveals a highly specialized genetic program of plasma cells. Blood 2003;101:4013–21.
59. Tarte K, Zhan F, De Vos J, Klein B, Shaughnessy J, Jr. Gene expression profiling of plasma cells and plasmablasts: toward a better understanding of the late stages of B-cell differentiation. Blood 2003;102:592–600.
60. Avery DT, Kalled SL, Ellyard JI, et al. BAFF selectively enhances the survival of plasmablasts generated from human memory B cells. J Clin Invest 2003;112:286–97.
61. Smith SH, Cancro MP. Cutting edge: B cell receptor signals regulate BLyS receptor levels in mature B cells and their immediate progenitors. J Immunol 2003;170:5820–3.
62. Treml LS, Carlesso G, Hoek KL, et al. TLR stimulation modifies BLyS receptor expression in follicular and marginal zone B cells. J Immunol 2007;178:7531–9.
63. Shu H-B, Johnson H. B cell maturation protein is a receptor for the tumor necrosis factor family member TALL-1. Proc Natl Acad Sci USA 2000;97:9156–61.
64. Hatzoglou A, Roussel J, Bourgeade M-F, et al. TNF receptor family member BCMA (B cell maturation) associates with TNF receptor-associated factor (TRAF) 1, TRAF2, and TRAF3 and activates NF-κB, Elk-1, c-Jun N-terminal kinase, and p38 mitogen-activated protein kinase. J Immunol 2000;165:1322–30.
65. Xu L-G, Shu H-B. TNFR-associated factor-3 is associated with BAFF-R and negatively regulates BAFF-R-mediated NF-κB activation and IL-10 production. J Immunol 2002;169:6883–9.
66. Hikida M, Johmura S, Hashimoto A, Takezaki M, Kurosaki T. Coupling between B cell receptor and phospholipase C-γ2 is essential for mature B cell development. J Exp Med 2003;198:581–9.
67. Claudio E, Brown K, Park S, Wang H, Siebenlist U. BAFF-induced NEMO-independent processing of NK-κB2 in maturing B cells. Nat Immunol 2002;3:958–65.
68. Kayagaki N, Yan M, Seshasayee D, et al. BAFF/BLyS receptor 3 binds the B cell survival factor BAFF ligand through a discrete surface loop and promotes processing of NF-κB2. Immunity 2002;17:515–24.
69. Hatada EN, Do RKG, Orlofsky A, et al. NF-κB1 p50 is required for BLyS attentuation of apoptosis but dispensible for processing of NF-κB2 p100 to p52 in quiescent mature B cells. J Immunol 2003;171:761–8.
70. Herrin BR, Justement LB. Expression of the adaptor protein hematopoietic Src homology 2 is up-regulated in response to stimuli that promote survival and differentiation of B cells. J Immunol 2006;176:4163–72.
71. Mecklenbräuker I, Kalled SL, Leitges M, Mackay F, Tarakhovsky A. Regulation of B-cell survival by BAFF-dependent PKCδ-mediated nuclear signalling. Nature 2004;431:456–61.
72. Hase H, Kanno Y, Kojima M, et al. BAFF/BLyS can potentiate B-cell selection with the B-cell co-receptor complex. Blood 2004;103:2257–65.

73. d'Arbonneau F, Pers J-O, Devauchelle V, Pennec Y, Saraux A, Youinou P. BAFF-induced changes in B cell antigen receptor-containing lipid rafts in Sjögren's syndrome. Arthritis Rheum 2006;54:115–26.
74. Thompson JS, Schneider P, Kalled SL, et al. BAFF binds to the tumor necrosis factor receptor-like molecule B cell maturation antigen and is important for maintaining the peripheral B cell population. J Exp Med 2000;192:129–35.
75. Do RKG, Hatada E, Lee H, Tourigny MR, Hilbert D, Chen-Kiang S. Attenuation of apoptosis underlies B lymphocyte stimulator enhancement of humoral immune response. J Exp Med 2000;192:953–64.
76. Batten M, Groom J, Cachero TG, et al. BAFF mediates survival of peripheral immature B lymphocytes. J Exp Med 2000;192:1453–65.
77. Harless SM, Lentz VM, Sah AP, et al. Competition for BLyS-mediated signaling through Bcmd/BR3 regulates peripheral B lymphocyte numbers. Curr Biol 2001;11:1986–9.
78. Hsu BL, Harless SM, Lindsley RC, Hilbert DM, Cancro MP. Cutting edge: BLyS enables survival of transitional and mature B cells through distinct mediators. J Immunol 2002;168:5993–6.
79. Mongini PKA, Inman JK, Han H, Fattah RJ, Abramson SB, Attur M. APRIL and BAFF promote increased viability of replicating human B2 cells via mechanism involving cyclooxygenase 2. J Immunol 2006;176:6736–51.
80. Woodland RT, Fox CJ, Schmidt MR, et al. Multiple signaling pathways promote B lymphocyte stimulator-dependent B-cell growth and survival. Blood 2008;111:750–60.
81. Craxton A, Draves KE, Gruppi A, Clark EA. BAFF regulates B cell survival by downregulating the BH3-only family member Bim via the ERK pathway. J Exp Med 2005;202:1363–74.
82. Oliver PM, Vass T, Kappler J, Marrack P. Loss of the proapoptotic protein, Bim, breaks B cell anergy. J Exp Med 2006;203:731–41.
83. Amanna IJ, Dingwall JP, Hayes CE. Enforced *bcl-xL* gene expression restored splenic B lymphocyte development in BAFF-R mutant mice. J Immunol 2003;170:4593–600.
84. Sasaki Y, Derudder E, Hobeika E, et al. Canonical NF-κB activity, dispensable for B cell development, replaces BAFF-receptor signals and promotes B cell proliferation upon activation. Immunity 2006;24:729–39.
85. Enzler T, Bonizzi G, Silverman GJ, et al. Alternative and classical NF-κB signaling retain autoreactive B cells in the splenic marginal zone and result in lupus-like disease. Immunity 2006;25:403–15.
86. Gardam S, Sierro F, Basten A, Mackay F, Brink R. TRAF2 and TRAF3 signal adapters act cooperatively to control the maturation and survival signals delivered to B cells by the BAFF receptor. Immunity 2008;28:391–401.
87. Xie P, Stunz LL, Larison KD, Yang B, Bishop GA. Tumor necrosis factor receptor-associated factor 3 is a critical regulator of B cell homeostasis in secondary lymphoid organs. Immunity 2007;27:253–67.
88. Patke A, Mecklenbräuker I, Erdjument-Bromage H, Tempst P, Tarakhovsky A. BAFF controls B cell metabolic fitness through a PKCβ- and Akt-dependent mechanism. J Exp Med 2006;203:2551–62.
89. Acosta-Rodríguez EV, Craxton A, Hendricks DW, et al. BAFF and LPS cooperate to induce B cells to become susceptible to CD95/Fas-mediated cell death. Eur J Immunol 2007;37:990–1000.
90. Rolink AG, Tschopp J, Schneider P, Melchers F. BAFF is a survival and maturation factor for mouse B cells. Eur J Immunol 2002;32:2004–10.
91. Tardivel A, Tinel A, Lens S, et al. The anti-apoptotic factor Bcl-2 can functionally substitute for the B cell survival but not for the marginal zone B cell differentiation activity of BAFF. Eur J Immunol 2004;34:509–18.
92. Yamada T, Zhang K, Yamada A, Zhu D, Saxon A. B lymphocyte stimulator activates p38 mitogen-activated protein kinase in human Ig class switch recombination. Am J Respir Cell Mol Biol 2005;32:388–94.
93. Castigli E, Wilson SA, Scott S, et al. TACI and BAFF-R mediate isotype switching in B cells. J Exp Med 2005;201:35–9.

94. Ettinger R, Sims GP, Robbins R, et al. IL-21 and BAFF/BLyS synergize in stimulating plasma cell differentiation from a unique population of human splenic memory B cells. J Immunol 2007;178:2872–82.
95. Xu W, Santini PA, Matthews AJ, et al. Viral double-stranded RNA triggers Ig class switching by activating upper respiratory mucosa B cells through an innate TLR3 pathway involving BAFF. J Immunol 2008;181:276–87.
96. Dall'Era M, Chakravarty E, Wallace D, et al. Reduced B lymphocyte and immunoglobulin levels after atacicept treatment in patients with systemic lupus erythematosus: results of a multicenter, phase Ib, double-blind, placebo-controlled, dose-escalating trial. Arthritis Rheum 2007;56:4142–50.
97. Tak PP, Thurlings RM, Rossier C, et al. Atacicept in patients with rheumatoid arthritis: results of a multicenter, phase Ib, double-blind, placebo-controlled, dose-escalating, single- and repeated-dose study. Arthritis Rheum 2008;58:61–72.
98. Badr G, Borhis G, Lefevre EA, et al. BAFF enhances chemotaxis of primary human B cells: a particular synergy between BAFF and CXCL13 on memory B cells. Blood 2008;111: 2744–54.
99. Ng LG, Ng C-H, Woehl B, et al. BAFF costimulation of Toll-like receptor-activated B-1 cells. Eur J Immunol 2006;36:1837–46.
100. Chang SK, Mihalcik SA, Jelinek DF. B lymphocyte stimulator regulates adaptive immune responses by directly promoting dendritic cell maturation. J Immunol 2008;180:7394–403.
101. Sutherland APR, Ng LG, Fletcher CA, et al. BAFF augments certain Th1-associated inflammatory responses. J Immunol 2005;174:5537–44.
102. López-Fraga M, Fernández R, Albar JP, Hahne M. Biologically active APRIL is secreted following intracellular processing in the Golgi apparatus by furin convertase. EMBO Rep 2001;2:945–51.
103. Pradet-Balade B, Medema JP, López-Fraga M, et al. An endogenous hybrid mRNA encodes TWE-PRIL, a functional cell surface TWEAK-APRIL fusion protein. EMBO J 2002;21:5711–20.
104. Hahne M, Kataoka T, Schröter M, et al. APRIL, a new ligand of the tumor necrosis factor family, stimulates tumor cell growth. J Exp Med 1998;188:1185–90.
105. Kelly K, Manos E, Jensen G, Nadauld L, Jones DA. APRIL/TRDL-1, a tumor necrosis factor-like ligand, stimulates cell death. Cancer Res 2000;60:1021–7.
106. Marsters SA, Yan M, Pitti RM, Haas PE, Dixit VM, Ashkenazi A. Interaction of the TNF homologues BLyS and APRIL with the receptor homologues BCMA and TACI. Curr Biol 2000;10:785–8.
107. Yu G, Boone T, Delaney J, et al. APRIL and TALL-1 and receptors BCMA and TACI: system for regulating humoral immunity. Nat Immunol 2000;1:252–6.
108. Wu Y, Bressette D, Carrell JA, et al. Tumor necrosis factor (TNF) receptor superfamily member TACI is a high affinity receptor for TNF family members APRIL and BLyS. J Biol Chem 2000;275:35478–85.
109. Rennert P, Schneider P, Cachero TG, et al. A soluble form of B cell maturation antigen, a receptor for the tumor necrosis factor family member APRIL, inhibits tumor cell growth. J Exp Med 2000;192:1677–83.
110. Day ES, Cachero TG, Qian F, et al. Selectivity of BAFF/BLyS and APRIL for binding to the TNF family receptors BAFFR/BR3 and BCMA. Biochemistry 2005;44: 1919–31.
111. Nagatani K, Itoh K, Nakajima K, et al. Rheumatoid arthritis fibroblast-like synoviocytes express BCMA and are stimulated by APRIL. Arthritis Rheum 2007;56: 3554–63.
112. Tan S-M, Xu D, Roschke V, et al. Local production of B lymphocyte stimulator protein and APRIL in arthritic joints of patients with inflammatory arthritis. Arthritis Rheum 2003;48:982–92.
113. Koyama T, Tsukamoto H, Masumoto K, et al. A novel polymorphism of the human *APRIL* gene is associated with systemic lupus erythematosus. Rheumatology 2003;42:980–5.

114. Lee YH, Ota F, Kim-Howard X, Kaufman KM, Nath SK. *APRIL* polymorphism and systemic lupus erythematosus (SLE) susceptibility. Rheumatology 2007;46:1274–6.
115. Roschke V, Sosnovtseva S, Ward CD, et al. BLyS and APRIL form biologically active heterotrimers that are expressed in patients with systemic immune-based rheumatic diseases. J Immunol 2002;169:4314–21.
116. Wallweber HJA, Compaan DM, Starovasnik MA, Hymowitz SG. The crystal structure of a proliferation-inducing ligand, APRIL. J Mol Biol 2004;343:283–90.
117. Bossen C, Ingold K, Tardivel A, et al. Interactions of tumor necrosis factor (TNF) and TNF receptor family members in the mouse and human. J Biol Chem 2006;281: 13964–71.
118. Craxton A, Magaletti D, Ryan EJ, Clark EA. Macrophage- and dendritic cell-dependent regulation of human B-cell proliferation requires the TNF family ligand BAFF. Blood 2003;101:4464–71.
119. Sakurai D, Hase H, Kanno Y, Kojima H, Okumura K, Kobata T. TACI regulates IgA production by APRIL in collaboration with HSPG. Blood 2007;109:2961–7.
120. He B, Xu W, Santini PA, et al. Intestinal bacteria trigger T cell-independent immunoglobulin A2 class switching by inducing eipthelial-cell secretion of the cytokine APRIL. Cell 2007;26:812–26.
121. Hendriks J, Planelles L, de Jong-Odding J, et al. Heparan sulfate proteoglycan binding promotes APRIL-induced tumor cell proliferation. Cell Death Differ 2005;12:637–48.
122. Ingold K, Zumsteg A, Tardivel A, et al. Identification of proteoglycans as the APRIL-specific binding partners. J Exp Med 2005;201:1375–83.
123. Belnoue E, Pihlgren M, McGaha TL, et al. APRIL is critical for plasmablast survival in the bone marrow and poorly expressed by early-life bone marrow stromal cells. Blood 2008;111:2755–64.
124. Varfolomeev E, Kischkel F, Martin F, et al. APRIL-deficient mice have normal immune system development. Mol Cell Biol 2004;24:997–1006.
125. Castigli E, Scott S, Dedeoglu F, et al. Impaired IgA class switching in APRIL-deficient mice. Proc Natl Acad Sci USA 2004;101:3903–8.
126. Stein JV, López-Fraga M, Elustondo FA, et al. APRIL modulates B and T cell immunity. J Clin Invest 2002;109:1587–98.
127. Ramanujam M, Wang X, Huang W, et al. Similarities and differences between selective and nonselective BAFF blockade in murine SLE. J Clin Invest 2006;116:724–34.
128. Bischof D, Elsawa SF, Mantchev G, et al. Selective activation of TACI by syndecan-2. Blood 2006;107:3235–42.
129. Patel DR, Wallweber HJA, Yin JP, et al. Engineering an APRIL-specific B-cell maturation antigen (BCMA). J Biol Chem 2004;279:16727–35.
130. Stohl W, Metyas S, Tan S-M, et al. Inverse association between circulating APRIL levels and serologic and clinical disease activity in patients with systemic lupus erythematosus. Ann Rheum Dis 2004;63:1096–1103.
131. Koyama T, Tsukamoto H, Miyagi Y, et al. Raised serum APRIL levels in patients with systemic lupus erythematosus. Ann Rheum Dis 2005;64:1065–7.
132. Gross JA, Dillon SR, Mudri S, et al. TACI-Ig neutralizes molecules critical for B cell development and autoimmune disease: impaired B cell maturation in mice lacking BLyS. Immunity 2001;15:289–302.
133. Schiemann B, Gommerman JL, Vora K, et al. An essential role for BAFF in the normal development of B cells through a BCMA-independent pathway. Science 2001;293: 2111–4.
134. Schneider P, Takatsuka H, Wilson A, et al. Maturation of marginal zone and follicular B cells requires B cell activating factor of the tumor necrosis factor family and is independent of B cell maturation antigen. J Exp Med 2001;194:1691–7.
135. Xu S, Lam D-P. B-cell maturation protein, which binds the tumor necrosis factor family members BAFF and APRIL, is dispensible for humoral immune responses. Mol Cell Biol 2001;21:4067–74.

136. O'Connor BP, Raman VS, Erickson LD, et al. BCMA is essential for the survival of long-lived bone marrow plasma cells. J Exp Med 2004;199:91–7.
137. Benson MJ, Dillon SR, Castigli E, et al. Cutting edge: the dependence of plasma cells and independence of memory B cells on BAFF and APRIL. J Immunol 2008;180:3655–9.
138. Yang M, Hase H, Legarda-Addison D, Varughese L, Seed B, Ting AT. BCMA, the receptor for APRIL and BAFF, induces antigen presentation in B cells. J Immunol 2005;175: 2814–24.
139. Shlomchik MJ, Madaio MP, Ni D, Trounstein M, Huszar D. The role of B cells in *lpr/lpr*-induced autoimmunity. J Exp Med 1994;180:1295–306.
140. Chan O, Shlomchik MJ. A new role for B cells in systemic autoimmunity: B cells promote spontaneous T cell activation in MRL-*lpr/lpr* mice. J Immunol 1998;160:51–9.
141. Chan OTM, Hannum LG, Haberman AM, Madaio MP, Shlomchik MJ. A novel mouse with B cells but lacking serum antibody reveals an antibody-independent role for B cells in murine lupus. J Exp Med 1999;189:1639–47.
142. von Bülow G-U, van Deursen JM, Bram RJ. Regulation of the T-independent humoral response by TACI. Immunity 2001;14:573–82.
143. Yan M, Wang H, Chan B, et al. Activation and accumulation of B cells in TACI-deficient mice. Nat Immunol 2001;2:638–43.
144. Seshasayee D, Valdez P, Yan M, Dixit VM, Tumas D, Grewal IS. Loss of TACI causes fatal lymphoproliferation and autoimmunity, establishing TACI as an inhibitory BLyS receptor. Immunity 2003;18:279–88.
145. Sakurai D, Kanno Y, Hase H, Kojima H, Okumura K, Kobata T. TACI attenuates antibody production costimulated by BAFF-R and CD40. Eur J Immunol 2007;37:110–8.
146. Castigli E, Wilson SA, Elkhal A, Ozcan E, Garibyan L, Geha RS. Transmembrane activator and calcium modulator and cyclophilin ligand interactor enhances CD40-driven plasma cell differentiation. J Allergy Clin Immunol 2007;120:885–91.
147. Katsenelson N, Kanswal S, Puig M, Mostowski H, Verthelyi D, Akkoyunlu M. Synthetic CpG oligodeoxynucleotides augment BAFF- and APRIL-mediated immunoglobulin secretion. Eur J Immunol 2007;37:1785–95.
148. Salzer U, Chapel HM, Webster ADB, et al. Mutations in *TNFRSF13B* encoding TACI are associated with common variable immunodeficiency in humans. Nat Genet 2005;37:820–8.
149. Castigli E, Wilson SA, Garibyan L, et al. TACI is mutant in common variable immunodeficiency and IgA deficiency. Nat Genet 2005;37:829–34.
150. Mantchev GT, Cortesao CS, Rebrovich M, Cascalho M, Bram RJ. TACI is required for efficient plasma cell differentiation in response to T-independent type 2 antigens. J Immunol 2007;179:2282–8.
151. Kanswal S, Katsenelson N, Selvapandiyan A, Bram RJ, Akkoyunlu M. Deficient TACI expression on B lymphocytes of newborn mice leads to defective Ig secretion in response to BAFF or APRIL. J Immunol 2008;181:976–90.
152. Diaz-de-Durana Y, Mantchev GT, Bram RJ, Franco A. TACI-BLyS signaling via B-cell-dendritic cell cooperation is required for naive $CD8^+$ T-cell priming in vivo. Blood 2006;107:594–601.
153. Sasaki Y, Casola S, Kutok JL, Rajewski K, Schmidt-Supprian M. TNF family member B cell-activating factor (BAFF) receptor-dependent and -independent roles for BAFF in B cell physiology. J Immunol 2004;173:2245–52.
154. Shulga-Morskaya S, Dobles M, Walsh ME, et al. B cell-activating factor belonging to the TNF family acts through separate receptors to support B cell survival and T cell-independent antibody formation. J Immunol 2004;173:2331–41.
155. Mayne CG, Amanna IJ, Nashold FE, Hayes CE. Systemic autoimmunity in BAFF-R-mutant A/WySnJ strain mice. Eur J Immunol 2008;38:587–98.
156. Miller JP, Stadanlick JE, Cancro MP. Space, selection, and surveillance: setting boundaries with BLyS. J Immunol 2006;176:6405–10.

157. Mackay F, Woodcock SA, Lawton P, et al. Mice transgenic for BAFF develop lymphocytic disorders along with autoimmune manifestations. J Exp Med 1999;190:1697–710.
158. Khare SD, Sarosi I, Xia X-Z, et al. Severe B cell hyperplasia and autoimmune disease in TALL-1 transgenic mice. Proc Natl Acad Sci USA 2000;97:3370–75.
159. Aït-Azzouzene D, Gavin AL, Skog P, Duong B, Nemazee D. Effect of cell:cell competition and BAFF expression on peripheral B cell tolerance and B-1 cell survival in transgenic mice expressing a low level of Igκ-reactive macroself antigen. Eur J Immunol 2006;36:985–96.
160. Morel L, Rudofsky UH, Longmate JA, Schiffenbauer J, Wakeland EK. Polygenic control of susceptibility to murine systemic lupus erythematosus. Immunity 1994;1:219–29.
161. Rozzo SJ, Allard JD, Choubey D, et al. Evidence for an interferon-inducible gene, *Ifi202*, in the susceptibility to systemic lupus. Immunity 2001;15:435–43.
162. Morel L, Mohan C, Yu Y, et al. Functional dissection of systemic lupus erythematosus using congenic mouse strains. J Immunol 1997;158:6019–28.
163. Mohan C, Alas E, Morel L, Yang P, Wakeland EK. Genetic dissection of SLE pathogenesis: *Sle1* on murine chromosome 1 leads to a selective loss of tolerance to H2A/H2B/DNA subnucleosomes. J Clin Invest 1998;101:1362–72.
164. Stohl W, Xu D, Kim KS, et al. BAFF overexpression and accelerated glomerular disease in mice with an incomplete genetic predisposition to systemic lupus erythematosus. Arthritis Rheum 2005;52:2080–91.
165. Groom J, Kalled SL, Cutler AH, et al. Association of BAFF/BLyS overexpression and altered B cell differentiation with Sjögren's syndrome. J Clin Invest 2002;109:59–68.
166. Zhang M, Ko K-H, Lam QLK, et al. Expression and function of TNF family member B cell-activating factor in the development of autoimmune arthritis. Int Immunol 2005;17:1081–92.
167. Zhang J, Roschke V, Baker KP, et al. Cutting edge: a role for B lymphocyte stimulator in systemic lupus erythematosus. J Immunol 2001;166:6–10.
168. Cheema GS, Roschke V, Hilbert DM, Stohl W. Elevated serum B lymphocyte stimulator levels in patients with systemic immune-based rheumatic diseases. Arthritis Rheum 2001;44:1313–9.
169. Stohl W, Metyas S, Tan S-M, et al. B lymphocyte stimulator overexpression in patients with systemic lupus erythematosus: longitudinal observations. Arthritis Rheum 2003;48:3475–86.
170. Petri M, Stohl W, Chatham W, et al. Association of plasma B lymphocyte stimulator levels and disease activity in systemic lupus erythematosus. Arthritis Rheum 2008;58:2453–9.
171. Collins CE, Gavin AL, Migone T-S, Hilbert DM, Nemazee D, Stohl W. B lymphocyte stimulator (BLyS) isoforms in systemic lupus erythematosus: disease activity correlates better with blood leukocyte BLyS mRNA levels than with plasma BLyS protein levels. Arthritis Res Ther 2006;8:R6.
172. Becker-Merok A, Nikolaisen C, Nossent HC. B-lymphocyte activating factor in systemic lupus erythematosus and rheumatoid arthritis in relation to autoantibody levels, disease measures and time. Lupus 2006;15:570–6.
173. Ju S, Zhang D, Wang Y, Ni H, Kong X, Zhong R. Correlation of the expression levels of BLyS and its receptors mRNA in patients with systemic lupus erythematosus. Clin Biochem 2006;39:1131–7.
174. Morimoto S, Nakano S, Watanabe T, et al. Expression of B-cell activating factor of the tumour necrosis factor family (BAFF) in T cells in active systemic lupus erythematosus: the role of BAFF in T cell-dependent B cell pathogenic autoantibody production. Rheumatology 2007;46:1083–6.
175. Seyler TM, Park YW, Takemura S, et al. BLyS and APRIL in rheumatoid arthritis. J Clin Invest 2005;115:3083–92.
176. Soto H, Hevezi P, Roth RB, et al. Gene array analysis comparison between rat collagen-induced arthritis and human rheumatoid arthritis. Scand J Immunol 2008;68:43–57.

177. Bosello S, Youinou P, Daridon C, et al. Concentrations of BAFF correlate with autoantibody levels, clinical disease activity, and response to treatment in early rheumatoid arthritis. J Rheumatol 2008;35:1256–64.
178. La DT, Collins CE, Yang H-T, Migone T-S, Stohl W. B lymphocyte stimulator expression in patients with rheumatoid arthritis treated with tumour necrosis factor α antagonists: differential effects between good and poor clinical responders. Ann Rheum Dis 2008;67:1132–8.
179. Cantaert T, Kolln J, Timmer T, et al. B lymphocyte autoimmunity in rheumatoid synovitis is independent of ectopic lymphoid neogenesis. J Immunol 2008;181:785–94.
180. Mariette X, Roux S, Zhang J, et al. The level of BLyS (BAFF) correlates with the titre of autoantibodies in human Sjögren's syndrome. Ann Rheum Dis 2003;62:168–71.
181. Jonsson MV, Szodoray P, Jellestad S, Jonsson R, Skarstein K. Association between circulating levels of the novel TNF family members APRIL and BAFF and lymphoid organization in primary Sjögren's syndrome. J Clin Immunol 2005;25:189–201.
182. Pers J-O, Daridon C, Devauchelle V, et al. BAFF overexpression is associated with autoantibody production in autoimmune diseases. Ann NY Acad Sci 2005;1050:34–9.
183. Pers J-O, d'Arbonneau F, Devauchelle-Pensec V, Saraux A, Pennec Y-L, Youinou P. Is periodontal disease mediated by salivary BAFF in Sjögren's syndrome? Arthritis Rheum 2005;52:2411–4.
184. Szodoray P, Jellestad S, Teague MO, Jonsson R. Attenuated apoptosis of B cell activating factor-expressing cells in primary Sjögren's syndrome. Lab Invest 2003;83:357–65.
185. Lavie F, Miceli-Richard C, Quillard J, Roux S, Leclerc P, Mariette X. Expression of BAFF (BLyS) in T cells infiltrating labial salivary glands from patients with Sjögren's syndrome. J Pathol 2004;202:496–502.
186. Ittah M, Miceli-Richard C, Gottenberg J-E, et al. B cell-activating factor of the tumor necrosis factor family (BAFF) is expressed under stimulation by interferon in salivary gland epithelial cells in primary Sjögren's syndrome. Arthritis Res Ther 2006;8:R51.
187. Ittah M, Miceli-Richard C, Gottenberg J-E, et al. Viruses induce high expression of BAFF by salivary gland epithelial cells through TLR- and type-I IFN-dependent and -independent pathways. Eur J Immunol 2008;38:1058–64.
188. Daridon C, Pers J-O, Devauchelle V, et al. Identification of transitional type II B cells in the salivary glands of patients with Sjögren's syndrome. Arthritis Rheum 2006;54:2280–8.
189. Daridon C, Devauchelle V, Hutin P, et al. Aberrant expression of BAFF by B lymphocytes infiltrating the salivary glands of patients with primary Sjögren's syndrome. Arthritis Rheum 2007;56:1134–44.
190. Matsushita T, Hasegawa M, Yanaba K, Kodera M, Takehara K, Sato S. Elevated serum BAFF levels in patients with systemic sclerosis: enhanced BAFF signaling in systemic sclerosis B lymphocytes. Arthritis Rheum 2006;54:192–201.
191. Krumbholz M, Specks U, Wick M, Kalled SL, Jenne D, Meinl E. BAFF is elevated in serum of patients with Wegener's granulomatosis. J Autoimmun 2005;25:298–302.
192. Sanders J-SF, Huitema MG, Kallenberg CGM, Stegeman CA. Plasma levels of soluble interleukin 2 receptor, soluble CD30, interleukin 10 and B cell activator of the tumor necrosis factor family during follow-up in vasculitis associated with proteinase 3-antineutrophil cytoplasmic antibodies: associations with disease activity and relapse. Ann Rheum Dis 2006;65:1484–9.
193. Fabris M, Quartuccio L, Sacco S, et al. B-Lymphocyte stimulator (BLyS) up-regulation in mixed cryoglobulinaemia syndrome and hepatitis-C virus infection. Rheumatology 2007;46:37–43.
194. Toubi E, Gordon S, Kessel A, et al. Elevated serum B-lymphocyte activating factor (BAFF) in chronic hepatitis C virus infection: association with autoimmunity. J Autoimmun 2006;27:134–9.
195. Sène D, Limal N, Ghillani-Dalbin P, Saadoun D, Piette J-C, Cacoub P. Hepatitis C virus-associated B-cell proliferation – the role of serum B lymphocyte stimulator (BLyS/BAFF). Rheumatology 2007;46:65–9.

196. Landau D-A, Rosenzwajg M, Saadoun D, Klatzmann D, Cacoub P. The B lymphocyte stimulator receptor-ligand system in hepatitis C virus-induced B-cell clonal disorders. Ann Rheum Dis 2008;68:337–44.
197. Gorelik L, Cutler AH, Thill G, et al. Cutting edge: BAFF regulates CD21/35 and CD23 expression independent of its B cell survival function. J Immunol 2004;172:762–6.
198. Lesley R, Xu Y, Kalled SL, et al. Reduced competitiveness of autoantigen-engaged B cells due to increased dependence on BAFF. Immunity 2004;20:441–53.
199. Thien M, Phan TG, Gardam S, et al. Excess BAFF rescues self-reactive B cells from peripheral deletion and allows them to enter forbidden follicular and marginal zone niches. Immunity 2004;20:785–98.
200. Jacob CO, Pricop L, Putterman C, et al. Paucity of clinical disease despite serological autoimmunity and kidney pathology in lupus-prone New Zealand Mixed 2328 mice deficient in BAFF. J Immunol 2006;177:2671–80.
201. Stohl W, Jacob N, Quinn WJ, III, et al. Global T cell dysregulation in non-autoimmune-prone mice promotes rapid development of BAFF-independent, systemic lupus erythematosus-like autoimmunity. J Immunol 2008;181:833–41.
202. Schiffer L, Bethunaickan R, Ramanujam M, et al. Activated renal macrophages are markers of disease onset and disease remission in lupus nephritis. J Immunol 2008;180: 1938–47.
203. Ju ZL, Shi GY, Zuo JX, Zhang JW, Sun J. Unexpected development of autoimmunity in BAFF-R-mutant MRL-*lpr* mice. Immunology 2007;120:281–9.
204. Ramanujam M, Wang X, Huang W, et al. Mechanism of action of transmembrane activator and calcium modulator ligand interactor-Ig in murine systemic lupus erythematosus. J Immunol 2004;173:3524–34.
205. Wang H, Marsters SA, Baker T, et al. TACI-ligand interactions are required for T cell activation and collagen-induced arthritis in mice. Nat Immunol 2001;2:632–7.
206. Matsushita T, Fujimoto M, Hasegawa M, et al. BAFF antagonist attenuates the development of skin fibrosis in tight-skin mice. J Invest Dermatol 2007;127:2772–80.
207. Gilbert JA, Kalled SL, Moorhead H, et al. Treatment of autoimmune hyperthyroidism in a murine model of Graves′ disease with tumor necrosis factor-family ligand inhibitors suggests a key role for B cell activating factor in disease pathology. Endocrinology 2006;147: 4561–8.
208. Neri P, Kumar S, Fulciniti MT, et al. Neutralizing B-cell-activating factor antibody improves survival and inhibits osteoclastogenesis in a severe combined immunodeficient human multiple myeloma model. Clin Cancer Res 2007;13:5903–9.
209. Baker KP, Edwards BM, Main SH, et al. Generation and characterization of LymphoStat-B, a human monoclonal antibody that antagonizes the bioactivities of B lymphocyte stimulator. Arthritis Rheum 2003;48:3253–65.
210. Halpern W, Lappin P, Zanardi T, et al. Chronic administration of belimumab, a BLyS antagonist, decreases tissue and peripheral blood B-lymphocyte populations in cynomolgus monkeys: pharmacokinetic, pharmacodynamic and toxicologic effects. Toxicol Sci 2006;91: 586–99.
211. Furie R, Stohl W, Ginzler EM, et al. Biologic activity and safety of belimumab, a neutralizing anti-B-lymphocyte stimulator (BLyS) monoclonal antibody: a phase I trial in patients with systemic lupus erythematosus. Arthritis Res Ther 2008;10:R109.
212. McKay J, Chwalinska-Sadowska H, Boling E, et al. Belimumab (BmAb), a fully human monoclonal antibody to B-lymphocyte stimulator (BLyS), combined with standard of care therapy reduces the signs and symptoms of rheumatoid arthritis in a heterogeneous subject population. Arthritis Rheum 2005;52:S710–11.
213. Stohl W, Chatham W, Weisman M, et al. Belimumab (BmAb), a novel fully human monoclonal antibody to B-lymphocyte stimulator (BLyS), selectively modulates B-cell subpopulations and immunoglobulins in a heterogeneous rheumatoid arthritis subject population. Arthritis Rheum 2005;52:S444.

214. Wallace DJ, Lisse J, Stohl W, et al. Belimumab (BmAb) reduces SLE disease activity and demonstrates durable bioactivity at 76 weeks. Arthritis Rheum 2006;54:S790.
215. Carter RH, Zhao H, Liu X, et al. Expression and occupancy of BAFF-R on B cells in systemic lupus erythematosus. Arthritis Rheum 2005;52:3943–54.
216. Carbonatto M, Yu P, Bertolino M, et al. Nonclinical safety, pharmacokinetics, and phamcodynamics of atacicept. Toxicol Sci 2008;105(1):200–10.
217. Nestorov I, Munafo A, Papasouliotis O, Visich J. Pharmacokinetics and biological activity of atacicept in patients with rheumatoid arthritis. J Clin Pharmacol 2008;48:406–417.
218. Vugmeyster Y, Seshasayee D, Chang W, et al. A soluble BAFF antagonist, BR3-Fc, decreases peripheral blood B cells and lymphoid tissue marginal zone and follicular B cells in cynomolgus monkeys. Am J Pathol 2006;168:476–89.
219. Fleischmann R, Wei N, Shaw M, et al. BR3-Fc phase I study: safety, pharmacokinetics (PK) and pharmacodynamic (PD) effects of a novel BR3-Fc fusion protein in patients with rheumatoid arthritis. Arthritis Rheum 2006;54:S229–30.
220. Shaw M, Del Giudice J, Trapp R, et al. The safety, pharmacokinetics (PK) and pharmacodynamic (PD) effects of repeated doses of BR3-Fc in patients with rheumatoid arthritis (RA). Arthritis Rheum 2007;56:S568–9.
221. Belouski SS, Rasmussen HE, Thomas JK, Ferbas J, Zack DJ. Changes in B cells and B cell subsets induced by BAFF neutralization in vivo. Arthritis Rheum 2007;56:S565.
222. Sabahi R, Owen T, Barnard J, et al. Immunologic effects of BAFF antagonism in the treatment of human SLE. Arthritis Rheum 2007;56:S566.
223. Edwards JCW, Szczepanski L, Szechinshi J, et al. Efficacy of B-cell-targeted therapy with rituximab in patients with rheumatoid arthritis. N Engl J Med 2004;350:2572–81.
224. Emery P, Fleischmann R, Filipowicz-Sosnowska A, et al. The efficacy and safety of rituximab in patients with active rheumatoid arthritis despite methotrexate treatment: results of a phase IIb randomized, double-blind, placebo-controlled, dose-ranging trial. Arthritis Rheum 2006;54:1390–400.
225. Cohen SB, Emery P, Greenwald MW, et al. Rituximab for rheumatoid arthritis refractory to anti-tumor necrosis factor therapy: results of a multicenter, randomized, double-blind, placebo-controlled, phase III trial evaluating primary efficacy and safety at twenty-four weeks. Arthritis Rheum 2006;54:2793–806.
226. Leandro MJ, Edwards JC, Cambridge G, Ehrenstein MR, Isenberg DA. An open study of B lymphocyte depletion in systemic lupus erythematosus. Arthritis Rheum 2002;46:2673–7.
227. Looney RJ, Anolik JH, Campbell D, et al. B cell depletion as a novel treatment for systemic lupus erythematosus: a phase I/II dose-escalation trial of rituximab. Arthritis Rheum 2004;50:2580–9.
228. Gunnarsson I, Sundelin B, Jónsdóttir T, Jacobson SH, Henriksson EW, van Vollenhoven RF. Histopathologic and clinical outcome of rituximab treatment in patients with cyclophosphamide-resistant proliferative lupus nephritis. Arthritis Rheum 2007;56:1263–72.
229. Albert D, Dunham J, Khan S, et al. Variability in the biological response to anti-CD20 B cell depletion in systemic lupus erythematosus. Ann Rheum Dis 2008;67:1724–31.
229a. Merrill JT, Neuwelt CM, Wallace DJ, et al. Efficacy and safety of rituximab in patients with moderately to severely active systemic lupus erythematosus: results from the randomized, double-blind phase II/III study EXPLORER. Arthritis Rheum 2008;58:4029–30.
230. Thaunat O, Patey N, Gautreau C, et al. B cell survival in intragraft tertiary lymphoid organs after rituximab therapy. Transplantation 2008;85:1648–53.
231. Gong Q, Ou Q, Ye S, et al. Importance of cellular microenvironment and circulatory dynamics in B cell immunotherapy. J Immunol 2005;174:817–26.
232. Vallerskog T, Heimbürger M, Gunnarsson I, et al. Differential effects on BAFF and APRIL levels in rituximab treated patients with systemic lupus erythematosus and rheumatoid arthritis. Arthritis Res Ther 2006;8:R167.

Chapter 9
The Role of BAFF and APRIL in Regulating Human B-Cell Behaviour: Implications for Disease Pathogenesis

Stuart G. Tangye and David A Fulcher

Abstract B cells require signals from multiple sources for their development from precursor cells in the bone marrow and differentiation into effector cells. BAFF and APRIL are members of the TNF superfamily of cytokines and have been identified as critical regulators of B-cell development and differentiation. Defects in the production of BAFF and APRIL, and/or expression of their receptors, have been associated with a diverse array of human diseases characterised by perturbed B-cell function and behaviour, including autoimmunity, malignancy, and immunodeficiency. This chapter will discuss the role of BAFF and APRIL in normal B-cell physiology as well as the emerging evidence of their involvement in the pathogenesis of these human immunopathologies.

Keywords Human B cells · Differentiation · Autoimmunity · Malignancy · Immunodeficiency

9.1 Introduction

The generation of the mature B-cell pool involves the step-wise development of haematopoietic stem cells into pro-B cells, which mature into pre-B cells and then immature B cells [1–3]. Immature B cells are then exported to the periphery as transitional B cells which undergo further selection and developmental events to yield mature B cells [4, 5]. When mature B cells encounter T-cell-dependent (TD) antigen (Ag), they differentiate into high-affinity effector cells, namely immunoglobulin (Ig)-secreting cells (ISCs) or plasma cells (PCs), as well as memory B cells [3, 6]. This process generally occurs within specialised structures in secondary lymphoid tissues called germinal centres (GCs) [3, 6] (Fig. 9.1). Thus, mature B cells are responsible for the generation of long-lived humoral immunity. The co-ordinated

S.G. Tangye (✉)
Immunology and Inflammation Group, Garvan Institute of Medical Research, Darlinghurst, NSW 2010, Australia

M.P. Cancro (ed.), *BLyS Ligands and Receptors*, Contemporary Immunology,
DOI 10.1007/978-1-60327-013-7_9,

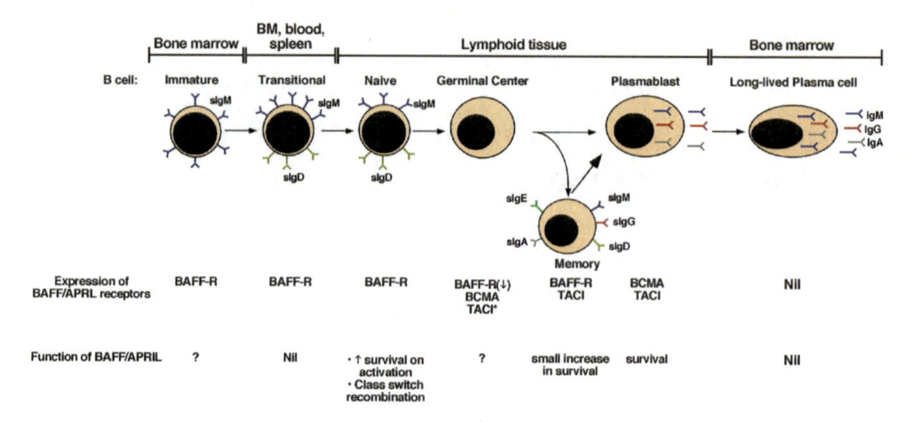

Fig. 9.1 Expression of receptors for BAFF and APRIL, and their functions, during human B-cell development and differentiation. The expression of BAFF-R, TACI, and BCMA at the different stages of B-cell development (immature → transitional B cell) and differentiation (mature B cell → GC → memory B cell/PC), as well as the function of BAFF/APRIL at distinct stages of human B-cell maturation are indicated. ∗ indicates uncertainty, since GC B cells have been reported to lack or express TACI

differentiation of B cells at these different stages of development and maturation is influenced by multiple factors, such as stromal cells and cytokines provided by the bone marrow (BM) microenvironment, strength of interactions with specific Ag, and cross-talk between B cells, Ag-specific T cells, follicular dendritic cells (FDC), and dendritic cells (DCs) in peripheral lymphoid tissues [1, 3, 6].

The differentiation of mature B cells into effector cells requires strict regulation so as to facilitate the generation of Ag-specific humoral immune responses whilst simultaneously avoiding the generation of autoantibodies. Receptor/ligand pairs of the tumour necrosis factor receptor (TNF-R)/TNF superfamily play critical roles during B-cell responses. The best characterised of these involves interactions between CD40 [7], CD27 [8], CD134 (OX40) [9], and TNF-R [10] on B cells and their respective ligands (CD40L, CD70, CD134L, TNF-α), usually on $CD4^+$ T cells, which promote B-cell proliferation, differentiation, and Ig secretion, while ligation of CD30 [11] and CD95 [12] negatively regulate B-cell behaviour. During the past 10 years, BAFF and APRIL, ligands of the TNF superfamily, have emerged as potent regulators of multiple functions of human and murine B cells. Here, we will review the role of BAFF, APRIL, and their respective receptors in B-cell activation during normal immune responses as well as in the pathogenesis of a variety of human diseases.

9.2 BAFF and APRIL: Ligands of the TNF Family

B-cell-activating factor belonging to the TNF family (BAFF) [13–16] and a proliferation-inducing ligand (APRIL) [17] were independently identified based on their homology to the TNF superfamily (reviewed in [18–20]). BAFF and APRIL are produced by hematopoietic cells such as monocytes, macrophages, DCs, astrocytes [21–24] and neutrophils [25, 26], as well as non-hematopoietic cells, namely epithelial cells present in the intestine and respiratory tract [27–29], and FDC in secondary lymphoid tissues [30]. Production of BAFF and APRIL by these cell types can be increased following stimulation with a broad range of cytokines (CD40L, IL-10, IFN-α, IFN-β, IFN-γ [21–24, 27], G-CSF [25, 31], the IL-7-related cytokine thymic stromal lymphopoietin (TSLP) [28, 29]) or ligands for specific Toll-like receptors (TLRs) [27–29]. Expression of both BAFF and APRIL can also be induced in human B cells following infection with Epstein Barr virus (EBV) [32] or dual stimulation through CD40 and the B-cell receptor (BCR) [33].

BAFF binds three receptors belonging to the TNF-R superfamily – BAFF receptor (BAFF-R/BR3) [34, 35], transmembrane activator of and calcium modulator and cyclophilin ligand (CAML) interactor (TACI), and B-cell maturation antigen (BCMA). APRIL does not interact with BAFF-R; however, it can bind to both TACI and BCMA [36–40]. Interestingly, heparin sulphate proteoglycan (HSPG) has been identified as a unique receptor for APRIL [41]. In mice, BAFF-R is expressed at low levels on early transitional B cells and up-regulated on late transitional, follicular, and marginal zone (MZ) B cells [42]; however, it is absent from pro-B and pre-B cells in the BM. TACI exhibits a similar expression profile to BAFF-R, being low/absent on murine splenic T1 B cells, induced at the follicular stage and further increased on late transitional and MZ B cells [26, 43]. In contrast to both BAFF-R and TACI, BCMA message can be detected in transitional B cells in murine spleen, but is then down-regulated at later stages of B-cell maturation [43]. While neither BAFF-R nor TACI were detected in murine BMPCs, these cells did contain mRNA for BCMA [44].

In humans, transitional, naïve, GC and memory B cells are all capable of binding soluble BAFF. Transitional and naïve B cells bind BAFF exclusively through BAFF-R, because TACI and BCMA are absent from these cells. On the other hand, memory B cells can interact with BAFF through either BAFF-R or TACI, but not BCMA [26, 30, 45–53]. Human GC B cells express BAFF-R, albeit at a reduced level compared to naïve and memory B cells, as well as BCMA [26, 30, 51]. Expression of TACI by human GC B cells is controversial, as it has been reported to be both absent [26, 30, 51] and present [54] from this B-cell subset. Human plasmablasts acquire expression of BCMA and TACI, yet down-regulate BAFF-R [30, 45, 55, 56]. Interestingly, PCs present in human tonsils retain expression of TACI and BCMA following their maturation from the plasmablast stage [52, 54], but then down-regulate expression of all known BAFF receptors once they undergo terminal differentiation and migrate to the BM [45, 46, 52]. Thus, it would appear that normal human B cells first express BAFF receptors at the transitional stage of development and remain capable of receiving BAFF-dependent signals at least until they terminally differentiate into PCs (see Fig. 9.1).

9.3 Functions of BAFF

9.3.1 B-Cell Survival and Proliferation

9.3.1.1 Murine B Cells

A predominant function of BAFF is in promoting and/or sustaining the survival of late transitional and mature murine B cells (reviewed in [18–20]). The mechanism by which BAFF exerts this effect is by altering the ratio between pro-survival and pro-apoptotic molecules. Expression of the anti-apoptotic genes A1, bcl-2, bcl-xL [43, 57], and Mcl-1 [44, 58] was increased following in vitro exposure of murine B cells to BAFF, while that of pro-apoptotic molecules Bak [57], Blk [59], and Bim [60, 61] was reduced. As a result of enhanced survival, BAFF could strongly increase proliferation of murine B cells induced by engagement of the BCR [13, 43, 62]. The source of BAFF that regulates homeostatic survival during B-cell development is believed to be from non-haematopoietic stromal cells present within BM and/or spleen [63]. In addition to its effects at the transitional stage of B-cell development, BAFF and APRIL can enhance survival of terminally differentiated murine PC [44].

Studies of mice lacking functional BAFF-R, TACI, or BCMA revealed that BAFF exerts its pro-survival effect during B-cell development predominantly through BAFF-R, because spleens from BAFF-deficient mice had a severe reduction in the number of transitional and mature B cells, while B-cell development in mice deficient in TACI [64, 65], BCMA [66–68], or both [69] was intact. These findings established the unique role played by the BAFF/BAFF-R signalling pathway in regulating B-cell survival and homeostasis. The impaired survival of developing B cells in mice deficient for BAFF or BAFF-R could be overcome by enforced expression of bcl-2 [70] or bcl-xL [71], supporting the proposal that BAFF maintains B-cell survival by modulating expression of anti-apoptotic molecules. This led to the proposal that signals delivered through BAFF-R, in concert with Ag-mediated engagement of the BCR, promote positive selection of transitional B cells and allow for their continued maturation [43, 72]. This was supported by the findings that signalling through the BCR increased binding of BAFF to murine B cells [73] and an intact BCR signal transduction pathway is required for acquisition of BAFF-responsiveness in developing B cells [74].

9.3.1.2 Human B Cells

In contrast to murine B cells, emerging data suggests that B cells from humans and non-human primates are less dependent on BAFF for their development and survival. Unlike murine B cells [72, 75], BAFF has minimal effect on the survival of subsets of human B cells, despite the expression of BAFF-R on these cells. BAFF only weakly promoted the survival of human memory B cells, but not transitional, naïve, or PCs [45, 46, 48, 53, 58, 76, 77] (Fig. 9.1). It is unclear whether the weak effect that BAFF has on the survival of memory B cells is mediated by BAFF-R, TACI, or both. However based on the dramatic consequences of BAFF-R deficiency in mice, it is plausible that BAFF-R has this function in human memory B cells. Consistent with these results were the findings that in vivo blockade of BAFF in cynomolgus monkeys had no effect on the numbers of putative transitional B cells and tissue PCs, while mature peripheral B cells were reduced <2-fold compared to control animals [78–80].

Despite these species differences in the ability of BAFF to support survival of resting human B cells, BAFF (and APRIL) can sustain the viability of (a) naïve B cells stimulated through the BCR, CD21, and IL-4R, which then acquire features of Ag-presenting cells [58, 77], and (b) human memory B cells stimulated with CD40L/IL-2/IL-10, which develop into plasmablasts [30, 45, 51, 55]. The latter finding parallels the ability of BAFF and APRIL to promote survival of plasmablasts arising from activated murine B cells in vivo [57, 62, 81, 82]. Interestingly, BAFF-R is down-regulated on plasmablasts, while expression of TACI is increased and BCMA induced [30, 45, 52, 55]. Furthermore, the effect of BAFF and APRIL on murine plasmablast survival was reduced in the absence of TACI or BCMA and completely abrogated in the absence of both receptors [62]. Thus, it is likely that the survival effect exerted by BAFF/APRIL on activated B cells as they differentiate into Ig-secreting cells is mediated by BCMA and/or TACI (Fig. 9.1). Recent findings have proposed that this is achieved by the induction of cyclo-oxygenase 2 and subsequent production of prostaglandin E2, which has known roles in promoting survival of multiple cell lineages [58]. It is likely that myeloid cells – monocytes, macrophages, DCs – in spleen and BM provide the BAFF/APRIL that regulates survival of human and murine plasmablasts [20, 23, 45, 63, 82]. Recently, a novel function of BAFF was revealed by its ability to preferentially promote chemotaxis of human memory B cells to CXCL13 [83]. Thus, BAFF may have a dual function on memory B cells by causing a mild increase in their survival [45], coupled with a greater migratory capacity to lymphoid homing chemokines [83]. The production of both CXCL13 and BAFF by cells within lymphoid follicles (i.e. stromal cells, FDC, myeloid cells), and the enhanced response of memory B cells to both ligands, may contribute to the rapid response of memory B cells, relative to naive B cells, that is characteristic of secondary responses to TD Ags.

9.3.2 Class Switch Recombination

9.3.2.1 Human B Cells

Following Ag stimulation, naïve B cells can undergo class switch recombination (CSR) to express and produce the downstream isotypes IgG, IgA, or IgE. This allows versatility in both function and distribution of the Ig molecule, while retaining Ag specificity [3]. CSR from Cμ to Cγ, Cα and Cε occurs in response to signals delivered through CD40/CD40L, usually in combination with specific cytokines, or TLR ligands [84–86], and is accompanied by characteristic molecular events such as induction of activation-induced cytidine deaminase, a DNA-editing enzyme necessary for CSR, and expression of Ig heavy chain germline transcripts and switch circles [84–86]. These events can be detected in human B cells when stimulated with BAFF or APRIL alone, and are greatly increased in B cells treated with BAFF or APRIL together with IL-4 or IL-10 [22, 29]. BAFF may further contribute to CSR by inducing B cells to secrete IL-10 [87], a known switch factor for production of IgG and IgA by human B cells [7]. Although CSR from Cμ to Ce was not observed when human B cells were exposed to BAFF or APRIL alone, switching to IgE did occur when combined with IL-4 [22] (Table 9.1). Interestingly, BAFF or APRIL alone were not sufficient to induce secretion of switched Ig isotypes by stimulated human B cells; rather, Ig secretion required additional signals, such as cross-linking of the BCR together with specific cytokines (IL-4, IL-10) [22, 29]. Notably, APRIL – more so than BAFF – favoured production of IgA2 by human B cells stimulated with IL-10 and the TLR5 ligand flagellin, while levels of IgA2 secretion were further enhanced by the combination of BAFF together with APRIL [29]. BAFF and APRIL also contribute to humoral immune responses by enhancing CSR, acquisition of expression of switched Ig isotypes IgG and IgA, as well as their secretion, by naïve B cells activated by EBV [32] or the TLR3 and TLR9 ligands dsRNA and CpG, respectively [88, 89] (Table 9.1).

The in vivo relevance of the in vitro findings of a role for BAFF and APRIL in regulating Ig CSR has been examined in the context of interactions between BAFF/APRIL-producing cells, such as DCs and epithelial cells, and B cells. It has been proposed that TLR ligands present in microorganisms stimulate innate immune cells (DCs) and non-haematopoietic cells (e.g. mucosal epithelial cells) to produce BAFF and APRIL. TLR-stimulated epithelial cells can also produce TSLP, which acts in a paracrine loop to augment TLR-induced production of BAFF by DCs. In addition to activating epithelial cells and DCs, microbial stimuli can activate B cells through corresponding TLRs (e.g. TLR3, TLR9), thereby directly initiating CSR. Thus, BAFF and APRIL, together with TLR ligands, activate CSR and, when combined with DC-derived cytokines including IL-10, co-operate to elicit secretion of switched Ig isotypes by responding B cells, resulting in the generation of an integrated humoral immune response [22, 28, 29, 85, 88, 89] (Table 9.1).

9.3.2.2 Murine B Cells

BAFF and APRIL also induce murine B cells to undergo CSR in vitro [90, 91]. Specifically, stimulation of murine B cells with BAFF or APRIL resulted in the secretion of IgG1 and IgA. In contrast, IL-4 was required for production of IgE by BAFF- or APRIL-stimulated murine B cells [90, 91]. Analyses of gene-targeted mice have revealed important and Ig isotype-specific roles for TACI and BAFF-R in this process. Both APRIL and BAFF failed to induce IgA secretion by naïve B cells from TACI$^{-/-}$ mice, but no such failure occurred in mice whose B cells

Table 9.1 Induction and amplification of class switch recombination and Ig secretion by BAFF and APRIL

Ig isotype	Induced by	Enhanced by	Secreted in response to:	Source of BAFF/APRIL	References
IgG1, IgG2	•BAFF, APRIL	•IL-4, IL-10	*Total IgG*: •BAFF/anti-Ig ± IL-15 • APRIL/anti-Ig ± IL-15	DCs stimulated with • CD40L • IFN-α • IFN-γ Monocytes stimulated with • LPS	[22, 29]
IgG3	• BAFF/IL-4 • APRIL/IL-4 • BAFF/IL-10 • APRIL/IL-10	ND			
IgG4	BAFF + IL-4 APRIL + IL-4	ND			
IgA1, IgA2	• BAFF • APRIL	• TGF-b	*Total IgA*: • BAFF/anti-Ig ± IL-15 • APRIL/anti-Ig ± IL-15	• IFN-α • IFN-γ	
IgE	• BAFF/IL-4 • APRIL/IL-4	ND	ND		
IgA2	• APRIL/IL-10	• TLR5 ligand (Flagellin)	• APRIL, IL-10, Flagellin	• Intestinal epithelial cells stimulated with TLR5 ligand ± TSLP	[29]
IgG1, G2, G3	• CpG (TLR9 ligand)	• IL-10	*Total IgG* •BAFF	• IFN-α stimulated DCs	[88]
IgG1, IgG2, IgG3, IgA1, IgA2, IgE	• EBV	• BAFF • APRIL	ND	• EBV-infected B cells	[32]
IgG1	• TLR3 ligand (poly I:C; ds RNA mimic)	• IL-10	*Total IgG and IgA* • poly I:C/IL10/BAFF	• TLR3-stimulated plasmacytoid and mucosal DCs	[89]

ND – not done

either expressed a non-functional BAFF-R or lacked BCMA [91]. This suggests that TACI is exclusively responsible for eliciting CSR to IgA in response to BAFF or APRIL, and is consistent with impaired switching to IgA in vivo in APRIL-deficient mice [90]. TACI-deficient B cells also failed to secrete IgG1 and IgE in response to APRIL plus IL-4, while the response of B cells lacking BCMA or a functional BAFF-R was intact [91]. Thus, TACI is also capable of mediating APRIL-induced switching to IgG1 and IgE. Despite the inability to respond to APRIL, TACI$^{-/-}$ B cells synthesised normal levels of IgG1 and IgE in response to stimulation with BAFF/IL-4 [91]. This study, therefore, demonstrated that in the absence of TACI, production of these isotypes can be compensated by the BAFF/BAFF-R signalling pathway, revealing redundancy in the abilities of both BAFF-R and TACI to mediate switching to IgG1 and IgE in the presence of corresponding ligands.

The relative roles of BAFF-R and TACI in inducing CSR in human B cells by BAFF and APRIL remain incompletely defined. Curiously, stimulation with the TLR9 agonist CpG strongly induces expression of TACI on human naïve B cells [52]. Since BAFF is capable of augmenting CSR induced in human B cells by CpG [88], it is tempting to speculate that this is achieved by BAFF interacting with TACI. This is supported by the finding that siRNA-mediated knockdown of TACI expression on human B cells abrogated the ability of APRIL to induce IgA production by these B cells; however, production of IgG was unaffected [92]. Taken together, it appears that TACI is required for APRIL-induced production of IgA, but not IgG, whereas BAFF-R contributes to BAFF-mediated CSR to IgG and IgA by human B cells. This is discussed further (see Section 9.6) in the context of humans with loss-of-function mutations in TACI whose B cells failed to secrete IgA in response to activation with either APRIL or BAFF [93].

9.4 Aberrant Expression of BAFF, APRIL, and BAFF Receptors in Human Disease

Dysregulated expression and/or function of BAFF, APRIL, and/or their receptors has been implicated in the pathogenesis of a diverse range of human diseases including autoimmunity, haematological malignancies, and immunodeficiencies.

9.4.1 Autoimmunity and Related Diseases

BAFF, APRIL, and BAFF/APRIL multimers are elevated in serum and synovial tissue of patients with autoimmune diseases such as systemic lupus erythematosus (SLE), rheumatoid arthritis (RA), Sjogren's syndrome (SjS) [94–102], scleroderma [103], systemic sclerosis [104, 105], atopic dermatitis [106], bullous pemphigoid [107, 108], and Wegener's granulomatosis [109]. BAFF has also been detected in neurological lesions of patients with multiple sclerosis (MS) [24]. In general, the increased levels of serum BAFF and APRIL were associated with increased levels of autoantibodies [95–97, 99, 101] or disease activity [105, 106]. A recent study also reported detection of BAFF in lesions of females with endometriosis, but neither in control endometrial tissue, nor patients with other gynaecological conditions such as adenomyosis and uterine fibrosis [110]. The presence of BAFF in endometriosis lesions was accompanied by infiltration of substantial numbers of BCMA-expressing plasmablasts, and substantially elevated levels of serum BAFF [110]. These findings suggest that an autoimmune/inflammatory component contributes to some of the pathological features of endometriosis.

The cell type responsible for the increased production of BAFF in some of these conditions has been identified. BAFF is produced by T cells and macrophages infiltrating inflamed salivary glands in SjS [111], and by astrocytes [24] and monocytes [112] in MS. In RA patients, macrophages exclusively produced BAFF, while DC secreted APRIL [101]. BAFF-expressing macrophages were also detected in lesions in endometriosis [110]. Neutrophils, which can be detected in inflamed synovium, may also contribute to elevated levels of BAFF in RA [101]. Several other cytokines, such as soluble CD40L, IL-4, IL-6, IFN-α, and IL-10, are elevated in autoimmune conditions [99, 113–116]. Furthermore, there is a large increase in the level of expression of CD40L on T cells and B cells from SLE patients [117]. Interestingly, most of these cytokines can augment production of BAFF by myeloid cells in vitro [21–24]. Thus, it is likely that myeloid cells exposed to CD40L, IFN-α, and IL-10 are responsible for the increased serum levels of BAFF in SLE.

In autoimmune diseases, the affected tissues are usually non-lymphoid, and are thus devoid of immune cells. However, inflamed synovial tissue and salivary glands from RA and SjS patients contains large numbers of infiltrating lymphocytes which form GC-like structures comprised of B cells, T cells, DCs, and FDC [118, 119]. The formation of these ectopic GC-like structures likely results from aberrant expression of molecules such as TNFa, lymphotoxin, CXCL13, CCL19 and CCL21, that are required for normal lymphoid neogenesis [80, 119, 120]. It was recently revealed that the incidence of ectopic GC in inflamed tissues in RA and SjS correlates with the levels of lymphotoxin, CCL19, BAFF, and APRIL [98, 99, 101, 115]. Thus, it is likely that BAFF and/or APRIL plays a critical role in the maintenance of ectopic GCs in autoimmune patients, by eliciting (i) pro-survival effects on autoreactive B cells and (ii) inducing them to produce class-switched pathogenic autoantibodies, which are characteristic of humoral autoimmune conditions. The ability of BAFF to enhance chemotaxis of memory B cells may also contribute to the infiltration of effector B cells to sites of autoimmune-mediated inflammation [83].

9.5 B-Cell Malignancies

Numerous studies have provided convincing evidence that BAFF/APRIL may contribute to malignancies of mature B cells (non-Hodgkin's lymphoma [NHL], chronic lymphocytic leukaemia [B-CLL], Hodgkin's lymphoma [HL]), plasmablasts (Waldenstrom's macroglobulinaemia [WM]) and PCs (multiple myeloma [MM]). Malignant B cells from these patients are all capable of binding soluble BAFF and, in some cases, APRIL [33, 47, 49, 50, 54, 56, 121–124]. However, the receptor expression profile of these malignant cells differed as NHL [123], B-CLL [49, 121], and WM [124] B cells co-express BAFF-R and TACI, while HL and MM cells preferentially expressed BCMA and TACI [47, 54, 56, 125]. In contrast to NHL, B-lymphomas in the central nervous system exhibited a heterogeneous phenotype, with variable expression of BAFF-R, TACI, and BCMA [24]. Malignant B cells isolated from patients with B-CLL [49, 121, 126], NHL [50, 122, 123], HL [54], MM [47, 56], and WM [124] were all capable of responding to the stimulatory effects of BAFF and/or APRIL in vitro, demonstrating expression of functional receptors.

A possible role of BAFF and APRIL in human B-cell malignancies may lie in their aberrant expression and production. First, expression of BAFF and APRIL is greater in B-CLL [49, 121, 127], NHL [33, 50], and MM [47, 56, 126] than in normal B cells. Second, levels of serum BAFF and APRIL are elevated in patients with NHL [122, 123], B-CLL [121, 127], and WM [124]. Notably, those NHL patients with the highest levels of serum BAFF/APRIL had reduced survival and poorer prognosis than those with lower levels [123, 128]. A complementary mechanism whereby BAFF may contribute to the development of some lymphomas came from the observation that EBV infection of primary human B cells induces expression of both BAFF and APRIL [32]. The significance of this is that several B-cell malignancies develop following immortalisation of normal human B cells by EBV [129]. Interestingly, patients with SjS are predisposed to B-cell malignancies, particularly NHL [99]. Because patients with NHL have increased levels of serum BAFF [122, 123], it is tempting to speculate that increased levels of serum BAFF in SjS contributes not only to the development of autoimmune manifestations of this syndrome, but also to B-cell malignancies frequently observed in these patients [99]. Third, the microenvironment of the malignancy may improve survival and persistence of malignant B cells. For instance, BAFF was detected in the BM of myeloma patients at higher levels than that observed for samples of normal BM [47]; cells infiltrating tissue-restricted NHL and HL express and secrete high levels of BAFF/APRIL which promotes the viability of the malignant B cells [54, 130]; nurse-like cells, which can be isolated from B-CLL patients, contain abundant amounts of BAFF and APRIL and improve the survival of B-CLL cells [126]; and tumour-infiltrating neutrophils are the main source of APRIL in cases of NHL [128].

BAFF protects normal B cells from apoptosis by modulating expression of members of the bcl-2 family of molecules (reviewed in [18, 19]). This is also likely to be the mechanism by which BAFF and APRIL preserve the viability of malignant B cells. In vitro exposure of B cells from patients with NHL, HL, B-CLL, or MM to BAFF or APRIL increased expression of the pro-survival proteins bcl-2, mcl-1, and bcl-xL and decreased the pro-apoptotic regulator bax [50, 54, 56]. Similarly, survival and expression of pro-survival proteins were reduced when malignant B cells were cultured in the presence of BAFF antagonists [33, 49, 54, 56]. Taken together, it is possible that aberrant and/or excessive production of BAFF or APRIL by malignant B cells themselves (i.e. autocrine) or by supporting cells present within the microenvironmental niche occupied by the malignant B cells (i.e. paracrine) may facilitate their growth and survival. Consequently, blocking interactions between BAFF and APRIL and their receptors may be a feasible therapeutic approach for treating some B-cell malignancies.

9.6 TACI Mutations in Immunodeficiency

The central role of the BAFF family of molecules in regulating B-cell survival and function has led naturally to an examination of their place in human humoral immunodeficiencies. This endeavour has concentrated on two poorly understood adult immunodeficiency disorders: common variable immunodeficiency (CVID) and IgA deficiency (IgAD). CVID is the most common adult immunodeficiency requiring treatment. It is a heterogeneous disorder manifest by hypogammaglobulinaemia resulting in recurrent upper and lower respiratory tract and gastrointestinal infections [131, 132]. Most cases occur sporadically, but 10–20% have a family history; kindreds with both autosomal dominant and recessive genetics have been described. About one-third of patients will have lymphoproliferation (lymphadenopathy, splenomegaly) as part of their presentation, and there is a heightened prevalence of autoimmunity. IgAD is a common immunodeficiency, accounting for about 1:600 patients in the Western world and is characterised by low to absent serum and mucosal concentrations of IgA. The condition is frequently asymptomatic, but those who develop infections typically display a similar spectrum to CVID [131, 132].

The last decade has seen the discovery of a number of well-defined genetic defects that account for CVID in very small numbers of patients. These include deficiency of ICOS in nine reported cases, most likely involving common ancestry [133, 134], and deficiency of CD19 in five other patients [135, 136]. There are also patients with "leaky" mutations in *Btk* [137, 138] and *SH2D1A* [139–141], who present similarly to CVID but are in fact atypical cases of X-linked agammaglobulinaemia and X-linked lymphoproliferative disease, respectively. On the other hand, the vast majority of CVID patients remain unexplained genetically and pathophysiologically, including those with a positive family history.

9.7 Genetics of TACI Mutation in Humoral Immunodeficiencies

In 2005, two groups independently examined cohorts of patients with CVID or IgAD for mutations in *TNFRSF13b*, which encodes TACI [142, 143]. Mutations were found in 5–10% of patients with familial and sporadic CVID (Fig. 9.1), and included homozygous and heterozygous cases. Of 162 CVID patients analysed by Salzer et al. [142], 3/27 familial and 10/135 sporadic cases had *TNFRSF13b* mutations; all patients had low IgG levels at diagnosis, whilst most also had low IgA and about half had low IgM levels. Pneumococcal antibody responses were generally poor. B-cell numbers were preserved, but there was a reduction in memory B cells. Lymphoproliferation and autoimmunity were frequent in these patients, although these features are also common in unselected cohorts of patients with CVID [131, 132]. These findings were largely supported by Castigli et al. [143], who documented *TNFRSF13b* mutations in 4/19 CVID patients (three homozygous and one compound heterozygote), all of whom had reduced IgG and IgA levels, normal IgM, and poor pneumococcal antibody responses. A positive family history was noted in three of these cases, the pedigree also showing family members harbouring identical TACI mutations that manifest phenotypically as either CVID or IgAD. Furthermore, one of the 16 patients with IgAD also harboured a TACI mutation, with perfect segregation with IgAD in that family. No TACI mutations were found in 50 control subjects.

The six mutations uncovered in these two studies either had, or were predicted to have, critical effects on TACI function. The most frequent mutation, C104R, altered a cysteine residue essential for protein folding and thus abolished BAFF binding. A181E changed a neutral amino acid to a charged residue in the transmembrane region, S144X and 204InsA abolished protein expression, R202H was predicted to alter TACI interaction with CAML, and S194X resulted in a truncated intracellular domain due to the introduction of a premature stop codon.

These genetic studies were expanded by both groups in follow-up studies published in 2007 [144, 145]. The first report included up to 852 CVID patients in cohorts spanning Sweden, Germany, and the USA [144]. *TNFRSF13b* was found to be a highly polymorphic gene, with nucleotide differences frequently detected in normal subjects, including the same mutations previously associated with CVID (Fig. 9.2). Only C104R and A181E were significantly associated with CVID, although in certain populations (e.g. Swedish patients for C104R and German patients for A181E), the association did not reach significance. 204InsA was rare, being found only in two CVID patients and no controls. The report also studied 474 Swedish IgAD patients, with only R202H being found more frequently than in the control population; however, family studies showed no segregation between this mutation and IgAD. A similar study of 212 CVID patients and 124 controls by Castigli et al. [145] largely confirmed these findings, with only C104R and A181E being significantly associated with CVID, although they reported a number of rare

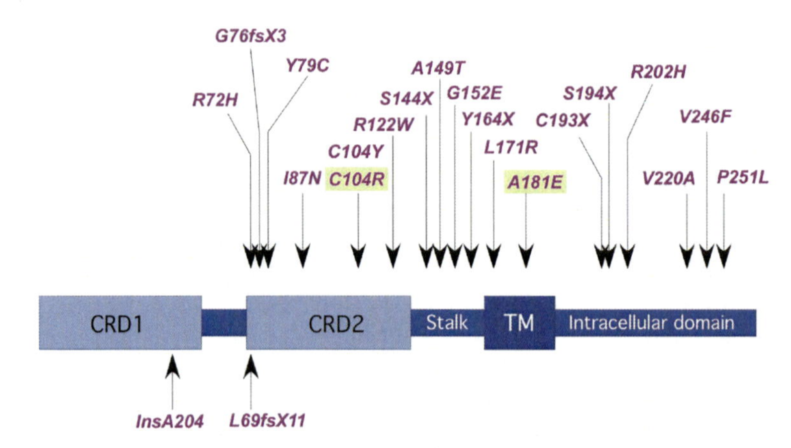

Fig. 9.2 Mutations within the TNFRSF13B gene that encodes TACI. Missense mutations affecting the open reading frame of TACI are shown. The two most closely associated with disease are highlighted

TACI alleles that were found only in CVID patients. All 34 patients with IgAD had wild-type *TNFRSF13b*, and taken together with the Swedish study, essentially excluded a role for TACI in this condition.

Most recently, another expanded CVID cohort (which included some patients previously reported in [144]) was studied [146]. *TNFRSF13b* mutations were found in 50/602 (6%) CVID patients compared with 12/589 (2%) of controls. Biallelic mutations were present in 11 subjects, all of whom had CVID, whereas none of the controls had biallelic mutations. C104R was significantly associated with CVID, but this was not found for A181E. Again, a number of rare *TNFRSF13b* mutations were only found in CVID patients. Six families of CVID patients with heterozygous *TNFRSF13b* mutations were studied, and three showed segregation with CVID whilst the remainder did not, suggesting that in the heterozygous state, the C104R mutation might be acting as a disease modifier rather than being directly causative.

Taken together, the conclusions from these studies are 3-fold. First, homozygous *TNFRSF13b* mutations have only been found in patients with CVID. Second, some rare heterozygous *TNFRSF13b* mutations have also so far only been found in CVID, leaving open the possibility that these may truly be causative in the heterozygous state. Third, there is a significant association between C104R and possibly A181E with CVID, albeit with incomplete penetrance, such that the mutation is more frequent in CVID but may also be found in healthy subjects.

9.8 Functional Implications of TNFRSF13b Mutations in B-Cell Responses

These genetic studies immediately posed questions as to the functional changes that accompany alterations in TACI structure that account for adult-on set loss of Ig production, the hallmark of CVID. Models to answer this question must not only explain the physiology of complete loss of TACI function, as is the case in patients with biallelic mutations, but also the weaker dominant negative effect in heterozygotes.

Complete loss of TACI function has been studied in at least two independently generated strains of gene-targeted mice [65, 147, 148]. TACI-deficient mice have normal B-cell development, yet B-cell numbers were elevated and proliferative responses were increased, with one strain showing a propensity for the development of B-cell lymphoma. These findings implied a negative regulatory role for TACI [65, 147, 148]. Such a role was supported by studies in which an agonistic anti-TACI mAb was used to stimulate human naïve B cells concomitantly activated with CD40L, BAFF, or anti-BAFF-R [149]. TACI activation consistently resulted in decreased proliferation and IgG production by the activated B cells. Furthermore, TACI expression on B cells increased during activation, leading the authors to speculate that TACI might function as a delayed brake on B-cell function analogous to the role of CTLA4 on T cells [149].

On the other hand, a positive role for TACI in type-II T-cell independent Ab responses has also been demonstrated in TACI$^{-/-}$ mice, which have poor pneumococcal vaccine responses and low IgA and IgM, but normal IgG, levels. There is also an established role for BAFF and APRIL in isotype switching by human and murine B cells [22, 28, 29, 89, 91], although redundancy in the BAFF/APRIL receptor system makes specific roles for each receptor in human B-cell differentiation difficult to isolate. Some progress has been made by Sakurai et al. [92], who explored the differential roles of the two main APRIL receptors on mature B cells, TACI, and HSPG. They first demonstrated that APRIL bound equally to each of these receptors, then examined responses of naïve B cells to stimulation with CD40L, BAFF, and APRIL in the presence of (i) siRNA to diminish TACI expression and/or (ii) heparitinase to eliminate HSPG binding. Engagement of TACI in isolation was insufficient to increase proliferation and Ig production by B cells. However, abrogation of TACI expression increased B-cell responses to BAFF, consistent with a negative regulatory role for TACI, as demonstrated in earlier experiments. It was shown further that either engagement of HSPG by APRIL, or BAFF-R by BAFF, was sufficient for IgG production and proliferation, but simultaneous stimulation of both TACI and HSPG was necessary for IgA production, such that loss of either abrogated this response (Fig. 9.3). Based on these studies, one would expect that in vivo loss of TACI function could result in impaired IgA production (although BAFF-R stimulation could potentially compensate for this loss), whilst there should be minimal effect on IgG production and B-cell proliferation, if not potential enhancement (in the case of

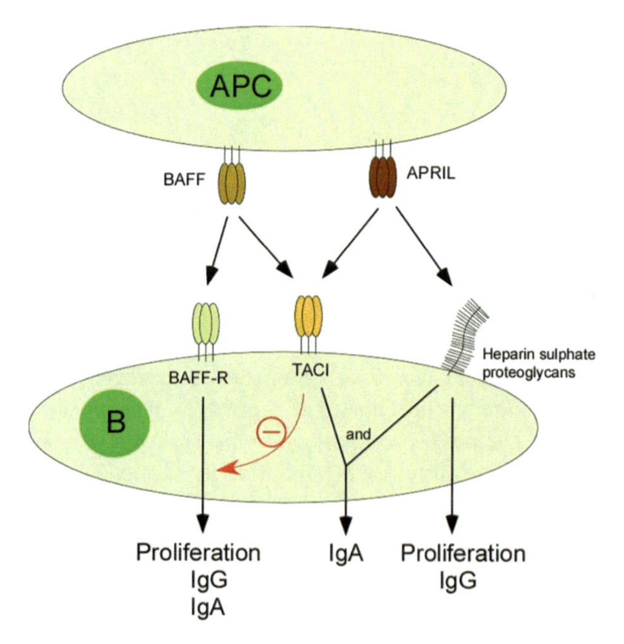

Fig. 9.3 Relative roles of BAFF-R, TACI, and HSPG in B-cell isotype switching. This model is based on experiments reported by Sakurai et al. [92]

BAFF) resulting from loss of negative regulation. This would be consistent with findings from studies of murine B cells which revealed that TACI is predominantly responsible for eliciting CSR to IgA in response to BAFF and APRIL [91].

In vitro studies of B-cell function in human homozygous TACI mutants have been limited. Original studies demonstrated that the C104R, but not A181E, mutation abolished BAFF binding when expressed in the human embryonic kidney cell line 293, although APRIL binding could not be studied due to the high expression of HSPG by these cells [143]. EBV-transformed B-cell lines from TACI mutant homozygotes (C104R, S144X) failed to bind APRIL in a staining system that neutralised HSPG binding. Not surprisingly, B cells isolated from TACI mutant homozygotes failed to respond to APRIL stimulation in terms of B-cell proliferation and CSR [142], again in the setting of neutralisation of HSPG. Thus, the lack of functional TACI expression seems to abrogate the positive effects of APRIL in the absence of HSPG, but the effect of homozygous mutations on the negative regulatory effects mediated by TACI remains to be determined, as do the functional changes in the setting of intact HSPG binding.

In heterozygotes, expression of a mutant surface protein need not necessarily alter cellular response to ligand, which could simply bind to wild-type receptors, notwithstanding the possibility of haploinsufficiency. However, Garibyan et al. [150] demonstrated that murine TACI assembled into trimers within the endoplasmic reticulum, independent of ligand binding, and that C104R mutant TACI was co-expressed in a trimeric receptor complex with wild-type TACI. Thus, in the heterozygous state, theoretically only one in eight trimers would be fully functional. Ligand binding studies demonstrated normal APRIL binding to EBV-transformed B-cell lines from TACI heterozygotes [142, 146], although others found binding was variable even between patients with the same mutation [151]. Garibyan et al. demonstrated normal BAFF binding to the hybrid mutant C104R/wt TACI heterotrimer [150], but there was a reduction in NFkB signalling in response to BAFF, consistent with a dominant negative effect. These findings are reminiscent of the effects of mutations in the gene encoding human CD95, which demonstrated a requirement for pre-association of CD95 monomers into a trimeric structure and that such assembly was compromised by heterozygous mutations in CD95 [152].

Given that a degree of receptor dysfunction accompanies the heterozygous state, one might therefore expect that functional studies in heterozygotes might be similar to homozygous deficiency. Such studies have generated conflicting results, which in general do not seem to reflect in vitro studies in normal B cells. Early reports demonstrated that B cells from TACI heterozygotes failed to produce IgG or IgA in response to APRIL, but also enigmatically failed to produce IgA in response to BAFF [143]. On the other hand, B cells from CVID patients themselves often display impaired function in vitro, as demonstrated by Zhang et al. who showed that whilst B-cell proliferation and isotype switching is variably impaired in C104R heterozygotes, this was not substantially different from CVID patients with wild-type TACI [151]. Furthermore, family members harbouring heterozygous C104R mutations but without CVID showed normal B-cell responses! This study [151] has important implications in terms of choosing appropriate controls for functional analysis, although it did not use purified B-cell cultures and relied on relatively insensitive proliferation assays.

Despite progress in the field of TACI mutations and CVID, the area continues to generate controversy. There seems little doubt that homozygous TACI mutations result in CVID [146], but there is also strong association in the heterozygous state particularly with the C104R and A181E mutations. Biochemical and functional data showing that the heterozygous state also impairs TACI signalling is consistent with this association [150, 151]. Although functional studies support a possible role for *TNFRSF13b* mutations in dysregulated CSR, they fall far short of explaining the CVID phenotype, given the redundancies in the BAFF/APRIL receptor system, since studies in normal B cells suggest that BAFF-R signalling generally should be able to compensate for loss of TACI function, at least in terms of signals provided by BAFF. Furthermore the functional effects of *TNFRSF13b* mutation in the setting of an intact HSPG receptor have not been studied. However the role of TACI in CSR in normal human B cells in vivo seems minor in comparison to BAFF-R and HSPG, studies of which show a non-redundant role only in IgA production, whilst the genetic association with IgAD has been largely refuted. If defects can be demonstrated, the link between TACI and CVID may lie in differential expression of BAFF versus APRIL in specific microenvironments, which may affect CSR or B-cell or PC longevity. Murine and human studies imply that TACI has predominantly a negative regulatory role in B-cell function; hence loss of function could result in B-cell over-activity, possibly explaining the propensity of affected CVID patients to develop lymphoproliferation, although it is unclear whether this is any more prevalent in CVID patients who have TACI mutations compared with those with wild-type TACI. Finally, the observation that B cells from CD40L-deficient patients with hyper-IgM syndrome fail to undergo CSR in vivo despite an intact BAFF/APRIL system (reviewed in [153]) would suggest either that the latter pathway is of relatively minor importance or that it is only relevant to B cells at a stage distal to CD40L activation. Further studies on specific B-cell subsets from heterozygous and homozygous C104R TACI mutant patients, along with carefully chosen controls, are needed to resolve these conflicting data.

9.9 Conclusions

It is well established that molecules expressed and produced by stromal cells, $CD4^+$ T cells, myeloid cells (e.g. DCs), and FDC are essential initially for the development of pluripotent stem cells into mature naïve B cells, and subsequently for their differentiation into effector cells, such as memory and Ig-secreting cells. The last decade has seen an explosion in our understanding of the fundamental roles of BAFF, APRIL, and the receptors BAFF-R, TACI, BCMA, and HSPG in these processes. It is now clear that specific interactions between these receptor/ligand pairs have important roles in B-cell development, CSR, and the maintenance of Ig-secreting cells. We have also gained an enormous appreciation that dysregulated signalling through these receptor/ligand complexes may underlie the development of a diverse array of human diseases, including autoimmunity, haematological malignancies, and immunodeficiency. However, these studies have also highlighted the complexities of the interactions between these molecules, and many questions and issues remain unresolved. These include the specific role of BCMA and HSPG; the apparent paradoxical ability of TACI to deliver both positive and negative signals to activated B cells; the signals that modify and regulate expression of receptors for BAFF and APRIL, as well as the production of these cytokines themselves; the exact mechanisms whereby aberrant interactions between these molecules contribute to the development of human diseases; and whether or not BAFF has the same role in the development of human B cells as it does in models of murine B-cell ontogeny. These unresolved issues, together with the targeting of the BAFF/APRIL/receptor axis as a means of treating autoimmunity and B-cell malignancy [78, 80, 154–156], will ensure a continued focus on the biology and regulation of function of members of the BAFF family on immune responses for at least the next decade.

References

1. Uckun FM. Regulation of human B-cell ontogeny. Blood 1990;76:1908–23.
2. Burrows PD, Cooper MD. B-cell development in man. Curr Opin Immunol 1993;5:201–6.
3. Banchereau J, Rousset F. Human B lymphocytes: phenotype, proliferation, and differentiation. Adv Immunol 1992;52:125–262.
4. Allman D, Lindsley RC, DeMuth W, Rudd K, Shinton SA, Hardy RR. Resolution of three nonproliferative immature splenic B cell subsets reveals multiple selection points during peripheral B cell maturation. J Immunol 2001;167:6834–40.
5. Carsetti R, Rosado MM, Wardmann H. Peripheral development of B cells in mouse and man. Immunol Rev 2004;197:179–91.
6. Liu YJ, Banchereau J. The paths and molecular controls of peripheral B-cell development. The Immunologist 1996;4:55–66.
7. Van Kooten C, Banchereau J. CD40-CD40 ligand: a multifunctional receptor-ligand pair. Adv Immunol 1996;61:1–77.
8. Agematsu K, Nagumo H, Oguchi Y, Nakazawa T, Fukushima K, Yasui K, Ito S, Kobata T, Morimoto C, Komiyama A. Generation of plasma cells from peripheral blood memory B cells: synergistic effect of interleukin-10 and CD27/CD70 interaction. Blood 1998;91:173–80.
9. Stuber E, Strober W. The T cell-B cell interaction via OX40-OX40L is necessary for the T cell-dependent humoral immune response. J Exp Med 1996;183:979–89.
10. Aversa G, Punnonen J, de Vries JE. The 26-kD transmembrane form of tumor necrosis factor alpha on activated CD4+ T cell clones provides a costimulatory signal for human B cell activation. J Exp Med 1993;177:1575–85.
11. Cerutti A, Schaffer A, Shah S, Zan H, Liou HC, Goodwin RG, Casali P. CD30 is a CD40-inducible molecule that negatively regulates CD40-mediated immunoglobulin class switching in non-antigen-selected human B cells. Immunity 1998;9:247–56.
12. Schattner EJ, Elkon KB, Yoo DH, Tumang J, Krammer PH, Crow MK, Friedman SM. CD40 ligation induces Apo-1/Fas expression on human B lymphocytes and facilitates apoptosis through the Apo-1/Fas pathway. J Exp Med 1995;182:1557–65.
13. Schneider P, MacKay F, Steiner V, Hofmann K, Bodmer JL, Holler N, Ambrose C, Lawton P, Bixler S, Acha-Orbea H, Valmori D, Romero P, Werner-Favre C, Zubler RH, Browning

JL, Tschopp J. BAFF, a novel ligand of the tumor necrosis factor family, stimulates B cell growth. J Exp Med 1999;189:1747–56.

14. Moore PA, Belvedere O, Orr A, Pieri K, LaFleur DW, Feng P, Soppet D, Charters M, Gentz R, Parmelee D, Li Y, Galperina O, Giri J, Roschke V, Nardelli B, Carrell J, Sosnovtseva S, Greenfield W, Ruben SM, Olsen HS, Fikes J, Hilbert DM. BLyS: member of the tumor necrosis factor family and B lymphocyte stimulator. Science 1999;285:260–3.
15. Shu HB, Hu WH, Johnson H. TALL-1 is a novel member of the TNF family that is down-regulated by mitogens. J Leukoc Biol 1999;65:680–3.
16. Mukhopadhyay A, Ni J, Zhai Y, Yu GL, Aggarwal BB. Identification and characterization of a novel cytokine, THANK, a TNF homologue that activates apoptosis, nuclear factor-kappaB, and c-Jun NH2-terminal kinase. J Biol Chem 1999;274:15978–81.
17. Hahne M, Kataoka T, Schroter M, Hofmann K, Irmler M, Bodmer JL, Schneider P, Bornand T, Holler N, French LE, Sordat B, Rimoldi D, Tschopp J. APRIL, a new ligand of the tumor necrosis factor family, stimulates tumor cell growth. J Exp Med 1998;188: 1185–90.
18. Mackay F, Browning JL. BAFF: a fundamental survival factor for B cells. Nat Rev Immunol 2002;2:465–75.
19. Mackay F, Schneider P, Rennert P, Browning JL. BAFF and APRIL: a tutorial on B cell survival. Ann Rev Immunol 2003;21:231–64.
20. Tangye SG, Bryant VL, Cuss AK, Good KL. BAFF, APRIL and human B cell disorders. Semin Immunol 2006;18:305–17.
21. Nardelli B, Belvedere O, Roschke V, Moore PA, Olsen HS, Migone TS, Sosnovtseva S, Carrell JA, Feng P, Giri JG, Hilbert DM. Synthesis and release of B-lymphocyte stimulator from myeloid cells. Blood 2001;97:198–204.
22. Litinskiy MB, Nardelli B, Hilbert DM, He B, Schaffer A, Casali P, Cerutti A. DCs induce CD40-independent immunoglobulin class switching through BLyS and APRIL. Nat Immunol 2002;3:822–9.
23. Craxton A, Magaletti D, Ryan EJ, Clark EA. Macrophage- and dendritic cell-dependent regulation of human B-cell proliferation requires the TNF family ligand BAFF. Blood 2003;101:4464–71.
24. Krumbholz M, Theil D, Derfuss T, Rosenwald A, Schrader F, Monoranu CM, Kalled SL, Hess DM, Serafini B, Aloisi F, Wekerle H, Hohlfeld R, Meinl E. BAFF is produced by astrocytes and up-regulated in multiple sclerosis lesions and primary central nervous system lymphoma. J Exp Med 2005;201:195–200.
25. Scapini P, Nardelli B, Nadali G, Calzetti F, Pizzolo G, Montecucco C, Cassatella MA. G-CSF-stimulated neutrophils are a prominent source of functional BLyS. J Exp Med 2003;197:297–302.
26. Ng LG, Sutherland AP, Newton R, Qian F, Cachero TG, Scott ML, Thompson JS, Wheway J, Chtanova T, Groom J, Sutton IJ, Xin C, Tangye SG, Kalled SL, Mackay F, Mackay CR. B cell-activating factor belonging to the TNF family (BAFF)-R is the principal BAFF receptor facilitating BAFF costimulation of circulating T and B cells. J Immunol 2004; 173:807–17.
27. Kato A, Truong-Tran AQ, Scott AL, Matsumoto K, Schleimer RP. Airway epithelial cells produce B cell-activating factor of TNF family by an IFN-beta-dependent mechanism. J Immunol 2006;177:7164–72.
28. Xu W, He B, Chiu A, Chadburn A, Shan M, Buldys M, Ding A, Knowles DM, Santini PA, Cerutti A. Epithelial cells trigger frontline immunoglobulin class switching through a pathway regulated by the inhibitor SLPI. Nat Immunol 2007;8:294–303.
29. He B, Xu W, Santini PA, Polydorides AD, Chiu A, Estrella J, Shan M, Chadburn A, Villanacci V, Plebani A, Knowles DM, Rescigno M, Cerutti A. Intestinal bacteria trigger T cell-independent immunoglobulin A(2) class switching by inducing epithelial-cell secretion of the cytokine APRIL. Immunity 2007;26:812–26.
30. Zhang X, Park CS, Yoon SO, Li L, Hsu YM, Ambrose C, Choi YS. BAFF supports human B cell differentiation in the lymphoid follicles through distinct receptors. Int Immunol 2005;17:779–88.

31. Scapini P, Carletto A, Nardelli B, Calzetti F, Roschke V, Merigo F, Tamassia N, Pieropan S, Biasi D, Sbarbati A, Sozzani S, Bambara L, Cassatella MA. Proinflammatory mediators elicit secretion of the intracellular B-lymphocyte stimulator pool (BLyS) that is stored in activated neutrophils: implications for inflammatory diseases. Blood 2005;105:830–7.
32. He B, Raab-Traub N, Casali P, Cerutti A. EBV-encoded latent membrane protein 1 cooperates with BAFF/BLyS and APRIL to induce T cell-independent Ig heavy chain class switching. J Immunol 2003;171:5215–24.
33. Fu L, Lin-Lee YC, Pham LV, Tamayo A, Youshimura L, Ford RJ. Constitutive NF-{kappa}B and NFAT activation leads to stimulation of The BLyS survival pathway in aggressive B cell lymphomas. Blood 2006;107:4540–8.
34. Thompson JS, Bixler SA, Qian F, Vora K, Scott ML, Cachero TG, Hession C, Schneider P, Sizing ID, Mullen C, Strauch K, Zafari M, Benjamin CD, Tschopp J, Browning JL, Ambrose C. BAFF-R, a newly identified TNF receptor that specifically interacts with BAFF. Science 2001;293:2108–11.
35. Yan M, Brady JR, Chan B, Lee WP, Hsu B, Harless S, Cancro M, Grewal IS, Dixit VM. Identification of a novel receptor for B lymphocyte stimulator that is mutated in a mouse strain with severe B cell deficiency. Curr Biol 2001;11:1547–52.
36. Gross JA, Johnston J, Mudri S, Enselman R, Dillon SR, Madden K, Xu W, Parrish-Novak J, Foster D, Lofton-Day C, Moore M, Littau A, Grossman A, Haugen H, Foley K, Blumberg H, Harrison K, Kindsvogel W, Clegg CH. TACI and BCMA are receptors for a TNF homologue implicated in B-cell autoimmune disease. Nature 2000;404:995–9.
37. Marsters SA, Yan M, Pitti RM, Haas PE, Dixit VM, Ashkenazi A. Interaction of the TNF homologues BLyS and APRIL with the TNF receptor homologues BCMA and TACI. Curr Biol 2000;10:785–8.
38. Yu G, Boone T, Delaney J, Hawkins N, Kelley M, Ramakrishnan M, McCabe S, Qiu W-R, Kornuc M, Xia X-Z, Guo J, Stolina M, Boyle WJ, Sarosi I, Hsu H, Senaldi G, Theill LE. APRIL and TALL-1 and receptors BCMA and TACI: system for regulating humoral immunity. Nature Immunol. 2000;1:252–6.
39. Xia XZ, Treanor J, Senaldi G, Khare SD, Boone T, Kelley M, Theill LE, Colombero A, Solovyev I, Lee F, McCabe S, Elliott R, Miner K, Hawkins N, Guo J, Stolina M, Yu G, Wang J, Delaney J, Meng SY, Boyle WJ, Hsu H. TACI is a TRAF-interacting receptor for TALL-1, a tumor necrosis factor family member involved in B cell regulation. J Exp Med 2000;192:137–43.
40. Thompson JS, Schneider P, Kalled SL, Wang L, Lefevre EA, Cachero TG, MacKay F, BixlerSA, Zafari M, Liu ZY, Woodcock SA, Qian F, Batten M, Madry C, Richard Y, Benjamin CD, Browning JL, Tsapis A, Tschopp J, Ambrose C. BAFF binds to the tumor necrosis factor receptor-like molecule B cell maturation antigen and is important for maintaining the peripheral B cell population. J Exp Med 2000;192:129–35.
41. Ingold K, Zumsteg A, Tardivel A, Huard B, Steiner QG, Cachero TG, Qiang F, Gorelik L, Kalled SL, Acha-Orbea H, Rennert PD, Tschopp J, Schneider P. Identification of proteoglycans as the APRIL-specific binding partners. J Exp Med 2005;201:1375–83.
42. Gorelik L, Cutler AH, Thill G, Miklasz SD, Shea DE, Ambrose C, Bixler SA, Su L, Scott ML, Kalled SL. Cutting Edge: BAFF regulates CD21/35 and CD23 expression independent of its B cell survival function. J Immunol 2004;172:762–6.
43. Hsu BL, Harless SM, Lindsley RC, Hilbert DM, Cancro MP. Cutting edge: BLyS enables survival of transitional and mature B cells through distinct mediators. J Immunol 2002;168:5993–6.
44. O'Connor BP, Raman VS, Erickson LD, Cook WJ, Weaver LK, Ahonen C, Lin L, Mantchev GT, Bram RJ, Noelle RJ. BCMA is essential for the survival of long-lived bone marrow plasma cells. J Exp Med 2004;199:91–7.
45. Avery DT, Kalled SL, Ellyard JI, Ambrose C, Bixler SA, Thien M, Brink R, Mackay F, Hodgkin PD, Tangye SG. BAFF selectively enhances the survival of plasmablasts generated from human memory B cells. J Clin Invest 2003;112:286–97.

46. Ellyard JI, Avery DT, Mackay CR, Tangye SG. Contribution of stromal cells to the migration, function and retention of plasma cells in human spleen: potential roles of CXCL12, IL-6 and CD54. Eur J Immunol 2005;35:699–708.
47. Novak AJ, Darce JR, Arendt BK, Harder B, Henderson K, Kindsvogel W, Gross JA, Greipp PR, Jelinek DF. Expression of BCMA, TACI, and BAFF-R in multiple myeloma: a mechanism for growth and survival. Blood 2004;103:689–94.
48. Cuss AK, Avery DT, Cannons JL, Yu LJ, Nichols KE, Shaw PJ, Tangye SG. Expansion of functionally immature transitional B cells is associated with human-immunodeficient states characterized by impaired humoral immunity. J Immunol 2006;176:1506–16.
49. Novak AJ, Bram RJ, Kay NE, Jelinek DF. Aberrant expression of B-lymphocyte stimulator by B chronic lymphocytic leukemia cells: a mechanism for survival. Blood 2002;100: 2973–9.
50. He B, Chadburn A, Jou E, Schattner EJ, Knowles DM, Cerutti A. Lymphoma B cells evade apoptosis through the TNF family members BAFF/BLyS and APRIL. J Immunol 2004;172:3268–79.
51. Darce JR, Arendt BK, Chang SK, Jelinek DF. Divergent effects of BAFF on human memory B cell differentiation into Ig-secreting cells. J Immunol 2007;178:5612–22.
52. Darce JR, Arendt BK, Wu X, Jelinek DF. Regulated expression of BAFF-binding receptors during human B cell differentiation. J Immunol 2007;179:7276–86.
53. Sims GP, Ettinger R, Shirota Y, Yarboro CH, Illei GG, Lipsky PE. Identification and characterization of circulating human transitional B cells. Blood 2005;105:4390–8.
54. Chiu A, Xu W, He B, Dillon SR, Gross JA, Sievers E, Qiao X, Santini P, Hyjek E, Lee JW, Cesarman E, Chadburn A, Knowles DM, Cerutti A. Hodgkin lymphoma cells express TACI and BCMA receptors and generate survival and proliferation signals in response to BAFF and APRIL. Blood 2007;109:729–39.
55. Avery DT, Ellyard JI, Mackay F, Corcoran LM, Hodgkin PD, Tangye SG. Increased expression of CD27 on activated human memory B cells correlates with their commitment to the plasma cell lineage. J Immunol 2005;174:4034–42.
56. Moreaux J, Legouffe E, Jourdan E, Quittet P, Reme T, Lugagne C, Moine P, Rossi JF, Klein B, Tarte K. BAFF and APRIL protect myeloma cells from apoptosis induced by interleukin 6 deprivation and dexamethasone. Blood 2004;103:3148–57.
57. Do RK, Hatada E, Lee H, Tourigny MR, Hilbert D, Chen-Kiang S. Attenuation of apoptosis underlies B lymphocyte stimulator enhancement of humoral immune response. J Exp Med 2000;192:953–64.
58. Mongini PK, Inman JK, Han H, Fattah RJ, Abramson SB, Attur M. APRIL and BAFF promote increased viability of replicating human B2 cells via mechanism involving cyclooxygenase 2. J Immunol 2006;176:6736–51.
59. Amanna IJ, Clise-Dwyer K, Nashold FE, Hoag KA, Hayes CE. Cutting edge: A/WySnJ transitional B cells overexpress the chromosome 15 proapoptotic Blk gene and succumb to premature apoptosis. J Immunol 2001;167:6069–72.
60. Craxton A, Draves KE, Gruppi A, Clark EA. BAFF regulates B cell survival by downregulating the BH3-only family member Bim via the ERK pathway. J Exp Med 2005;202:1363–74.
61. Lesley R, Xu Y, Kalled SL, Hess DM, Schwab SR, Shu HB, Cyster JG. Reduced Competitiveness of Autoantigen-Engaged B Cells due to Increased Dependence on BAFF. Immunity 2004;20:441–53.
62. Bossen C, Cachero TG, Tardivel A, Ingold K, Willen L, Dobles M, Scott ML, Maquelin A, Belnoue E, Siegrist CA, Chevrier S, Acha-Orbea H, Leung H, Mackay F, Tschopp J, Schneider P. TACI, unlike BAFF-R, is solely activated by oligomeric BAFF and APRIL to support survival of activated B cells and plasmablasts. Blood 2008;111: 1004–12.
63. Gorelik L, Gilbride K, Dobles M, Kalled SL, Zandman D, Scott ML. Normal B cell homeostasis requires B cell activation factor production by radiation-resistant cells. J Exp Med 2003;198:937–45.

64. von Bulow G-U, van Deursen JM, Bram RJ. Regulation of the T-independent humoral response by TACI. Immunity 2001;14:573–82.
65. Yan M, Wang H, Chan B, Roose-Girma M, Erickson S, Baker T, Tumas D, Grewal IS, Dixit VM. Activation and accumulation of B cells in TACI-deficient mice. Nat Immunol 2001;2:638–43.
66. Xu S, Lam KP. B-cell maturation protein, which binds the tumor necrosis factor family members BAFF and APRIL, is dispensable for humoral immune responses. Molec Cell Biol 2001;21:4067–74.
67. Schneider P, Takatsuka H, Wilson A, Mackay F, Tardivel A, Lens S, Cachero TG, Finke D, Beermann F, Tschopp J. Maturation of marginal zone and follicular B cells requires B cell activating factor of the tumor necrosis factor family and is independent of B cell maturation antigen. J Exp Med 2001;194:1691–7.
68. Schiemann B, Gommerman JL, Vora K, Cachero TG, Shulga-Morskaya S, Dobles M, Frew E, Scott ML. An essential role for BAFF in the normal development of B cells through a BCMA-independent pathway. Science 2001;293:2111–4.
69. Shulga-Morskaya S, Dobles M, Walsh ME, Ng LG, MacKay F, Rao SP, Kalled SL, Scott ML. B cell-activating factor belonging to the TNF family acts through separate receptors to support B cell survival and T cell-independent antibody formation. J Immunol 2004;173:2331–41.
70. Sasaki Y, Casola S, Kutok JL, Rajewsky K, Schmidt-Supprian M. TNF family member B cell-activating factor (BAFF) receptor-dependent and -independent roles for BAFF in B cell physiology. J Immunol 2004;173:2245–52.
71. Amanna IJ, Dingwall JP, Hayes CE. Enforced bcl-xL gene expression restored splenic B lymphocyte development in BAFF-R mutant mice. J Immunol 2003;170:4593–600.
72. Batten M, Groom J, Cachero TG, Qian F, Schneider P, Tschopp J, Browning JL, Mackay F. BAFF mediates survival of peripheral immature B lymphocytes. J Exp Med 2000;192: 1453–66.
73. Harless-Smith S, Cancro MP. Cutting edge: B cell receptor signals regulate BLyS receptor levels in mature B cells and their immediate progenitors. J Immunol 2003;170:5820–3.
74. Walmsley MJ, Ooi SK, Reynolds LF, Smith SH, Ruf S, Mathiot A, Vanes L, Williams DA, Cancro MP, Tybulewicz VL. Critical roles for Rac1 and Rac2 GTPases in B cell development and signaling. Science 2003;302:459–62.
75. Rolink AG, Tschopp J, Schneider P, Melchers F. BAFF is a survival and maturation factor for mouse B cells. Eur J Immunol 2002;32:2004–10.
76. Malaspina A, Moir S, Ho J, Wang W, Howell ML, O'Shea MA, Roby GA, Rehm CA, Mican JM, Chun TW, Fauci AS. Appearance of immature/transitional B cells in HIV-infected individuals with advanced disease: correlation with increased IL-7. Proc Natl Acad Sci USA 2006;103:2262–7.
77. Mongini PK, Inman JK, Han H, Kalled SL, Fattah RJ, McCormick S. Innate immunity and human B cell clonal expansion: effects on the recirculating B2 subpopulation. J Immunol 2005;175:6143–54.
78. Baker KP, Edwards BM, Main SH, Choi GH, Wager RE, Halpern WG, Lappin PB, Riccobene T, Abramian D, Sekut L, Sturm B, Poortman C, Minter RR, Dobson CL, Williams E, Carmen S, Smith R, Roschke V, Hilbert DM, Vaughan TJ, Albert VR. Generation and characterization of LymphoStat-B, a human monoclonal antibody that antagonizes the bioactivities of B lymphocyte stimulator. Arthritis Rheum 2003;48:3253–65.
79. Vugmeyster Y, Seshasayee D, Chang W, Storn A, Howell K, Sa S, Nelson T, Martin F, Grewal I, Gilkerson E, Wu B, Thompson J, Ehrenfels BN, Ren S, Song A, Gelzleichter TR, Danilenko DM. A soluble BAFF antagonist, BR3-Fc, decreases peripheral blood B cells and lymphoid tissue marginal zone and follicular B cells in cynomolgus monkeys. Am J Pathol 2006;168:476–89.
80. Martin F, Chan AC. B cell immunobiology in disease: evolving concepts from the clinic. Annu Rev Immunol 2006;24:467–96.

81. Balazs M, Martin F, Zhou T, Kearney J. Blood dendritic cells interact with splenic marginal zone B cells to initiate T-independent immune responses. Immunity 2002;17:341–52.
82. Belnoue E, Pihlgren M, McGaha TL, Tougne C, Rochat AF, Bossen C, Schneider P, Huard B, Lambert PH, Siegrist CA. APRIL is critical for plasmablast survival in the bone marrow and poorly expressed by early-life bone marrow stromal cells. Blood 2008;111: 2755–64.
83. Badr G, Borhis G, Lefevre EA, Chaoul N, Deshayes F, Dessirier V, Lapree G, Tsapis A, Richard Y. BAFF enhances chemotaxis of primary human B cells: a particular synergy between BAFF and CXCL13 on memory B cells. Blood 2008;111:2744–54.
84. Honjo T, Kinoshita K, Muramatsu M. Molecular mechanism of class switch recombination: linkage with somatic hypermutation. Annu Rev Immunol 2002;20:165–96.
85. Cerutti A. The regulation of IgA class switching. Nat Rev Immunol 2008;8:421–34.
86. Geha RS, Jabara HH, Brodeur SR. The regulation of immunoglobulin E class-switch recombination. Nat Rev Immunol 2003;3:721–32.
87. Xu LG, Wu M, Hu J, Zhai Z, Shu HB. Identification of downstream genes up-regulated by the tumor necrosis factor family member TALL-1. J Leukoc Biol 2002;72:410–6.
88. He B, Qiao X, Cerutti A. CpG DNA induces IgG class switch DNA recombination by activating human B cells through an innate pathway that requires TLR9 and cooperates with IL-10. J Immunol 2004;173:4479–91.
89. Xu W, Santini PA, Matthews AJ, Chiu A, Plebani A, He B, Chen K, Cerutti A. Viral double-stranded RNA triggers Ig class switching by activating upper respiratory mucosa B cells through an innate TLR3 pathway involving BAFF. J Immunol 2008;181:276–87.
90. Castigli E, Scott S, Dedeoglu F, Bryce P, Jabara H, Bhan AK, Mizoguchi E, Geha RS. Impaired IgA class switching in APRIL-deficient mice. Proc Natl Acad Sci USA 2004;101:3903–8.
91. Castigli E, Wilson SA, Scott S, Dedeoglu F, Xu S, Lam KP, Bram RJ, Jabara H, Geha RS. TACI and BAFF-R mediate isotype switching in B cells. J Exp Med 2005;201:35–9.
92. Sakurai D, Hase H, Kanno Y, Kojima H, Okumura K, Kobata T. TACI regulates IgA production by APRIL in collaboration with HSPG. Blood 2007;109:2961–7.
93. Castigli E, Wilson SA, Garibyan L, Rachid R, Bonilla F, Schneider L, Geha RS. TACI is mutant in common variable immunodeficiency and IgA deficiency. Nat Genet 2005;37: 829–34.
94. Groom J, Kalled SL, Cutler AH, Olson C, Woodcock SA, Schneider P, Tschopp J, Cachero TG, Batten M, Wheway J, Mauri D, Cavill D, Gordon TP, Mackay CR, Mackay F. Association of BAFF/BLyS overexpression and altered B cell differentiation with Sjogren's syndrome. J Clin Invest 2002;109:59–68.
95. Cheema GS, Roschke V, Hilbert DM, Stohl W. Elevated serum B lymphocyte stimulator levels in patients with systemic immune-based rheumatic diseases. Arthritis Rheum 2001;44:1313–9.
96. Zhang J, Roschke V, Baker KP, Wang Z, Alarcon GS, Fessler BJ, Bastian H, Kimberly RP, Zhou T. Cutting edge: a role for B lymphocyte stimulator in systemic lupus erythematosus. J Immunol 2001;166:6–10.
97. Mariette X, Roux S, Zhang J, Bengoufa D, Lavie F, Zhou T, Kimberly R. The level of BLyS (BAFF) correlates with the titre of autoantibodies in human Sjogren's syndrome. Ann Rheum Dis 2003;62:168–71.
98. Jonsson MV, Szodoray P, Jellestad S, Jonsson R, Skarstein K. Association between circulating levels of the novel TNF family members APRIL and BAFF and lymphoid organization in primary Sjogren's syndrome. J Clin Immunol 2005;25:189–201.
99. Szodoray P, Jonsson R. The BAFF/APRIL system in systemic autoimmune diseases with a special emphasis on Sjogren's syndrome. Scand J Immunol 2005;62:421–8.
100. Koyama T, Tsukamoto H, Miyagi Y, Himeji D, Otsuka J, Miyagawa H, Harada M, Horiuchi T. Raised serum APRIL levels in patients with systemic lupus erythematosus. Ann Rheum Dis 2005;64:1065–7.

101. Seyler TM, Park YW, Takemura S, Bram RJ, Kurtin PJ, Goronzy JJ, Weyand CM. BLyS and APRIL in rheumatoid arthritis. J Clin Invest 2005;115:3083–92.
102. Roschke V, Sosnovtseva S, Ward CD, Hong JS, Smith R, Albert V, Stohl W, Baker KP, Ullrich S, Nardelli B, Hilbert DM, Migone TS. BLyS and APRIL form biologically active heterotrimers that are expressed in patients with systemic immune-based rheumatic diseases. J Immunol 2002;169:4314–21.
103. Matsushita T, Hasegawa M, Matsushita Y, Echigo T, Wayaku T, Horikawa M, Ogawa F, Takehara K, Sato S. Elevated serum BAFF levels in patients with localized scleroderma in contrast to other organ-specific autoimmune diseases. Exp Dermatol 2007;16:87–93.
104. Matsushita T, Hasegawa M, Yanaba K, Kodera M, Takehara K, Sato S. Elevated serum BAFF levels in patients with systemic sclerosis: enhanced BAFF signaling in systemic sclerosis B lymphocytes. Arthritis Rheum 2006;54:192–201.
105. Matsushita T, Fujimoto M, Hasegawa M, Tanaka C, Kumada S, Ogawa F, Takehara K, Sato S. Elevated serum APRIL levels in patients with systemic sclerosis: distinct profiles of systemic sclerosis categorized by APRIL and BAFF. J Rheumatol 2007;34:2056–62.
106. Matsushita T, Fujimoto M, Echigo T, Matsushita Y, Shimada Y, Hasegawa M, Takehara K, Sato S. Elevated serum levels of APRIL, but not BAFF, in patients with atopic dermatitis. Exp Dermatol 2008;17:197–202.
107. Asashima N, Fujimoto M, Watanabe R, Nakashima H, Yazawa N, Okochi H, Tamaki K. Serum levels of BAFF are increased in bullous pemphigoid but not in pemphigus vulgaris. Br J Dermatol 2006;155:330–6.
108. Watanabe R, Fujimoto M, Yazawa N, Nakashima H, Asashima N, Kuwano Y, Tada Y, Maruyama N, Okochi H, Tamaki K. Increased serum levels of a proliferation-inducing ligand in patients with bullous pemphigoid. J Dermatol Sci 2007;46:53–60.
109. Krumbholz M, Specks U, Wick M, Kalled SL, Jenne D, Meinl E. BAFF is elevated in serum of patients with Wegener's granulomatosis. J Autoimmun 2005;25:298–302.
110. Hever A, Roth RB, Hevezi P, Marin ME, Acosta JA, Acosta H, Rojas J, Herrera R, Grigoriadis D, White E, Conlon PJ, Maki RA, Zlotnik A. Human endometriosis is associated with plasma cells and overexpression of B lymphocyte stimulator. Proc Natl Acad Sci USA 2007;104:12451–6.
111. Lavie F, Miceli-Richard C, Quillard J, Roux S, Leclerc P, Mariette X. Expression of BAFF (BLyS) in T cells infiltrating labial salivary glands from patients with Sjogren's syndrome. J Pathol 2004;202:496–502.
112. Thangarajh M, Gomes A, Masterman T, Hillert J, Hjelmstrom P. Expression of B-cell-activating factor of the TNF family (BAFF) and its receptors in multiple sclerosis. J Neuroimmunol 2004;152:183–90.
113. Banchereau J, Pascual V, Palucka AK. Autoimmunity through cytokine-induced dendritic cell activation. Immunity 2004;20:539–50.
114. Llorente L, Richaud-Patin Y. The role of interleukin-10 in systemic lupus erythematosus. J Autoimmun 2003;20:287–9.
115. Szodoray P, Alex P, Jonsson MV, Knowlton N, Dozmorov I, Nakken B, Delaleu N, Jonsson R, Centola M. Distinct profiles of Sjogren's syndrome patients with ectopic salivary gland germinal centers revealed by serum cytokines and BAFF. Clin Immunol 2005;117:168–76.
116. Llorente L, Richaud-Patin Y, Fior R, Alcocer-Varela J, Wijdenes J, Fourrier BM, Galanaud P, Emilie D. In vivo production of interleukin-10 by non-T cells in rheumatoid arthritis, Sjogren's syndrome, and systemic lupus erythematosus. A potential mechanism of B lymphocyte hyperactivity and autoimmunity. Arthritis Rheum 1994;37:1647–55.
117. Desai-Mehta A, Lu L, Ramsey-Goldman R, Datta SK. Hyperexpression of CD40 ligand by B and T cells in human lupus and its role in pathogenic autoantibody production. J Clin Invest 1996;97:2063–73.
118. Berek C, Kim HJ. B-cell activation and development within chronically inflamed synovium in rheumatoid and reactive arthritis. Semin Immunol 1997;9:261–8.

119. Takemura S, Braun A, Crowson C, Kurtin PJ, Cofield RH, O'Fallon WM, Goronzy JJ, Weyand CM. Lymphoid neogenesis in rheumatoid synovitis. J Immunol 2001;167:1072–80.
120. Ansel KM, Cyster JG. Chemokines in lymphopoiesis and lymphoid organ development. Curr Opin Immunol 2001;13:172–9.
121. Kern C, Cornuel JF, Billard C, Tang R, Rouillard D, Stenou V, Defrance T, Ajchenbaum-Cymbalista F, Simonin PY, Feldblum S, Kolb JP. Involvement of BAFF and APRIL in the resistance to apoptosis of B-CLL through an autocrine pathway. Blood 2004;103:679–88.
122. Briones J, Timmerman JM, Hilbert DM, Levy R. BLyS and BLyS receptor expression in non-Hodgkin's lymphoma. Exp Hematol 2002;30:135–41.
123. Novak AJ, Grote DM, Stenson M, Ziesmer SC, Witzig TE, Habermann TM, Harder B, Ristow KM, Bram RJ, Jelinek DF, Gross JA, Ansell SM. Expression of BLyS and its receptors in B-cell non-Hodgkin lymphoma: correlation with disease activity and patient outcome. Blood 2004;104:2247–53.
124. Elsawa SF, Novak AJ, Grote DM, Ziesmer SC, Witzig TE, Kyle RA, Dillon SR, Harder B, Gross JA, Ansell SM. B-lymphocyte stimulator (BLyS) stimulates immunoglobulin production and malignant B-cell growth in Waldenstrom's macroglobulinemia. Blood 2006;107:2882–8.
125. Moreaux J, Cremer FW, Reme T, Raab M, Mahtouk K, Kaukel P, Pantesco V, De Vos J, Jourdan E, Jauch A, Legouffe E, Moos M, Fiol G, Goldschmidt H, Rossi JF, Hose D, Klein B. The level of TACI gene expression in myeloma cells is associated with a signature of microenvironment dependence versus a plasmablastic signature. Blood 2005;106:1021–30.
126. Nishio M, Endo T, Tsukada N, Ohata J, Kitada S, Reed JC, Zvaifler NJ, Kipps TJ. Nurselike cells express BAFF and APRIL, which can promote survival of chronic lymphocytic leukemia cells via a paracrine pathway distinct from that of SDF-1alpha. Blood 2005;106:1012–20.
127. Planelles L, Carvalho-Pinto CE, Hardenberg G, Smaniotto S, Savino W, Gomez-Caro R, Alvarez-Mon M, de Jong J, Eldering E, Martinez AC, Medema JP, Hahne M. APRIL promotes B-1 cell-associated neoplasm. Cancer Cell 2004;6:399–408.
128. Schwaller J, Schneider P, Mhawech-Fauceglia P, McKee T, Myit S, Matthes T, Tschopp J, Donze O, Le Gal FA, Huard B. Neutrophil-derived APRIL concentrated in tumor lesions by proteoglycans correlates with human B-cell lymphoma aggressiveness. Blood 2007;109:331–8.
129. Kuppers, R. B cells under influence: transformation of B cells by Epstein-Barr virus. Nature Rev Immunol 2003;3:801–12.
130. Ogden CA, Pound JD, Batth BK, Owens S, Johannessen I, Wood K, Gregory CD. Enhanced apoptotic cell clearance capacity and B cell survival factor production by IL-10-activated macrophages: implications for Burkitt's lymphoma. J Immunol 2005;174:3015–23.
131. Cunningham-Rundles C, Bodian C. Common variable immunodeficiency: clinical and immunological features of 248 patients. Clin Immunol 1999;92:34–48.
132. Di Renzo M, Pasqui AL, Auteri A. Common variable immunodeficiency: a review. Clin Exp Med 2004;3:211–7.
133. Salzer U, Maul-Pavicic A, Cunningham-Rundles C, Urschel S, Belohradsky BH, Litzman J, Holm A, Franco JL, Plebani A, Hammarstrom L, Skrabl A, Schwinger W, Grimbacher B. ICOS deficiency in patients with common variable immunodeficiency. Clin Immunol 2004;113:234–40.
134. Warnatz K, Bossaller L, Salzer U, Skrabl-Baumgartner A, Schwinger W, van der Burg M, van Dongen JJM, Orlowska-Volk M, Knoth R, Durandy A, Draeger R, Schlesier M, Peter HH, Grimbacher B. Human ICOS deficiency abrogates the germinal center reaction and provides a monogenic model for common variable immunodeficiency. Blood 2006;107:3045–52.
135. Kanegane H, Agematsu K, Futatani T, Sira MM, Suga K, Sekiguchi T, van Zelm MC, Miyawaki T. Novel mutations in a Japanese patient with CD19 deficiency. Genes Immun 2007;8:663–70.

136. van Zelm MC, Reisli I, van der Burg M, Castano D, van Noesel CJM, van Tol MJD, Woellner C, Grimbacher B, Patino PJ, van Dongen JJM, Franco JL. An antibody-deficiency syndrome due to mutations in the CD19 gene. [see comment]. N Engl J Med 2006;354:1901–12.
137. Kanegane H. Tsukada S, Iwata T, Futatani T, Nomura K, Yamamoto J, Yoshida T, Agematsu K, Komiyama A, Miyawaki T. Detection of Bruton's tyrosine kinase mutations in hypogammaglobulinaemic males registered as common variable immunodeficiency (CVID) in the Japanese Immunodeficiency Registry. Clin Exp Immunol 2000;120:512–7.
138. Weston SA, Prasad ML, Mullighan CG, Chapel H, Benson EM. Assessment of male CVID patients for mutations in the Btk gene: how many have been misdiagnosed? Clin Exp Immunol 2001;124:465–9.
139. Eastwood D, Gilmour KC, Nistala K, Meaney C, Chapel H, Sherrell Z, Webster AD, Davies EG, Jones A, Gaspar HB. Prevalence of SAP gene defects in male patients diagnosed with common variable immunodeficiency. Clin Exp Immunol 2004;137:584–8.
140. Morra M, Silander O, Calpe S, Choi M, Oettgen H, Myers L, Etzioni A, Buckley R, and Terhorst C. Alterations of the X-linked lymphoproliferative disease gene SH2D1A in common variable immunodeficiency syndrome. Blood 2001;98:1321–25.
141. Soresina A, Lougaris V, Giliani S, Cardinale F, Armenio L, Cattalini M, Notarangelo LD, Plebani A. Mutations of the X-linked lymphoproliferative disease gene SH2D1A mimicking common variable immunodeficiency. Eur J Pediatr 2002;161:656–9.
142. Salzer U, Chapel HM, Webster AD, Pan-Hammarstrom Q, Schmitt-Graeff A, Schlesier M, Peter HH, Rockstroh JK, Schneider P, Schaffer AA, Hammarstrom L, Grimbacher B. Mutations in TNFRSF13B encoding TACI are associated with common variable immunodeficiency in humans. Nat Genet 2005;37:820–8.
143. Castigli E, Wilson SA, Gariby L, Rachid R, Bonilla F, Schneider L, Geha RS. TACI is mutant in common variable immunodeficiency and IgA deficiency. Nat Genet 2005;37:829–34.
144. Pan-Hammarstrom Q, Salzer U, Du L, Bjorkander J, Cunningham-Rundles C, Nelson DL, Bacchelli C, Gaspar HB, Offer S, Behrens TW, Grimbacher B, Hammarstrom L. Reexamining the role of TACI coding variants in common variable immunodeficiency and selective IgA deficiency. Nat Genet 2007;39:429–30.
145. Castigli E, Wilson S, Garibyan L, Rachid R, Bonilla F, Schneider L, Morra M, Curran J, Geha R. Reexamining the role of TACI coding variants in common variable immunodeficiency and selective IgA deficiency. Nat Genet 2007;39:430–1.
146. Salzer U, Bacchelli C, Buckridge S, Pan-Hammarström Q, Jennings S, Lougaris V, Hagena T, Birmelin J, Plebani A, Webster ADB, Peter H-H, Suez D, Chapel H, Maclean-Tooke A, Spickett GP, Anover-Sombke S, Ochs HD, Urschel S, Belohradsky BH, Kumararatne DS, Lawrence TC, Holm AM, Franco JL, Schulze I, Schneider P, Hammarström L, Thrasher AJ, Gaspar HB, Grimbacher B.Relevance of biallelic versus monoallelic TNFRSF13B mutations in distinguishing disease causing from disease modifying TNFRSF13B variants in common variable immunodeficiency. 2008;113:1967–76.
147. von Bulow GU, van JM Deursen, Bram RJ. Regulation of the T-independent humoral response by TACI. Immunity 2001;14:573–82.
148. Seshasayee D, Valdez P, Yan M, Dixit VM, Tumas D, Grewal IS. Loss of TACI causes fatal lymphoproliferation and autoimmunity, establishing TACI as an inhibitory BLyS receptor. Immunity 2003;18:279–88.
149. Sakurai D, Kanno Y, Hase H, Kojima H, Okumura K, Kobata T. TACI attenuates antibody production costimulated by BAFF-R and CD40. Eur J Immunol 2007;37:110–8.
150. Garibyan L, Lobito AA, Siegel RM, Call ME, Wucherpfennig KW, Geha RS. Dominant-negative effect of the heterozygous C104R TACI mutation in common variable immunodeficiency (CVID). J Clin Invest 2007;117:1550–7.
151. Zhang L, Radigan L, Salzer U, Behrens TW, Grimbacher B, Diaz G, Bussel J, Cunningham-Rundles C. Transmembrane activator and calcium-modulating cyclophilin ligand interactor mutations in common variable immunodeficiency: clinical and immunologic outcomes in heterozygotes. J Allergy Clin Immunol 2007;120:1178–85.

152. Siegel RM, Frederiksen JK, Zacharias DA, Chan FK, Johnson M, Lynch D, Tsien RY, Lenardo MJ. Fas preassociation required for apoptosis signaling and dominant inhibition by pathogenic mutations. Science 2000;288:2354–7.
153. Durandy A, Taubenheim N, Peron S, Fischer A. Pathophysiology of B-cell intrinsic immunoglobulin class switch recombination deficiencies. Adv Immunol 2007;94:275–306.
154. Nimmanapalli R, Lyu MA, Du M, Keating MJ, Rosenblum MG, Gandhi V. The growth factor fusion construct containing B-lymphocyte stimulator (BLyS) and the toxin rGel induces apoptosis specifically in BAFF-R-positive CLL cells. Blood 2007;109:2557–64.
155. Lyu MA, Cheung LH, Hittelman WN, Marks JW, Aguiar RC, Rosenblum MG. The rGel/BLyS fusion toxin specifically targets malignant B cells expressing the BLyS receptors BAFF-R, TACI, and BCMA. Mol Cancer Ther 2007;6:460–70.
156. Lin WY, Gong Q, Seshasayee D, Lin Z, Ou Q, Ye S, Suto E, Shu J, Lee WP, Lee CW, Fuh G, Leabman M, Iyer S, Howell K, Gelzleichter T, Beyer J, Danilenko D, Yeh S, DeForge LE, Ebens A, Thompson JS, Ambrose C, Balazs M, Starovasnik MA, Martin F. Anti-BR3 antibodies: a new class of B-cell immunotherapy combining cellular depletion and survival blockade. Blood 2007;110:3959–67.

Chapter 10
Translation of BAFF Inhibition from Mouse to Non-human Primate and Human

Lachy McLean, Dhaya Seshasayee, Susan L. Kalled, and Flavius Martin

Abstract The identification of BAFF as a fundamental B cell survival factor in mouse and man, its over-expression in certain autoimmune disease patient populations, and the discovery of three cognate receptors, stimulated interest in understanding the role of BAFF in the pathogenesis of autoimmunity, and designing novel therapeutics to blunt B cell participation in disease pathogenesis via blockade of this pathway. Positive clinical trial results with B cell depletion have demonstrated beyond doubt the pathogenic B cell component in human rheumatoid arthritis [1, 2]. This current chapter focuses on data obtained in mice (both normal and disease models), non-human primates and human clinical trials with several large molecule inhibitors of the BAFF B cell survival pathway. The focus is on normal, non-diseased animals as well as rheumatology diseases and disease models although implications are clear for all other diseases with a B cell pathogenic component. Three strategies for blocking BAFF/receptor interactions have been used in mice and primates, 1) blocking antibodies directed specifically against BAFF, 2) receptor/receptor fragment- Fc fusion proteins (receptor:Fc) to act as decoys, and 3) depleting mAb directed against BR3, the BAFF receptor responsible for transducing a survival signal and which is expressed on all mature B cells. The first two strategies target B cells indirectly by depriving them of a survival signal, while the third option utilizes an additional direct approach through antibody mediated cell killing. We described the experience with each of these types of molecules first in mice, then monkeys and human. Finally, we discuss the differences observed between species regarding in vivo blockade of this pathway and potential implications for human disease therapy.

Keywords BAFF · BAFF-R · BR3 · TACI · Belimumab · LymphoStat- B · Atacicept · Rheumatoid arthritis · Systemic lupus erythematosus · Sjögren syndrome

---

F. Martin (✉)
Department of Immunology, Genentech Inc., 1 DNA Way, MS34, 12-286, South San Francisco, CA 94080, USA

M.P. Cancro (ed.), *BLyS Ligands and Receptors,* Contemporary Immunology,
DOI 10.1007/978-1-60327-013-7_10,

10.1 Blocking BAFF in Mice

Early clinical trial data and case reports of B-cell-depleting anti-CD20 therapy in rheumatoid arthritis (RA) and systemic lupus erythematosus (SLE) were suggestive of an important role for B cells in these autoimmune diseases [3]; therefore, several laboratories proceeded to test the impact of BAFF or BAFF/APRIL blockade on B-cell survival and function in normal mice, as well as efficacy of BAFF inhibition in animal models of disease.

Three receptor:Fc proteins have been used in normal mice and murine disease models to block BAFF alone (BR3-Fc) or BAFF plus APRIL (BCMA-Fc, TACI-Fc). It is unfortunate that these decoy receptors have not been evaluated in side-by-side experiments to understand the functional equivalence and impact of B-cell survival and biology in vivo. Nevertheless, data from several laboratories are available for comparison.

10.1.1 BAFF and BAFF/APRIL Blockade in Normal Mice

In normal mice, all receptor:Fc proteins have been shown to reduce, although not eliminate, antigen-specific immune responses to a variety of T-cell-dependent and independent antigens [4–6]. Pelletier et al. [7]. evaluated directly BR3-Fc and BCMA-Fc in normal mice to examine the effect on splenic B-cell survival at a single high (300 μg), intermediate (100 μg), and low (10 μg) dose. The data showed that after 7 days, a 50% reduction in the number of splenic B cells was obtained with both molecules; however, 10× more BCMA-Fc than BR3-Fc was needed to achieve this, and this is likely due to different BAFF-binding properties inherent in each molecule, including affinity [8]. While no such single-dose study has been described for TACI-Fc, one report showed that multi-dose treatment of normal mice for 2 weeks resulted in a 52% and 75% decline in the frequency of mesenteric lymph node and splenic B cells, respectively [9]. Independent studies also showed that while germinal centers (GCs) could still form in BR3-Fc (S. Kalled, Biogen Idec, unpublished data), BCMA-Fc [6], and TACI-Fc-treated mice [10], their size was reduced and kinetics was altered when compared to controls.

To understand more fully the impact of BAFF blockade on the kinetics of B-cell loss and recovery, we undertook a study, in whereby normal mice received four injections (2.5 or 10 mg/kg) of BR3-Fc, and blood B cells were monitored weekly. As seen in Fig. 10.1, both dose groups exhibited a similar loss of $B220^+$ B cells with a 57–67% decline in frequency after 1 week, reaching a nadir after 6 weeks with an 84–88% reduction in frequency. Continuous dosing with BR3-Fc did not lead to a further decline (D. Hess, Biogen Idec, unpublished data), indicating that a subset of circulating B cells do not require BAFF for survival, which correlates with findings in BAFF knockout animals [10]. Clear B-cell recovery could be observed beginning

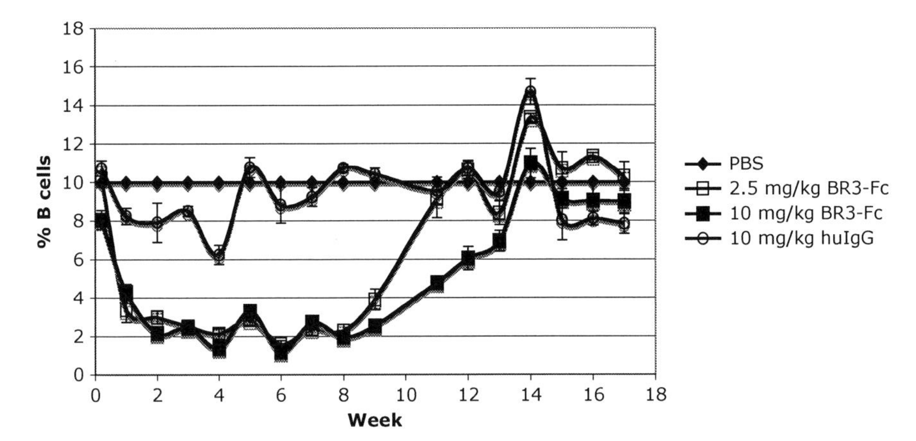

Fig. 10.1 Kinetics of B-cell loss and recovery with BR3-Fc therapy in mice. Female BALB/c mice, 6–8-weeks old (5/group) received BR3-Fc, control human IgG or saline, i.p. once a week for 4 weeks. Blood B cells were monitored weekly via flow cytometry analyses. Results were normalized to the PBS controls for each time point. The percentage of B220+ cells was averaged for each group, and then calculated as the percentage of the B cells in the saline group at each time point

at week 9 for both dose groups, with full recovery obtained by week 12 and 14 for the 2.5 and 10 mg/kg dose groups, respectively, and correlated with drug clearance (D. Hess, Biogen Idec, unpublished data). A single, 100-μg dose (~5 mg/kg) of BR3-Fc was also found to reduce splenic B-cell numbers by ~40% in human CD20 transgenic mice just 4 days post-treatment [11].

While there is no published report describing the effect of BAFF inhibition on B cells using a blocking anti-BAFF mAb in normal mice, such a reagent is being tested currently in clinical trials and will be discussed below. There has been, however, a body of work describing anti-BR3 therapy in mice. In normal mice, anti-BR3 therapy results in a depletion of mature B cells in lymphoid tissues by a dual mechanism-of-action, namely blockade of the BR3/BAFF survival signal plus Fc-mediated killing [12]. In an important study, BR3-Fc and anti-BR3 treatments were compared and assessed for their ability to eliminate B cells in blood and peripheral lymphoid organs, including an examination of follicular (FO) and marginal zone (MZ) B cells, germinal centers, and splenic plasma cells. These studies clearly illustrated the superior ability of anti-BR3 therapy to eliminate B cells compared to BR3-Fc in both kinetics and extent of B-cell depletion, with anti-BR3 treatment resulting in 90–99% reduction in B-cell populations in all compartments. These data were also of interest for demonstrating that the BR3-Fc molecule had a greater overall impact on tissue-resident B-cell populations when compared to circulating blood B cells (marginal zone > follicular > blood). Both molecules were found to exhibit similar reductions in the number of splenic plasma cells and germinal centers within the Peyer's patches.

10.1.2 Impact of BAFF and BAFF/APRIL Blockade in Models of Human Disease

While the data in normal mice indicated that peripheral B-cell reduction could be achieved quickly with full recovery following drug clearance and inhibit antigen-specific immune responses, the primary interest of investigators has been the therapeutic potential of BAFF or BAFF/APRIL blockade in autoimmune indications. Indeed, BR3-Fc, TACI-Fc, and an anti-BAFF mAb are currently being tested in clinical trials in RA and SLE patient populations (see below). Furthermore, since the year 2000, there have been ten reports describing the use of receptor:Fc fusion proteins, and one describing the use of a blocking BAFF mAb therapy, in animal models of disease. These include models of RA, SLE, multiple sclerosis (MS), chronic graft-versus-host disease (cGVHD), asthma, systemic sclerosis, and multiple myeloma (Table 10.1). Overall, BAFF inhibition, or BAFF plus APRIL inhibition in rodent models has provided impressive efficacy data that created excitement among investigators that the BAFF or BAFF/APRIL pathway could provide a novel target for modulating B cells in autoimmune disorders. How well these data translated to clinical success will be discussed in a forthcoming section of this chapter.

Rheumatoid arthritis patients have, in recent years, gained a wider selection of drugs for treatment, including the B-cell-depleting drug rituximab (Genentech, Biogen Idec). Unfortunately, not all patients are served by the new arsenal of drugs, and therefore, novel therapeutics with distinct mechanisms-of-action are desirable. TACI-Fc and BR3-Fc were both tested in the collagen-induced arthritis (CIA) mouse model, one that resembles human RA and requires both T-cell and B-cell involvement for disease pathogenesis. In this model mice typically are immunized with collagen on days 0 and 21, and disease is characterized by inflammation of the synovium and joint tissue, with cartilage and bone erosion, as well as development of anti-collagen antibodies. Two independent laboratories examined the impact of TACI-Fc in this model, and both showed efficacy when compared to control-treated animals. Wang et al. [13] treated mice continuously for 6 weeks, beginning

Table 10.1 Summary of BAFF antagonists tested in animal models of disease

Disease: model	Molecule tested	References
RA: CIA	TACI-Fc	[9, 13]
SLE: (NZBxNZW)F1	TACI-Fc	[14]
	BR3-Fc	[15, 16, 18]
	anti-BR3 mAb	[18]
cGVHD: parent→un-irradiated F1	BCMA-Fc, BR3-Fc	[21]
MS: EAE	BCMA-Fc	[19]
Asthma: OVA-induced	TACI-Fc	[46]
Systemic sclerosis: tight-skin mouse	BR3-Fc	[20]
Multiple myeloma: HuSCID	TACI-Fc, BR3-Fc	[23]
	Anti-BAFF	[22]

3 days after the second collagen immunization. In contrast to control-treated animals, TACI-Fc-treated mice exhibited a marked inhibition in disease progression and reduced total disease score, which included minimal joint inflammation and pathologic changes. A marked reduction in anti-collagen antibody titers was also observed for TACI-Fc-treated mice compared to controls when examined out to 48 days after the second collagen immunization. In a separate report Gross et al. [9] provided TACI-Fc treatment either in a prophylactic mode, beginning 7 days prior to the second collagen immunization, or in a therapeutic mode, beginning 7 days after the second collagen immunization. Mice were treated continuously for 3 weeks. In the prophylactic study disease onset was delayed, fewer animals developed disease, and inhibition of an anti-collagen antibody response was observed in the TACI-Fc-treated group when compared to controls. In the therapeutic study, ~50% of the animals exhibited joint inflammation at the start of treatment, and TACI-Fc therapy resulted in a reduced progression of disease with fewer animals exhibiting established disease, diminished disease scores and a reduced anti-collagen antibody response compared to controls. While the data described above indicate that blocking both BAFF and APRIL ameliorates disease in the CIA model, blocking BAFF alone with BR3-Fc gives similar results (H. Hess, Biogen Idec, unpublished data).

A complex animal model of autoimmunity is one that is reflective of human SLE-nephritis. (NZB × NZW)F1 mice develop spontaneous, lupus nephritis as they age with corresponding splenomegaly, lymphadenopathy, autoantibodies, and increased levels of p52 in their B cells. Results for both BR3-Fc and TACI-Fc have been reported in this SLE model and both have been shown to provide impressive efficacy with just a short treatment course, indicating that blocking BAFF alone is sufficient for ameliorating disease. The earliest study involved BAFF/APRIL blockade with TACI-Fc. Gross et al. [14] reported that in an 18-week study, treatment of prenephritic mice for 5 weeks with 100 μg/dose (3×/week) of TACI-Fc delayed the onset of nephritis and increased survival. Treatment with a lower dose (20 μg) had an obvious impact on disease compared to control-treated mice, but the effect was less robust than the higher dosing strategy. Neither dose regimen reduced anti-dsDNA autoantibody titers, although both significantly reduced the frequency of peripheral blood B cells, 53% and 32% for the 100 μg and 20 μg doses, respectively, when measured 8 weeks after the start of treatment. The reduction in B cells remained in the 100-μg dose group 13 weeks after the start of treatment, and recovered by week 17. In a separate report NZB/W mice were treated for 5 weeks (100 μg 3×/week) with BR3-Fc [15]. Like TACI-Fc there was an increase in survival, 100% over the 35-week course of the study, and a reduced frequency of mice developed severe nephritis. In contrast to what was reported for TACI-Fc, in this study, autoantibody titers were markedly diminished, although no decline in peripheral blood B cells was observed. The reason for the disparities in regards to B-cell loss and autoantibody responses is unclear, but they could be related to affinity for BAFF and/or pharmacokinetic properties of each molecule. A continuous dosing strategy (5.5 months) with BR3-Fc in the NZB/W model resulted in marked B-cell declines in blood and spleen, as well as decreases in the total number of anti-dsDNA antibody titers (F. Martin, Genentech, unpublished data). Interestingly, splenic anti-dsDNA antibody-producing plasma cells and total plasma cells were reduced compared to controls, and this reduction could be seen as early as 2 weeks after treatment (F. Martin, Genentech, unpublished data). A third laboratory using an adenovirus expression system to deliver BR3-Fc to NZB/W mice also observed efficacy as determined by survival and progression of proteinuria, as well as a decline in the number of mature B cells and activated T cells and delayed appearance of anti-dsDNA antibodies [16]. However, unlike the data described above with BR3-Fc dosing, a decline in plasma cells was not observed [16], which may be due to differences in systemic availability of BR3-Fc in each study and/or the point in time splenic plasma cells were enumerated (2 weeks vs. 16–18 weeks after commencing treatment). TACI-Fc delivered by the same adenoviral system was only detected in the sera for a short period of time (7–10 days), nevertheless, a beneficial effect on survival and development of proteinuria could still be obtained [17].

An obvious question to pose for a BAFF-modulating therapy is whether B cells in an autoimmune host are as sensitive to BAFF inhibition as B cells in a nonautoimmune environment. Lin et al. [18] addressed this question by examining blood and splenic B cells in NZB/W and C57Bl/6 mice after BR3-Fc and anti-BR3 mAb treatment. Interestingly, it was found that while B cell loss was observed in the blood and spleen of NZB/W mice, the extent was not as great as that observed in normal mice for either molecule. When comparing effects of BR3-Fc and anti-BR3 mAb therapy in NZB/W mice, as in normal animals, anti-BR3 mAb treatment had a greater impact on B-cell survival than BR3-Fc.

In addition to models of RA and SLE, BAFF and BAFF/APRIL blockade have been shown to provide efficacy in animal models of human MS [19], systemic sclerosis [20], cGVHD [21], and multiple myeloma [22, 23]. While the results described in these animal models are of interest, and the corresponding human diseases of importance, we have chosen to concentrate on RA and SLE for this review because of the availability of clinical trial data in these indications, allowing a more complete comparison of animal model data with clinical experience and a discussion on the complexities of translating rodent data to non-human primate and then to man.

10.2 Comparison of BAFF Antagonists in Non-human Primates

Several approaches described in mice have been developed to antagonize BAFF function as potential therapeutic options. Studies to test efficacies of these various approaches in inhibiting B-cell survival and function in non-human primates are reviewed below (Table 10.2).

The first of these approaches involved the development of a neutralizing antibody to BAFF by Human Genome Sciences, belimumab (LymphoStat-B). This is a humanized mAb that blocks the binding of BAFF to its three receptors BR3, TACI, and BCMA with high affinity, and can inhibit human BAFF-stimulated (i) survival and proliferation of B cells in vitro and (ii) B-cell hyperplasia and immunoglobulin

Table 10.2 Summary of BAFF antagonists tested in non-human primates

BAFF antagonist	Dose tested (mg/kg)	Company	References
Belimumab (anti-BAFF)	5, 15, 50	HGS/GSK	[24, 25]
Briobacept (BR3-Fc)	2, 20	Genentech/BIIB	[26]
Atacicept (TACI-Fc)	0.4, 2, 10	ZymoGenetics/Serono	[69]
Anti-BR3	20	Genentech/BIIB	[18]

production in vivo in mice [24]. In the first of two published studies to analyze effects of this mAb on non-human primate B cells in vivo [24], cynomolgus monkeys were administered three doses (5, 15, or 50 mg/kg) once a week for 4 weeks. Necropsy of these animals on day 29 revealed no changes in serum Ig levels and circulating peripheral B-cell numbers, while mature B cells ($CD20^+CD21^+$) in lymphoid tissues such as spleen and mesenteric lymph nodes were decreased by 40–50% in all treatment groups (relative to the vehicle group) with no observable dose dependency. This decrease was maintained during an additional 4-week treatment-free recovery period, perhaps because, LymphoStat-B, with a half-life of ~2 weeks, was still present at appreciable levels. In a follow-up study to test if long-term treatment would affect circulating B cells [25], cynomolgus monkeys were dosed at 5, 15, or 50 mg/kg every 2 weeks for 26 weeks, followed by a 34-week recovery period. B-cell populations were assessed in blood and tissues at 13 and 26 weeks of treatment, and at the end of the recovery period. While peripheral blood B cells were unaffected after 4 weeks of treatment in the previous study, 13 weeks of treatment resulted in a significant decrease of circulating mature B cells by 40% relative to baseline across all treatment groups. Blood B cells numbers continued to drop with treatment, and after 26 weeks, mature B cells were at 70% of baseline values in belimumab-treated monkeys. This decrease in B-cell populations was also observed in spleen and mesenteric lymph nodes (70–90% decrease relative to vehicle group). Interestingly, no statistically significant effects of treatment were detected on serum Igs throughout the length of this study. As before, there was no dose-dependent effect, and recovery of B cells was observed in blood and tissues by week 60.

Soluble receptor fusion proteins provide an alternate indirect method for targeting BAFF-dependent B cells, and Vugmeyster et al. studied effects of BAFF blockade on B-cell subsets in cynomolgus monkeys administered a human BR3-Fc fusion protein [26]. Animals were dosed with vehicle or BR3-Fc (Briobacept), in 2 or 20 mg/kg treatment groups, weekly for 18 weeks and followed for a further 23 weeks for recovery. Tracking circulating B-cell subsets in blood throughout the study indicated that naïve B cells ($CD20^+CD21^+CD27^-$) started dropping significantly by week 2 of treatment, and reached a maximum depletion of ~50% (relative to vehicle group) by week 18 in both treatment groups. Circulating memory B cells ($CD20^+CD21^+CD27^+$) were unaffected throughout the study, in contrast to what was observed in lymphoid tissues. An interim necropsy of vehicle- and 20 mg/kg-treated monkeys at week 13 of treatment, revealed an ~50% decrease in naïve B cells (relative to vehicle) and ~60% decrease in marginal zone B cells ($CD20^{+}CD21^{hi}$) in spleen and lymph nodes. A full necropsy of all groups at week 18 of treatment indicated two important distinctions from observations at week 13: a significant decrease of 50% in memory B cells in lymph nodes observed for the first time, and a continued drop in marginal zone B cells in spleen/lymph nodes resulting in a maximal depletion of 70–80%. Naïve B cells remained at ~50% of vehicle-treated monkeys in the spleen, lymph nodes, and bone marrow. Germinal center B cells and plasma cells in tissues (as assayed by immunohistochemistry) were unaffected by treatment. No dose-related effects were observed across the 2 and 20 mg/kg groups, and there was full recovery of all B-cell subsets by week 41 at the end of the recovery period. Effects of BAFF blockade on antigen-specific immunoglobulin responses were also assayed in this study. The observed decrease in various B-cell subsets in blood and tissue due to treatment with BR3-Fc fusion protein resulted in a modest but significant decrease in tetanus toxoid-specific IgG titers induced during a memory response by a recall immunization. The detailed analyses of various B-cell subsets performed in this study clearly revealed a differential dependence of these subsets on BAFF for survival, with naïve and marginal zone B cells being the most affected, followed by memory B cells, with germinal center B cells/plasma cells being the least affected.

A soluble receptor fusion protein composed of a second BAFF receptor, TACI, (TACI-Fc, Atacicept) was also tested in cynomolgus monkeys as a means to target B cells indirectly by depriving them of BAFF-induced survival signals. One important distinction, however, between TACI-Fc and BR3-Fc is that while BR3 binds only BAFF, TACI is also a receptor for a related ligand, APRIL, which has been shown to potentially affect long-lived plasma-cell survival (through BCMA) [27]. Thus, while BR3-Fc would block only BAFF function, TACI-Fc would antagonize both BAFF and APRIL. In the study conducted by Serono International and ZymoGenetics [69], cynomolgus monkeys were treated with three doses of TACI-Fc (0.4, 2 and 10 mg/kg) every third day for 39 weeks, followed by a 25-week recovery period. The more frequent dosing of this molecule in comparison to the anti-BAFF mAb and BR3-Fc was clearly a result of its lower half-life (1 week for TACI-Fc versus ~2 weeks for anti-BAFF mAb and BR3-Fc). Analyses of circulating mature B-cell numbers ($CD20^{+}IgD^{+}$) at weeks 8, 12, 24, and 39 revealed a maximal depletion of ~50% (relative to baseline) in the 2 mg/kg and 20 mg/kg groups (no efficacy in the 0.4 mg/kg group) by week 12 through dosing, followed by full recovery by week 64. Analyses of serum Ig levels revealed decreased levels of IgG (20–40%), IgM (50–75%), and IgA (40–60%) relative to baseline across all treatment groups from week 8 through the end of dosing at week 39. Full recovery was observed for all Igs by end of the recovery period. This drop in total serum Igs was observed only for TACI-Fc in contrast to anti-BAFF mAb and BR3-Fc molecules, suggesting a potential role for APRIL in regulating Ig production in primates.

A third class of BAFF antagonists comprises a blocking antibody directed against BR3. This novel approach of B-cell-directed therapy would combine the depleting capacity of a therapeutic mAb with deprivation of BAFF-induced survival signals through BR3. The anti-BR3 mAb was shown to inhibit BAFF-mediated B-cell survival and proliferation in vitro [18] and was tested for effects on B cells in cynomolgus monkeys in vivo. In this report, monkeys were administered anti-BR3 mAb at 20 mg/kg weekly for 8 weeks. $CD20^+$ blood B cells decreased by approximately 50% after 2 weeks of treatment with anti-BR3 mAb, with a maximal drop to 75% after 1 month relative to baseline. In comparison, in monkeys treated with BR3-Fc (20 mg/kg weekly), there was a 10% decrease in 2 weeks, and 35% decrease in 4 weeks. Necropsy after 4 weeks of treatment revealed decreases relative to vehicle in mature B cells in the spleen (70% decrease for anti-BR3 mAb versus 50% for BR3-Fc), naïve B cells in the lymph nodes (60% decrease for anti-BR3 mAb versus 80% for BR3-Fc), memory B cells in lymph nodes (50% decrease for both anti-BR3 mAb and BR3-Fc), marginal zone B cells in lymph nodes (70% decrease for both anti-BR3 mAb and BR3-Fc) and a 30% decrease in lymph node germinal center B cells for anti-BR3 mAb. Interestingly, there was no change in total serum Ig levels for anti-BR3 mAb over 3 months, similar to anti-BAFF mAb and BR3-Fc, both of which target only BAFF-induced signals. Analyses of anti-BR3 mAb treatment effects on cynomolgus B cells clearly indicate that reduction of naive B cells in blood and lymph nodes (both follicular and marginal zone subsets) was faster and more profound compared with memory B cells, suggesting a differential sensitivity for BAFF blockade for these subsets in primates. In non-human primates, similar to mice, anti-BR3 mAb-mediated B-cell depletion was faster than BR3-Fc. However, the maximal B-cell reduction obtained after 1 month of treatment was approximately 80% in certain subsets (naive, follicular, and marginal zone) and less in others (memory, germinal center, and plasma cells). This reduction was superior to previously reported anti-BAFF reagents (anti-BAFF mAb, BR3-Fc, TACI-Fc), showing that anti-BR3 mAb, in addition to blocking survival, is also depleting B cells in non-human primates (Fig. 10.2).

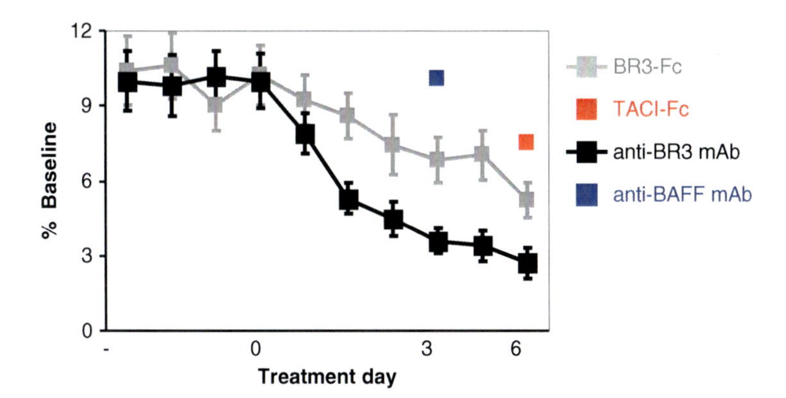

Fig. 10.2 Comparison of BAFF antagonists in mediating peripheral B-cell depletion in cynomolgus monkeys. Kinetics of peripheral B-cell reduction in 8 weeks in cynomolgus monkeys treated with the indicated BAFF antagonists are shown. Values were calculated for each monkey as a percentage of its baseline and averaged for each treatment group at each time point

10.3 Clinical Aspects of BAFF Inhibition

10.3.1 BAFF Overexpression in Autoimmune Disorders in Man

Many disease states are thought to have some immunological component mediated by B cells (Table 10.3) and therefore may be dependent on BAFF to some degree. These include autoimmune diseases like RA, SLE, and Sjögren syndrome, and hematological malignancies of the B lymphocyte series (multiple myeloma, chronic lymphocytic leukemia, and certain lymphoma types). B cells participate in autoimmune processes by antigen presentation, by the provision of co-stimulatory signals required for T-cell functions, by secretion of pro-inflammatory cytokines such as tumor necrosis factor-α (TNF-α), interleukins 1 and 6, and chemokines, and by the production of autoantibodies. Here we will mainly focus on human autoimmune disease for which there is available clinical data from human in vivo studies with BAFF antagonists.

Table 10.3 Examples of human immune disorders with B-cell-mediated pathology[a]

Disorder	Evidence of B-Cell Involvement
Allergic rhinitis, conjunctivitis, urticaria	IgE against allergens
Asthma	Anti-IgE effective
Bullous pemphigoid	Pathogenic autoantibodies
Grave's Disease (autoimmune hyperthyroidism)neonatal disease[b]	Pathogenic autoantibodies,
Hyperacute allograft rejection	Preformed antibody against graft
Immune thrombocytopenic purpura	Pathogenic autoantibodies
Myasthenia gravis	Pathogenic autoantibodies; Ig removal[c] effective; neonatal disease
Multiple sclerosis	Anti-CD20 effective[d]
Pemphigus vulgaris	Pathogenic autoantibodies
Primary antiphospholipid syndrome	Pathogenic autoantibodies; Ig removal effective
Rheumatoid arthritis	Anti-CD20 effective
Sjögren syndrome	Pathogenic autoantibodies
Systemic Lupus Erythematosus	Pathogenic autoantibodies; neonatal disease; anti-BAFF effective
Thrombotic thrombocytopenic purpura	Pathogenic autoantibodies

[a] Comprehensive descriptions of the clinical features and management of these disorders can be found in recent reference textbooks of Rheumatology or Internal Medicine
[b] Neonatal disease reflects transfer of IgG autoantibodies across the placenta in pregnancy
[c] Therapeutic removal of circulating Ig by plasmapheresis
[d] Preliminary evidence

10.3.1.1 Rheumatoid Arthritis (RA)

RA affects approximately 1% of the population and is more common in women. A typical clinical presentation is a middle aged (50s) woman with multiple painful, swollen joints including the hands, wrist, and feet. The synovial membrane that lines joints and tendon sheaths is expanded and infiltrated by inflammatory cells, resulting in pain, swelling, and stiffness. In the longer term, the adjacent bone and cartilage may be damaged irreversibly, associated with deformity of the joints and progressive disability. Systemic features include fatigue and anemia. The spectrum of severity spans a wide range; at the more severe end, RA carries mortality comparable to severe ischemic heart disease and lymphoma.

While the pathogenesis of RA is commonly believed to be an autoimmune response triggered by an environmental agent in a host predisposed by her genetic background (HLA-DR, PTPN22, STAT4, and others) the underlying etiology remains unclear as there is no consistent evidence for particular triggering agents or autoantigenic targets. Approximately 70% of patients have rheumatoid factor, and IgM autoantibodies directed against the Fc portion of IgG, and autoantibodies against a range of citrullinated proteins (anti-CCP), which are relatively specific for RA.

Therapy depends on the severity in a given patient and consists of patient education, physical and occupational therapy, symptomatic relief with analgesics and non-steroidal anti-inflammatory drugs, and immunosupressive therapy. The latter includes small molecule "disease modifying anti-rheumatic drugs" (DMARDs) such as methatrexate, leflunomide, and sulphasalazine, and biological agents including TNFα antagonists (etanercept, infliximab, adalimumab), the CTLA-4 fusion construct abatacept, and the B-cell depleting anti-CD20 monoclonal antibody rituximab. In recent years the use of biologicals, either alone or in combination with methatrexate and other DMARDs, has significantly improved both symptom control and reduced joint damage.

BAFF levels are elevated in RA, and the level correlates with disease activity as assessed clinically (DAS28 score) and by acute phase reactant levels (ESR, C-reactive protein), and with levels of rheumatoid factor [28, 29]. The level of BAFF is higher in the synovial fluid from inflamed joints of RA patients, consistent with local production within the synovium [28]. Whether autoantibodies such as rheumatoid factor and anti-CCP contribute directly to RA pathogenesis remains questionable, but the clinical efficacy of rituximab and of other anti-CD20 monoclonal antibodies has removed any doubt about the importance of B cells in RA. Moreover, part of the benefit of TNF sequestration in RA may be mediated by the effects of TNF and lymphotoxin depletion on germinal center B cells [30]. Small molecule drugs directed against B cells are also in development, including antagonists of the kinases *Syk* and *Btk*.

10.3.1.2 Systemic Lupus Erythematosus (SLE)

SLE is an autoimmune disease with a very wide range of clinical features, affecting around 1 in 1,000 people. It has a marked female predominance (9:1) and is more common in certain ethnic groups including African-American and Hispanic. Although certain features are more frequent (arthritis, skin rash, photosensitivity, glomerulonephritis, inflammation of the pleural linings, and immune cytopenias), almost every organ in the body can be affected; SLE has replaced syphilis as "the great mimic." This poses huge challenges to patients and their physicians, and for the design of clinical trials to test new drugs for SLE.

The cardinal immunologic feature is the production of autoantibodies against a range of nuclear components, including native dsDNA. It is generally believed that autoantibodies are directly pathogenic in SLE. Inflammation results from the deposition of immune complexes comprising autoantibody and nuclear antigens, for example in the small blood vessels of the dermis of the skin, the synovium, and the glomerulus of the kidney. As with RA, a variety of microbial agents have been studied as potential triggers.

Circulating BAFF levels are elevated in over 50% of SLE patients, and show some degree of correlation to levels of anti-dsDNA autoantibodies [31, 32]. The relationship between BAFF levels and clinical disease activity is less clear. Although corticosteroid therapy reduces BAFF levels in SLE, there is no correlation with specific organ system involvement. In part this is a reflection of the heterogenous nature of SLE and the difficulties in measuring disease activity.

Current treatments for SLE are either symptomatic or consist of non-specific immunosupression (corticosteroids, DMARDs, and cytotoxic drugs such as cyclophosphamide). Existing therapies often prove inadequate. There is significant toxicity from the long-term use of corticosteroids (for example cataracts, metabolic side effects including hyperlipidemia, weight gain, cardiovascular disease, and osteoporosis), and ovarian failure from cytotoxic agents is a particularly distressing issue given the predominance of young female patients affected by SLE.

It is over 40 years since a new drug was approved to treat SLE. Preliminary reports suggest efficacy from B-cell depletion with anti-CD20 [33, 34] and with the BAFF antagonist belimumab (discussed below). Phase III clinical trials in SLE are underway with both anti-CD20 and BAFF antagonists.

10.3.1.3 Sjögren Syndrome (SS)

In Sjögren syndrome, there is dysfunction of the exocrine secretory glands, manifesting as dryness (sicca) of the eyes, mouth, vagina, and occasionally other surfaces of the body [35]. Glandular dysfunction is associated with infiltration of the glands by B lymphocytes and with the production of autoantibodies [36] including functional autoantibodies directly involved in reducing secretions [37]. Sjögren patients have marked polyclonal increases in IgG level and an approximately 16-fold increase in B-cell non-Hodgkin's lymphoma compared with the general population [38]. In contrast to RA, anti-TNF therapies have not shown efficacy in primary SS.

BAFF is produced within the affected glands by T cells, macrophages, and ductal epithelial cells [39], and elevated BAFF levels in Sjögren patients correlate with autoantibody levels [40]. There are preliminary reports suggesting clinical efficacy of B-cell depletion in Sjögren Syndrome with rituximab [41–43] and with the anti-CD22 monoclonal antibody epratuzumab [44], but none using BAFF antagonism in human Sjögren's.

10.3.1.4 Other Disorders

IgE-mediated allergic syndromes include urticaria, allergic rhinitis and conjunctivitis, certain drug and food reactions, and contribute to most cases of allergic asthma. Depletion of IgE by the monoclonal anti-IgE antibody omalizumab is beneficial for moderate-to-severe asthma [45]. Depletion of BAFF and APRIL with a TACI-Fc fusion construct was beneficial in a mouse model of asthma [46] but no human studies have been reported to date.

It could be argued that there is a potential role for BAFF antagonism in treating virtually any disease in which B cells have been implicated. Other important considerations include the safety and effectiveness of existing standard-of-care treatments; whether there are well-defined and easily measured endpoints to allow robust assessment of efficacy; the ability to recruit enough patients for timely undertaking of clinical trials; the framework for regulatory approval for a new drug in a given indication; risk–benefit and pharmacoeconomic considerations; and the perceived likelihood that the pharmaceutical or biotechnology company sponsoring the clinical studies will recover the growing costs associated with conducting the studies.

10.3.2 Clinical Experience with BAFF Antagonists

There are four BAFF antagonists (Table 10.4) for which there is human clinical data in the public domain: belimumab (*LymphoStat-B*^{TM}), atacicept (TACI-Ig), briobacept (BR3-Fc), and AMG-623 (Anthera/Amgen).

Table 10.4 BAFF antagonists with human in vivo experience

Agent	Type	Company	Development Stage[a]	Indications[a]
AMG-623	Fusion protein	Anthera/Amgen	Phase I/II	SLE
Atacicept	Fc-R fusion	Zymogenetics/Serono	Phase III	SLE, LN
Belimumab	mAb	HGS/GSK	Phase III	SLE
Briobacept	Fc-R fusion	Genentech/BIIB	Phase Ib	Rheumatology

[a] Based on information in the public domain, January 2008.
Abbreviations: BIIB, Biogen Idec; HGS, Human Genome Sciences; GSK, GlaxoSmithKline; LN, Lupus nephritis; mAb, monoclonal antibody; Fc-R, Fc-receptor

10.3.2.1 Belimumab

Belimumab is a fully human monoclonal antibody against BAFF. It has been generally safe and well tolerated in phase I and II studies involving RA and SLE patients, with no clinically significant excess of infections seen in patients receiving belimumab compared to those receiving placebo. The phase I pharmacokinetics study in mild active SLE patients indicated a half-life of 13–17 days [47]. Following a single dose of belimumab, reductions in B cells were noted between 42 and 105 days later.

With repeated dosing in both RA and SLE patients, belimumab produces a gradual reduction in peripheral blood B cells, reaching a plateau after several months of therapy (in contrast to the rapid depletion seen with anti-CD20 mAb treatments) of an approximately 50% reduction. The decrease is largely confined to the $CD27^-$ naïve B-cell subset, whereas the $CD27^+$ memory subset increased acutely and remained elevated for several months of ongoing treatment. Activated $CD69^+$ B cell numbers are reduced. There is no data available on the effect of belimumab (or of other BAFF antagonists) on tissue B cells in humans. In efficacy studies reported to date, belimumab has been administered by intravenous infusion, and in recent clinical trials, has been administered using three initial doses at 14 day intervals followed by a regular once-a-month regimen.

In the Phase II RA study, 283 patients who had active arthritis despite standard therapy received either placebo or one of three doses of belimumab (1, 4, or 10 mg/kg). After 24 weeks, belimumab achieved a 20% or more reduction in the American College of Rheumatology composite disease activity score (ACR20) in 29% of treated patients overall, as compared to 16% of placebo-treated patients [48]. The autoantibody rheumatoid factor (IgM anti-IgGFc) was significantly lowered by belimumab, and modest reductions were produced in total Ig levels (IgG by 6% compared to baseline, IgA 10%, IgM 16%, and IgE 25%), with few subjects dropping below the lower limit of normal. There was no clear dose–response relationship.

Although the reduction in RA activity was statistically superior to placebo with certain belimumab dose groupings, the overall magnitude of response seemed less than that seen in RA patients treated with competing biological therapies such as TNF antagonists, anti-CD20 B-cell depleting antibodies, and abatacept. No subsequent RA studies were reported.

A 1-year phase II SLE study enrolled 449 patients assigned to treatment with placebo or with belimumab, at monthly IV infusions of 1, 4, or 10 mg/kg. The two primary endpoints were reduction in the Systemic Lupus Disease Assessment Index (SLEDAI) score and a reduction in the incidence of flares of lupus activity. Neither of these endpoints was achieved. However, a retrospective analysis indicated that belimumab had greater benefit then placebo in 71% of subjects who had positive anti-nuclear antibody serology at the start of the study. When data from the three belimumab treatment groups was combined, the mean SLEDAI score improved from 9.9 down to 7, whereas the placebo-treated patients improved from 9.9 to 8.8. Improvement also occurred in a physician global assessment and in the SF-36, a quality of life measure [49]. Levels of anti-dsDNA autoantibodies decreased. As in RA, there were small reductions in Ig levels, and safety follow-up studies have shown no excess of infections or other adverse events. Some have suggested that the magnitude of clinical benefit is small relative to other SLE therapies.

Two large double-blind placebo-controlled randomized clinical trials are under way (BLISS-52 and BLISS-76). The primary endpoint for both studies is the proportion of patients achieving a composite outcome approved by regulatory agencies and consisting of (i) a $\geq$ 4-point reduction in baseline SLEDAI score, (ii) a physician global assessment that improves or is stable (not worsening by more than 0.3 on a

10-point scale), and (iii) no lupus flares, as assessed using the British Isles Lupus Assessment Group (BILAG) instrument.

There is no information on the efficacy of belimumab for treating glomerulonephritis in SLE. Clinical trials generally evaluate severe kidney disease in separate studies because the poor prognosis of untreated lupus nephritis dictates background standard-of-care immunosupression (usually high-dose corticosteroids and either cyclophosphamide or mycophenolate) that would be inappropriate for most SLE patients without severe kidney disease. Phase II and III SLE studies generally also exclude patients with serious central nervous system disease or other manifestations needing urgent intervention, and patients at elevated risk of infections or who have other major co-morbidities. This may reduce the applicability of the findings to the SLE population seen in routine clinical practice, but this is a limitation common to all pre-registration clinical trials.

10.3.2.2 Atacicept

Early phase clinical trial data for atacicept (TACI-Ig) is available in both RA and SLE [50, 51]. Atacicept shows non-linear PK properties characterized by greater than dose-proportional increases in free drug concentration, a finding ascribed to saturation of the atacicept: BAFF/APRIL complex. The terminal elimination half-life is unusually long for a human Fc fusion receptor construct, at 25–63 days. The bioavailability was 28–40% following subcutaneous injection [52]. Atacicept was detected in synovial fluid of four RA patients following SC administration, confirming that it reaches this pathogenically relevant compartment [51].

The effect of atacicept on the total peripheral blood B-cell count is similar to the other BAFF antagonists, with an initial transient elevation in total B cells (mainly comprising memory B cells) followed over several weeks by a more gradual decline in total B-cell count by approximately 40% of baseline level. Most of this decrease was due to effects on the $CD19^{+}IgD^{+}CD27^{-}$ mature B-cell population [51]. The effects of atacicept on memory and naïve B-cell subsets have not been reported in detail.

As in preclinical models (discussed above), Atacicept has a greater impact on Ig levels than other BAFF antagonists (which do not affect APRIL) [51, 52]. In the RA Phase Ib study, total serum IgM showed a rapid dose-dependent decrease in the first two weeks after treatment, declining to approximately half the baseline IgM level. Total IgG level was reduced by a median of 21% and IgA by 37%, with similar changes seen in SLE patients. Ig levels were returning toward baseline at the end of the 6-month follow-up period, and in most patients did not go below the lower limit of normal. Whether this greater pharmacodynamic effect will translate into greater clinical efficacy will require further study, as will the safety implications.

Preliminary data were suggestive of clinical efficacy in RA, with 6 of 19 patients (32%) achieving an ACR20 response during 3 months of therapy. The DAS (a continuous measure of RA activity which is more sensitive to change than the categorical ACR scores) decreased from 6.4 ± 1.3 to 5.1 ± 1.4 in the patients receiving the highest doses (420 mg subcutaneous every 2 weeks) [51]. A possible clinical signal was also noted in SLE, although the phase Ib study contained too few patients to assess this with certainty [50]. There have been no clinical safety signals of concern reported for atacicept. Local injection site symptoms were reported by 24 of 73 subjects in the RA Phase Ib study following SC injections, but these were of a mild nature. In particular there has been no increase in infections, but as with all early phase trials these were of short duration and only powered to detect clinical events that have a high frequency.

Atacicept has also been studied in early phase clinical trials for B-cell malignancies including multiple myeloma, chronic lymphocytic leukemia, Waldenstrom's macroglobulinemia, and non-Hodgkin's lymphoma. Phase II/III-controlled clinical trials in extra-renal SLE and in lupus nephritis are now being done, with the nephritis patients receiving mycophenolate mofetil and corticosteroids as background standard-of-care immunosuppression. The studies are proceeding under a Special Protocol Assessment process with the US Food and Drug Administration, an arrangement which allows sponsors and the FDA to reach a binding advance agreement on study designs, endpoints, and analyses that will form the basis of an efficacy claim in a regulatory submission – highly desirable in the face of the many challenges in designing SLE trials.

10.3.2.3 Briobacept

Single and multiple dose studies of briobacept (BR3-Fc) have been done in RA patients, which showed a half-life of 10–12 days and <50% bioavailability when administered via the SC route [53]. Single IV doses of up to 10 mg/kg were generally safe and well tolerated. In the repeat dose phase Ib study, there was a numerical excess of herpes zoster reactivations (two events among nine patients in the highest dose group, receiving 300 mg/week SC for nine doses). Other clinically significant events were not more common among the briobacept-treated patients compared to the placebo group, and there were no atypical or opportunistic infections. Briobacept reduced $CD19^+$ B-cell levels by a median of ~55% (Fig. 10.3). In many subjects, B-cell reductions persisted for at least 10 months after the last dose, long outlasting any measurable exposure to the drug. As with belimumab, B-cell reductions were largely limited to the $CD19^+CD27^-$ naïve population (by 60–70% of baseline level), while $CD19^+CD27^+$ memory cell numbers increased approximately twofold during dosing and gradually returned to baseline levels. This data confirms the generally benign safety profile of BAFF antagonism in humans. The extended pharmacodynamic effect is intriguing, and raises the possibility that relatively infrequent dosing may be needed to maintain B-cell depletion.

10.3.2.4 AMG-623

AMG-623 is a fusion molecule containing the Fc portion of IgG and a peptide sequence with high affinity for BAFF, although no details are publicly available on its structure. A phase I clinical study in SLE indicated a reduction in the naïve peripheral blood B-cell subset and an increase in memory cells [54], as

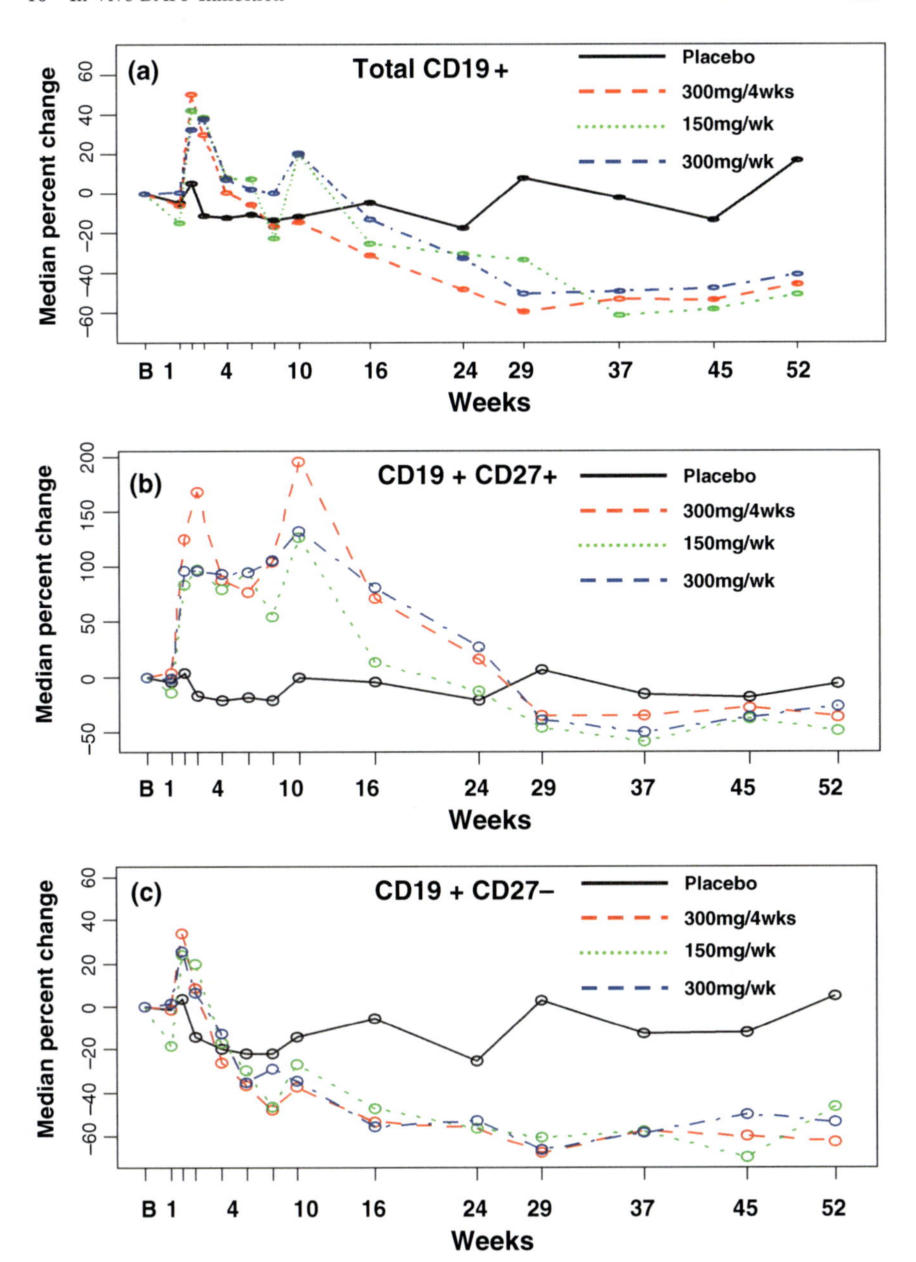

Fig. 10.3 Pharmacodynamic effect of BAFF inhibition in humans. Peripheral blood total CD19$^+$ B cells (**a**), memory (CD19$^+$CD27$^+$) subset (**b**), and naïve (CD19$^+$CD27$^-$) subset (**c**) following treatment of RA patients with briobacept (BR3-Fc) using three SC dose regimens, in a phase-Ib clinical trial. *Black* – placebo; *red* – 300 mg every 4 weeks; *green* – 150 mg once a week; *blue* – 300 mg once a week. B – Baseline, defined as the mean of screening and randomization values. Flow cytometry analyses gated on mononuclear cells; all counts expressed relative to baseline

noted with the belimumab and briobacept. Preclinical data on the related molecule AMG-523 indicated activity against multiple myeloma cell lines and myeloma cells from patients, suggesting possible utility for BAFF antagonists in this plasma cell malignancy.

10.3.2.5 Combination Therapy

The combination of BAFF inhibition with B-cell-depleting therapy has the potential to produce enhanced B-cell depletion and improve clinical efficacy. Although treatment with anti-CD20 mAb produces rapid and near-complete depletion of peripheral blood B cells in RA patients, tissue B cells are generally less affected – for example the B-cell population in the inflamed synovium of RA patients [55, 56]. B-cell depletion by anti-CD20 mAb also appears less reliable in SLE patients than in RA [57], and in MRL/lpr autoimmune-prone mice transgenic for human CD20, resistance to depletion by anti-CD20 mAb increases as the disease progresses [58]. In mice, marginal zone and germinal center B cells resist depletion by anti-CD20 mAb, but this can be overcome by BAFF inhibition [59]. Human studies of BAFF antagonism in combination with other B-cell therapies have not yet been reported.

Circulating BAFF levels rise markedly after B-cell depletion with anti-CD20 mAb, and this may contribute significantly to the survival and/or regeneration of B-cell populations capable of triggering clinical relapse [60]. Another potential combination approach would be to use a BAFF inhibitor at intervals after treatment with anti-CD20 mAb, to delay the return of pathogenic B cells and extend the clinical response to anti-CD20 mAb. No human studies of this sequential approach have been reported.

10.4 From Mice to Humans

As reviewed in this chapter, in vivo BAFF blockade has quantitatively distinct outcomes in humans, non-human primates, and mice. First, primates appear to be different from mice in two respects: (1) the kinetics of peripheral blood B-cell reduction is significantly slower in primates than mice, and (2) decrease in some non-human primate subsets – follicular naive cells and marginal zone B cells, do not reach the almost complete depletion observed in mice even after 12–26 weeks of treatment, although they appear to be relatively more sensitive to inhibition when compared with other B-cell subsets. Second, subtle differences appear to exist even between non-human primates and human rheumatology patients regarding the kinetics of peripheral blood memory B cell ($CD20^{+}CD27^{+}$) numbers during BAFF blockade. Memory B cells decrease in non-human primates despite being more resistant to BAFF blockade than naïve B cells. Conversely, memory B cells increase, at least temporarily, in blood of human patients treated with BAFF blockers. One point to consider while comparing memory B-cell depletion in mice and primates is the relative dependence of memory B cell on BAFF during generation/re-activation of memory versus maintenance of memory. During BAFF antagonist treatment all three processes can be tested in mice [6], while it is usually only maintenance of memory that is assayed in primates.

These differences between rodents and primates raise two questions that deserve further discussion and experimental investigation: (1) from a basic science perspective – what are the molecular and cellular differences between rodent and primate BAFF/B-cell systems, and (2) from a clinical perspective – what is the best way to optimize BAFF inhibition as a clinical treatment?

Possible causes for these different outcomes of BAFF blockade in rodents and primates are species-specific differences in B-cell subset composition and their BAFF-dependence or molecular differences in the BAFF/receptor system. As discussed, studies in BAFF- and BR3 knockout mice revealed a specific B-cell subset (B1 B cells) that is BAFF independent. One simple explanation for primate BAFF blockade response would be that there are proportionally more of these cells in humans. Opposite to most naïve re-circulating B cells, mouse B1 B cells have been shown to survive independent of BAFF through self-renewal, have a different repertoire, cell cycle characteristics, and depend on surface Ig signaling [61, 62]. However, identical cells do not exist in humans, but it is likely that some B1 B-cell characteristics instead of the entire phenotype are present in human B cells. Limited half-life studies in humans [63] show that human memory B cells divide at a fivefold higher rate than naïve cells. This raises the possibility that opposite to naïve B-cell differentiation from bone marrow precursors, the maintenance of a B-cell compartment that can divide is less BAFF dependent. This would apply well for memory cells, which were shown to divide more than naïve cells and to depend on additional signals like the TLR pathway for survival and activation. The more prominent role of the TLR and surface Ig pathways for survival in such a compartment would fit well with these signals being driven not through BR3/NF-κB2 but by TLR/Ig/NF-κB1 pathways. Additional survival signals through TNFR-family molecules (CD40, TACI, etc.) have been also described and could differ in quantity between rodents and primates. Of particular interest is the fact that in some mouse models of autoimmune diseases, B-cell reduction induced by B-cell immunotherapy is less than in normal mice. This could be due to excess activation and survival signals in this inflammatory environment that interfere with BAFF dependence. NF-κB2-independent/NF-κB1-dependent B-cell development has been demonstrated in mouse B-cell development and human B-cell lymphomagenesis suggesting that a variety of signals, if present and able to upregulate either pathway, would support B-cell development and maintenance [64, 65]. In addition, comparative analyses of B-cell depletion by BAFF antagonists in NZBxW autoimmune versus normal mice revealed that kinetics of depletion was slower, and extent of depletion less profound in autoimmune mice [18]. These results would suggest a lesser dependence of certain B-cell subsets on BAFF in autoimmune states, in comparison to non-disease situations.

Despite the fact that the BAFF/receptor system is well conserved between mice and primates, some properties could still be different. For example the phenotype of TACI knockout mice is dominated by autoimmunity with small aspects of immunodeficiency for T-independent immune responses. The human phenotype of TACI deficient patients is primarily immunodeficient (CVID) although autoimmunity has been associated as well [66, 67]. A detailed description of human BR3-deficient patients is not available at this time but preliminary data on one patient suggests some differences when compared with BR3 knockout mice [68]. Additionally, it is possible that APRIL/TACI-BCMA signaling could have a more prominent role in humans than mice, at least for specific B-cell subsets such as plasma cells. As seen in monkeys treated with Atacicept, serum Ig levels (including IgG, IgM, and IgA) dropped significantly with treatment, a phenomenon that has not been observed in monkeys treated with reagents that antagonize BAFF alone, or other B-cell therapeutics such as rituximab. This suggests a dependence of plasma cells in primates on BAFF and APRIL, as opposed to BAFF alone. Finally, the primate system might be rich in BAFF/APRIL heterotrimers, or BAFF 60-mers, making the interpretation of the blocking abilities of various therapeutic molecules not as straight forward as compared to BAFF trimers.

Although a few biological unknowns are left in this pathway, clinical tests are ongoing, seeking to identify the best disease or disease subset, and therapeutic regime in order to confer patient benefit. Due to the heterogeneous nature of human diseases, this question will likely require additional clinical data in subsets of patients well defined through molecular diagnostic techniques. Comparative data in rodents and non-human primates will continue to help make decisions in setting up these clinical tests.

Acknowledgments The authors would like to thank Melissa Starovasnik, Bruce Goldberg, Thomas Gelzleichter, Kathy Howell, Development Sciences FACS laboratory, Translational Immunology group, and the Genentech/Biogen-IDEC teams for their contributions to the BR3-Fc/anti-BR3 projects.

References

1. Martin F, Chan AC. B cell immunobiology in disease: evolving concepts from the clinic. Annu Rev Immunol 2006;24:467–96.
2. Edwards JC, Szczepanski L, Szechinski J, et al. Efficacy of B-cell-targeted therapy with rituximab in patients with rheumatoid arthritis. N Engl J Med 2004;350:2572–81.
3. Sanz I, Anolik JH, Looney RJ. B cell depletion therapy in autoimmune diseases. Front Biosci 2007;12:2546–67.
4. Yan M, Brady JR, Chan B, et al. Identification of a novel receptor for B lymphocyte stimulator that is mutated in a mouse strain with severe B cell deficiency. Curr Biol 2001;11:1547–52.
5. Xia XZ, Treanor J, Senaldi G, et al. TACI is a TRAF-interacting receptor for TALL-1, a tumor necrosis factor family member involved in B cell regulation. J Exp Med 2000;192:137–43.
6. Vora KA, Wang LC, Rao SP, et al. Cutting edge: germinal centers formed in the absence of B cell-activating factor belonging to the TNF family exhibit impaired maturation and function. J Immunol 2003;171:547–51.
7. Pelletier M, Thompson JS, Qian F, et al. Comparison of soluble decoy IgG fusion proteins of BAFF-R and BCMA as antagonists for BAFF. J Biol Chem 2003;278:33127–33.
8. Day ES, Cachero TG, Qian F, et al. Selectivity of BAFF/BLyS and APRIL for binding to the TNF family receptors BAFFR/BR3 and BCMA. Biochemistry 2005;44:1919–31.
9. Gross JA, Dillon SR, Mudri S, et al. TACI-Ig neutralizes molecules critical for B cell development and autoimmune disease. Impaired B cell maturation in mice lacking BLyS. Immunity 2001;15:289–302.

10. Schiemann B, Gommerman JL, Vora K, et al. An essential role for BAFF in the normal development of B cells through a BCMA-independent pathway. Science 2001;293:2111–4.
11. Gong Q, Ou Q, Ye S, et al. Importance of cellular microenvironment and circulatory dynamics in B cell immunotherapy. J Immunol 2005;174:817–26.
12. Lin Z, Dan-Rong Y, Xiao-Qing T, et al. Preliminary clinical measurement of the expression of B-cell activating factor in Chinese systemic lupus erythematosus patients. J Clin Lab Anal 2007;21:183–7.
13. Wang H, Marsters SA, Baker T, et al. TACI-ligand interactions are required for T cell activation and collagen-induced arthritis in mice. Nat Immunol 2001;2:632–7.
14. Gross JA, Johnston J, Mudri S, et al. TACI and BCMA are receptors for a TNF homologue implicated in B-cell autoimmune disease. Nature 2000;404:995–9.
15. Kayagaki N, Yan M, Seshasayee D, et al. BAFF/BLyS receptor 3 binds the B cell survival factor BAFF ligand through a discrete surface loop and promotes processing of NF-kappaB2. Immunity 2002;17:515–24.
16. Ramanujam M, Wang X, Huang W, et al. Similarities and differences between selective and nonselective BAFF blockade in murine SLE. J Clin Invest 2006;116:724–34.
17. Ramanujam M, Wang X, Huang W, et al. Mechanism of action of transmembrane activator and calcium modulator ligand interactor-Ig in murine systemic lupus erythematosus. J Immunol 2004;173:3524–34.
18. Lin WY, Gong Q, Seshasayee D, et al. Anti-BR3 antibodies: a new class of B-cell immunotherapy combining cellular depletion and survival blockade. Blood 2007;110: 3959–67.
19. Huntington ND, Tomioka R, Clavarino C, et al. A BAFF antagonist suppresses experimental autoimmune encephalomyelitis by targeting cell-mediated and humoral immune responses. Int Immunol 2006;18:1473–85.
20. Matsushita T, Fujimoto M, Hasegawa M, et al. BAFF antagonist attenuates the development of skin fibrosis in tight-skin mice. J Invest Dermatol 2007;127:2772–80.
21. Kalled SL, Ambrose C, Hsu YM. The biochemistry and biology of BAFF, APRIL and their receptors. Curr Dir Autoimmun 2005;8:206–42.
22. Neri P, Kumar S, Fulciniti MT, et al. Neutralizing B-cell activating factor antibody improves survival and inhibits osteoclastogenesis in a severe combined immunodeficient human multiple myeloma model. Clin Cancer Res 2007;13:5903–9.
23. Yaccoby S, Pennisi A, Li X, et al. Atacicept (TACI-Ig) inhibits growth of TACI(high) primary myeloma cells in SCID-hu mice and in coculture with osteoclasts. Leukemia 2008 Feb;22(2): 406–13.
24. Baker KP, Edwards BM, Main SH, et al. Generation and characterization of LymphoStat-B, a human monoclonal antibody that antagonizes the bioactivities of B lymphocyte stimulator. Arthritis Rheum 2003;48:3253–65.
25. Halpern WG, Lappin P, Zanardi T, et al. Chronic administration of belimumab, a BLyS antagonist, decreases tissue and peripheral blood B-lymphocyte populations in cynomolgus monkeys: pharmacokinetic, pharmacodynamic, and toxicologic effects. Toxicol Sci 2006;91: 586–99.
26. Vugmeyster Y, Seshasayee D, Chang W, et al. A soluble BAFF antagonist, BR3-Fc, decreases peripheral blood B cells and lymphoid tissue marginal zone and follicular B cells in cynomolgus monkeys. Am J Pathol 2006;168:476–89.
27. O'Connor BP, Raman VS, Erickson LD, et al. BCMA is essential for the survival of long-lived bone marrow plasma cells. J Exp Med 2004;199:91–8.
28. Cheema GS, Roschke V, Hilbert DM, et al. Elevated serum B lymphocyte stimulator levels in patients with systemic immune-based rheumatic diseases. Arthritis & Rheum 2001;44: 1313–9.
29. Stohl W, Xu D, Kim KS, et al. BAFF overexpression and accelerated glomerular disease in mice with an incomplete genetic predisposition to systemic lupus erythematosus. Arthritis Rheum 2005;52:2080–91.

30. Anolik JH, Barnard J, Owen T, et al. Delayed memory B cell recovery in peripheral blood and lymphoid tissue in systemic lupus erythematosus after B cell depletion therapy. Arthritis Rheum 2007;56:3044–56.
31. Zhang J, Roschke V, Baker KP, et al. Cutting edge: a role for B lymphocyte stimulator in systemic lupus erythematosus. Journal of Immunology 2001;166:6–10.
32. Stohl W, Metyas S, Tan SM, et al. B lymphocyte stimulator overexpression in patients with systemic lupus erythematosus: longitudinal observations. Arthritis Rheum 2003;48:3475–86.
33. Looney RJ, Anolik J, Sanz I. Treatment of SLE with anti-CD20 monoclonal antibody. Curr Dir Autoimmun 2005;8:193–205.
34. Leandro MJ, Cambridge G, Edwards JC, et al. B-cell depletion in the treatment of patients with systemic lupus erythematosus: a longitudinal analysis of 24 patients. Rheumatology 2005;44:1542–5.
35. Hansen A, Lipsky PE, Dorner T. Immunopathogenesis of primary Sjogren's syndrome: implications for disease management and therapy. Curr Opin Rheumatol 2005;17:558–65.
36. Looney RJ, Looney RJ. Will targeting B cells be the answer for Sjogren's syndrome? [comment]. Arthritis Rheum 2007;56:1371–7.
37. Dawson LJ, Fox PC, Smith PM, Dawson LJ, Fox PC, Smith PM. Sjogrens syndrome–the non-apoptotic model of glandular hypofunction. Rheumatology 2006;45:792–8.
38. Theander E, Henriksson G, Ljungberg O, et al. Lymphoma and other malignancies in primary Sjogren's syndrome: a cohort study on cancer incidence and lymphoma predictors. [see comment]. Ann Rheum Dis 2006;65:796–803.
39. Ittah M, Miceli-Richard C, Eric Gottenberg J, et al. B cell-activating factor of the tumor necrosis factor family (BAFF) is expressed under stimulation by interferon in salivary gland epithelial cells in primary Sjogren's syndrome. Arthritis Res Ther 2006;8:R51.
40. Mariette X, Roux S, Zhang J, et al. The level of BLyS (BAFF) correlates with the titre of autoantibodies in human Sjogren's syndrome. Ann Rheum Dis 2003;62:168–71.
41. Pijpe J, van Imhoff GW, Spijkervet FK, et al. Rituximab treatment in patients with primary Sjogren's syndrome: an open-label phase II study. Arthritis Rheum 2005;52:2740–50.
42. Devauchelle-Pensec V, Pennec Y, Morvan J, et al. Improvement of Sjogren's syndrome after two infusions of rituximab (anti-CD20). Arthritis Rheum 2007;57:310–7.
43. Seror R, Sordet C, Guillevin L, et al. Tolerance and efficacy of rituximab and changes in serum B cell biomarkers in patients with systemic complications of primary Sjogren's syndrome. Ann Rheum Dis 2007;66:351–7.
44. Steinfeld SD, Tant L, Burmester GR, et al. Epratuzumab (humanised anti-CD22 antibody) in primary Sjogren's syndrome: an open-label phase I/II study. Arthritis Res Ther 2006;8:R129.
45. Strunk RC, Bloomberg GR, Strunk RC, Bloomberg GR. Omalizumab for asthma. [see comment]. N Engl J Med 2006;354:2689–95.
46. Moon EY, Ryu SK, Moon E-Y, Ryu S-K. TACI:Fc scavenging B cell activating factor (BAFF) alleviates ovalbumin-induced bronchial asthma in mice. Exp Mol Med 2007;39:343–52.
47. Furie R, Stohl W, Ginzler E, et al. Safety, pharmacokinetic and pharmacodynamic results of a phase 1 single and double dose-escalation study of Lymphostat-B (human monoclonal antibody to BLyS) in SLE patients. Arthritis Rheum 2003;48:S377.
48. McKay J, Chwalinska-Sadowska H, Boling E, et al. Belimumab (BmAb), a fully human monoclonal antibody to B-Lymphocyte Stimulator (BLyS), combined with standard of care therapy reduces the signs and symptoms of rheumatoid arthritis in a heterogenous subject population. Arthritis Rheum 2005;52:S710–1.
49. Wallace DJ, Lisse J, Stohl W, et al. Belimumab (BmAb) reduces SLE disease activity and demonstrates durable bioactivity at 76 weeks. Arthritis Rheum 2006;54:S790.
50. Dall'Era M, Chakravarty E, Wallace D, et al. Reduced B lymphocyte and immunoglobulin levels after atacicept treatment in patients with systemic lupus erythematosus: Results of a multicenter, phase ib, double-blind, placebo-controlled, dose-escalating trial. Arthritis Rheum 2007;56:4142–50.

51. Tak PP, Thurlings RM, Rossier C, et al. Atacicept in patients with rheumatoid arthritis. Results of a multicenter, phase Ib, double-blind, placebo-controlled, dose-escalating, single- and repeated-dose study. Arthritis Rheum 2008;58:61–72.
52. Nestorov I, Pena Rossi C, Munafo A, et al. Pharmacokinetics and biological activity of atacicept after intravenous and subcutaneous administration to SLE patients. Ann Rheum Dis 2007;66:S57 (abstract).
53. Shaw M, Del Guidice J, Trapp R, et al. The safety, pharmacokinetic (PK) and pharmacodynamic (PD) effects of repeated doses of BR3-Fc in patients with rheumatoid arthritis (RA). Arthritis Rheum 2007;56:S568–9.
54. Sabahi R, Owen T, Barnard J, et al. Immunologic effects of BAFF antagonism in the treatment of human SLE. Arthritis Rheum 2007;56:S566.
55. Kavanaugh A, Rosengren S, Lee SL, et al. Assessment of rituximab's immunomodulatory synovial effects (the ARISE trial). I: clinical and synovial biomarker results. Ann Rheum Dis 2008 Mar;67(3):402–8.
56. Vos K, Thurlings RM, Wijbrandts CA, et al. Early effects of rituximab on the synovial cell infiltrate in patients with rheumatoid arthritis. Arthritis Rheum 2007;56:772–8.
57. Looney RJ, Anolik J, Sanz I. B lymphocytes in systemic lupus erythematosus: lessons from therapy targeting B cells. Lupus 2004;13:381–90.
58. Ahuja A, Shupe J, Dunn R, et al. Depletion of B cells in murine lupus: efficacy and resistance. J Immunol 2007;179:3351–61.
59. Gong Q, Ou Q, Ye S, et al. Importance of cellular microenvironment and circulatory dynamics in B cell immunotherapy. J Immunol 2005;174:817–26.
60. Cambridge G, Stohl W, Leandro MJ, et al. Circulating levels of B lymphocyte stimulator in patients with rheumatoid arthritis following rituximab treatment: relationships with B cell depletion, circulating antibodies, and clinical relapse. Arthritis Rheum 2006;54:723–32.
61. Piatelli MJ, Tanguay D, Rothstein TL, Chiles TC. Cell cycle control mechanisms in B-1 and B-2 lymphoid subsets. Immunol Res 2003;27:31–52.
62. Rowley B, Tang L, Shinton S, Hayakawa K, Hardy RR. Autoreactive B-1 B cells: constraints on natural autoantibody B cell antigen receptors. J Autoimmun 2007;29:236–45.
63. Macallan DC, Wallace DL, Zhang Y, et al. B-cell kinetics in humans: rapid turnover of peripheral blood memory cells. Blood 2005;105:3633–40.
64. Sasaki Y, Derudder E, Hobeika E, et al. Canonical NF-kappaB activity, dispensable for B cell development, replaces BAFF-receptor signals and promotes B cell proliferation upon activation. Immunity 2006;24:729–39.
65. Sasaki Y, Schmidt-Supprian M, Derudder E, Rajewsky K. Role of NFkappaB signaling in normal and malignant B cell development. Adv Exp Med Biol 2007;596:149–54.
66. Castigli E, Wilson SA, Garibyan L, et al. TACI is mutant in common variable immunodeficiency and IgA deficiency. Nat Genet 2005;37:829–34.
67. Zhang L, Radigan L, Salzer U, et al. Transmembrane activator and calcium-modulating cyclophilin ligand interactor mutations in common variable immunodeficiency: clinical and immunologic outcomes in heterozygotes. J Allergy Clin Immunol 2007;120:1178–85.
68. Lee WI, Huang JL, Kuo ML, et al. Analysis of genetic defects in patients with the common variable immunodeficiency phenotype in a single Taiwanese tertiary care hospital. Ann Allergy Asthma Immunol 2007;99:433–42.
69. Peano S, Ponce R, Bertolino M, et al. Nonclinical safety, pharmacokinetics, and pharmacodynamics of TACI-Ig: a soluble receptor fusion protein antagonist of BLyS (B Lymphocyte Stimulator) nad APRIL (A Proliferation-Inducing Ligand). Arthritis Rheum 2005;52(9 suppl):S285.

Chapter 11
BLyS/BR3 Receptor Signaling in the Biology and Pathophysiology of Aggressive B-Cell Lymphomas

Lingchen Fu, Lan V. Pham, Yen-chiu Lin-Lee, Archito T. Tamayo, and Richard J. Ford

Abstract Lymphoid neoplasms usually display various genetic alterations that dysregulate normal immune functions, primarily through abnormal cellular growth and survival mechanisms that expand and immortalize a single neoplastic clone, to the detriment of the indigenous normal immune system, and eventually the host. In addition to the nonrandom genetic translocations, other important cellular abnormalities are also involved. The abnormalities in cellular regulation generally occur in one or more of the following four areas: cell death, cell cycle progression, NF-κB activation, or critical signal transduction pathways, usually from cell surface growth and survival receptors. Recent studies of BLyS/BAFF family members, BLyS and APRIL, have shown their importance in non-Hodgkin's B-cell lymphoma (NHL-B), chronic lymphoid leukemia (CLL), and plasma cell myeloma cell growth and survival. In this chapter, we will discuss how BLyS is involved in mechanisms of tumor cell proliferation and survival in these malignant lymphoid cells. Potential therapeutic approaches to these lymphoid malignances through inhibition of BLyS signaling pathways are also discussed.

Keywords BLyS · BAFF · NHL-B · LBCL · MCL · NF-κB · NF-AT

11.1 LBCL and MCL: Representative Aggressive NHL-Bs

Non-Hodgkin's lymphomas (NHLs) are a common, heterogeneous group of primarily B lymphoid cell tumors (NHL-B) of the human immune (lymphoid) system. About 85% of NHLs are derived from the B-cell lineage (NHL-B) that ranks fifth in cancer incidence and mortality, among the most common cancers in the USA (∼63,000 new cases expected in 2007). The incidence of NHL has increased 84%

R.J. Ford (✉)
Department of Hematopathology, The University of Texas M.D. Anderson Cancer Center, Houston, Texas, USA

M.P. Cancro (ed.), *BLyS Ligands and Receptors*, Contemporary Immunology,
DOI 10.1007/978-1-60327-013-7_11,

from 1975 to 2004, an average annual percentage increase of 2.8%, thought to be due to a combination of environmental and genetic factors [1]. NHL-B are very heterogeneous tumors, with both indolent and aggressive lymphoid neoplasms comprising the 14 histotypes listed in the current (but soon to be updated) WHO classification. Large B-cell lymphoma and mantle cell lymphoma are two common types of aggressive NHL-B [2] that are generally representative of aggressive NHL-B.

11.2 Large B-Cell Lymphoma

Large B-cell lymphoma (LBCL) is the most common lymphoma, accounting for 30–40% of lymphoid neoplasms [3, 4]. LBCL cells are large transformed lymphocytes (Fig. 11.1a, *left panel*) that have been further divided into morphologic variants: centroblastic, immunoblastic, T-cell/histiocyte-rich, and plasmablastic

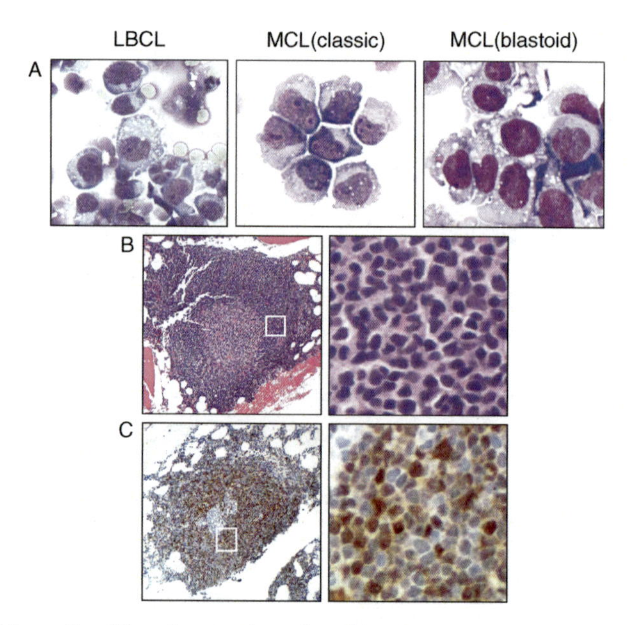

Fig. 11.1 (**a**) Large B-cell lymphoma and mantle cell lymphoma. *Left panel* shows clinical Hematoxylin and Eosin (H&E)-stained specimen with large B-cell lymphoma (400×), *middle panel* is H&E-stained classic mantle cell lymphoma cell line (MINO) (600×), and *right panel* is H&E-stained blastoid variant mantle cell lymphoma cell line (JMP1) (400×). (**b**) A bone marrow biopsy sample from a patient with typical mantle cell lymphoma demonstrates a germinal center surrounded by typical MCL cells, stained with H&E (100×) (*left panel*). And a higher magnification of section was shown in *right panel* (400×) (**c**) A bone marrow biopsy sample from the same patient demonstrates a germinal center surrounded by typical MCL cells, stained with cyclin D1 by immunohistochemistry (*left panel*) (100×). And a higher magnification of section was shown in *right panel* (400×)

[3, 4], but such morphologic variants generally do not have clinical or survival significance.

LBCLs have been proposed to originate from germinal center (GC) or post-GC B cells, due to the evidence of immunoglobulin gene somatic hypermutation (SHM) and expression of GC-associated genes [4–6]. Germinal center-like LBCL (GCB-LBCL), activated B-cell-like LBCL (ABC-LBCL), and primary mediastinal LBCL (PM-BCL) have been defined as the three major subgroups of LBCL, identified by genotyping using gene expression profiling and immunophenotyping [4]. GCB-LBCL appears to be derived from germinal center B cells, while ABC-LBCL have been proposed as being derived from a post-germinal center B cell and PMBCL may be derived from a thymic B cell, but shows similarities to both GC and ABC-LBCLs [4]. This type of classification is by no means absolute, as overlap in characteristics is common, and future studies will likely produce a more significant classification.

Several genetic abnormalities have been identified in subsets of LBCL due to nonrandom chromosomal translocations occurring during SHM. The three most frequently deregulated genes in LBCL are BCL6, BCL2, and cMYC [4, 6, 7]. The deregulation of these genes is due to chromosomal translocation that brings the target gene (oncogene) under inappropriate enforced expression, usually of an immunoglobulin regulatory element (IgH enhancer), resulting in constitutive over-expression of these genes. Abnormal constitutive expression of BCL6 may lead to maturation arrest, increased or hyperproliferation, and promote persistence of malignant clones, by inhibiting the p53-mediated apoptotic response to DNA damage [8]. The t(14;18) translocation, is present in approximately 15% of LBCLs that most commonly associated with BCL2 oncogene deregulation. Overexpression of the BCL-2 gene promotes the lymphoma cell escaping from apoptosis in the NHL-B pathophysiology. The cMYC oncogene, a powerful transcription factor associated with Burkitt's lymphoma, is deregulated in approximately 15% of LBCLs, which is most commonly associated with a t(8;14) chromosomal translocation. Constitutive expression of cMYC contributes to but is not exclusively responsible for abnormal proliferation of lymphoma cells in NHL-like Burkitt's lymphoma and some LBCL [7, 9].

Furthermore, about 20% of LBCLs have been reported as having FAS death domain (DD) mutations [10]. Loss of FAS leads to affinity maturation defects and failure to negatively select autoreactive B cells, which in some cases, may lead to development of a malignant clonal expansion [11]. The tumor suppressor gene Tsp53 (p53) has been associated with poor outcomes in LBCL patients [12, 13]. Previous studies suggested that up to 14% of LBCL patients had increased c-Rel expression, a component of the important transcriptional factor NF-κB family, that might be associated with increased NF-κB activity [14–16] (see below).

The cure rates of each LBCL subgroup in the pre-Rituxin (anti-CD20 antibody) treatment era were reported as being significantly different, with ABC-LBCL, GCB-LBCL, and PMBCL having 5-year survival rates of 30%, 59%, and 64%, respectively [17, 18]. Since Rituxan (rituximab) has been added to the standard CHOP(R-CHOP) chemotherapy regimen, these differences appear to have largely disappeared, and new approaches and tumor markers will likely be needed for identifying prognostically significant subgroups of LBCL [19–22].

The CHOP regimen (cyclophosphamide, doxorubicin, vincristine, and prednisone), which now includes Rituxan (R-CHOP), is now considered to be standard therapy. A more intensive strategy, however, involves high-dose chemotherapy (HDT) followed by autologous stem-cell transplantation (ASCT) [23]. Rituxan is a recombinant humanized monoclonal antibody against the pan-B-cell marker CD20 [20, 21], which is added to CHOP(R-CHOP) to treat LBCL patients with BCL-2 overexpression [22]. Based on gene expression profiling, several novel therapeutic approaches have been discovered, which target on PKC, PDE4B, NF-κB, NFAT, CD40, BLyS, Bcl-2, and Bcl-6 [8, 24–33].

11.3 Mantle-Cell Lymphoma

Mantle-cell lymphoma (MCL) is a subtype of B-cell non-Hodgkin's lymphomas (B-NHLs), immunophenotypically resembling B cells in GC mantle zone (Fig. 11.1b). MCL accounts for approximately 6% of NHL-B, but is currently increasing among the NHL-B histotypes [34, 35]. CD5 is expressed in most cases [36], suggesting that MCL may be derived from the B1 subset of the B-cell lineage, active in the inherent immune system in body cavities in infants/children, as opposed to the more prevalent B2 cells of the adaptive immune [37]. There are two main cytologic variants of MCL. One is the typical subtype with small to medium-sized neoplastic cells (Fig. 11.1a, *middle panel*); the other is the blastoid subtype that contains larger, morphologically transformed appearing as lymphoblast-like tumor cells [34, 38] (Fig. 11.1a, *right panel*). Based on various studies, the blastoid variant of MCL is associated with a more aggressive clinical behavior and greater drug resistance [39–41].

The chromosomal translocation t(11;14)(q13;q32) that can be detected in all but rare cases of MCL [34, 36], the juxtaposition of the immunoglobulin heavy chain (IgH) joining region on chromosome 14 to a chromosomal locus on 11q13 designated CCND1/bcl-1 (B-cell lymphoma/leukemia 1, Cyclin D1), thus bringing CCND1 under the control of the IgH enhancer, which leads to deregulation and overexpression of cyclin D1 (Fig. 11.1c), an important molecule in G1 phase progression in the cell cycle [41, 42]. Beside the initial description of the translocation t(11;14), several other oncogenic genetic events have been found in MCL [10]. It was reported that the homozygous deletions of the cyclin-dependent kinase (CDK) inhibitor $p16^{INK4a}$ on chromosome 9p21 were detected in some aggressive MCL cases [43, 44]. $p16^{INK4a}$ acts as an inhibitor of CDK4 and -6 and thus maintains the Rb protein in its active, antiproliferative state [45]. Deletion of $p16^{INK4a}$ and increased levels of cyclin D1 may therefore cooperate in promoting G1/S-phase transition in MCL cells by increasing the intracellular amount of active cyclin D1/CDK complexes [46]. The genomic locus $p16^{INK4a}$ also encodes for an alternative spliced transcript, $p14^{ARF}$, therefore homozygous deletions of this locus can affect both $p14^{ARF}$ and $p16^{INK4a}$ expression in MCL. Inactivation of this locus leads to simultaneous inhibition of the cell cycle regulatory and the p53 pathways. Deletions of the INK4a/ARF locus can be detected in approximately 20% of MCL cases, and the disruption of both pathways seems to be associated with a more aggressive clinical behavior than inactivation of the p53 pathway alone [47–50].

Mantle cell lymphoma is an aggressive lymphoma with the poorest long-term survival of any NHL-B subtype that is usually present in an advanced (widely disseminated) clinical stage. MCL has low sensitivity to most of conventional chemotherapies. CHOP produces a median survival of only 3 years and no curative option is available in advanced disease [51]. There are several approaches to treatment for MCL including chemotherapy plus monoclonal antibody therapy; high-dose therapy supported by autologous stem-cell transplantation; allogeneic bone marrow transplantation; the use of thalidomide plus rituximab; as well as newer with targeting agents like the protease inhibitor bortezomib (velcade), the cyclin-dependent kinase modulator flavopiridol, the mTOR signal transduction inhibitor temsirolimus (CCI 779), and radioimmunotherapy [52–55]. Potential idiotype vaccines directed against an antigen derived from the patients' own tumor has also begun to be explored in B-cell lymphomas including MCL [56, 57].

In the near future, further defining a MCL molecular signature that identifies alterations in gene sets, the expression of which can be associated with clinical parameters, will likely impact treatment decisions. Molecular profiling may enable physicians to identify subsets of patients with more indolent or aggressive MCL or subsets of genes that predict treatment responses [58–63].

11.4 Abnormal BLyS Expression in NHL-Bs

B-lymphocyte stimulator (BLyS; also known as BAFF, TALL-1, THANK, zTNF4, and TNFSF13B) is one of the newer members of tumor necrosis factor (TNF) family that was identified in 1999 by several groups of researchers and was rapidly recognized as a potent cell-survival factor for most B cells [64–66]. BlyS-deficient mice lack a mature B-cell compartment (follicular and marginal zone B lymphocytes), suggesting that the expression of this gene is necessary to maintain mature B-cell development [67]. Phenotypes present in BLyS transgenic mice included increased mature B cells in the periphery, enlarged lymph nodes and spleens, and hypergammaglobulinemia with autoimmune-like manifestations [68–70]. Since it is well established that risk for B-cell malignancy is higher in patients with autoimmune diseases such as SLE [71], and abnormal apoptosis control is known to be one of the key mechanisms of oncogenesis, BLyS was predicted to be a contributor in B-cell malignancy pathogenesis [72]. This prediction was confirmed by studies from several groups: TNF deficiency in BAFF-Tg mice led to a surprisingly high incidence of B-cell lymphomas (35%) [67]. Relatively high levels of BLyS was found in the serum of patients with NHL-Bs, and the serum levels of BLyS were significantly elevated in patients with follicular lymphomas ($n = 17$, mean $\pm$ SD; 13.4 ± 5.6 ng/mL) compared to that in healthy donors ($n = 15$, 4.6 ± 0.7 ng/mL; p 0.0001) [73]. Later, a subset of patients with LBCL was found having elevated serum BLyS levels (26.84 ng/mL (mean, 33.52 ± 4.18; range, undetectable to 161 ng/mL; $n = 56$), compared to those in healthy controls (17.95 ng/mL (mean, 18.24 ± 4.27; range, 7.62–32.41 ng/mL; $n = 6$). Furthermore, the BLyS level correlated significantly with overall patient clinical outcome, in that surviving patients had low BLyS levels (40 ng/mL) at diagnosis. Patients with high serum BLyS had a high mortality (75%), while 51% of patients with low serum BLyS levels also died, but the median OS was 39 months compared with 78 months ($P = 0.05$). Elevated pretreatment BLyS levels were also found to correlate with other clinical features including constitutional symptoms (fever, sweating, or weight loss) and an elevated serum LDH. Further studies showed that in vitro exogenous BLyS stimulation promoted the NHL-B-cell survival [73]. Although elevated serum BLyS level is consistently observed in patients with NHL-Bs, the underlying mechanism leading to this elevation is not clear.

As described by several groups of investigators, BLyS is a type-II transmembrane protein that is released in a soluble form after proteolytic processing [64–66]. It is expressed by monocytes, macrophages, and dendritic cells, as well as growth factor-stimulated neutrophils. Cytokines such as IL-10, IFN-γ, and IFN-α induce BLyS expression in monocytes, macrophages, and dendritic cells, while CD40 ligand stimulates BLyS secretion from dendritic cells [74–80]. However, BLyS expression is not found in normal Go peripheral blood B cells [64]. Studies from other groups as well as our laboratory have shown that BLyS and its receptors are constitutively expressed by several types of NHL-B cells including LBCL and MCL [73, 81–84]. These results indicate that BLyS promotes NHL-B-cell survival through an autocrine pathway. Various studies have also shown that inhibiting BLyS protein using decoy receptors or blocking antibodies, restores the apoptotic process in malignant B cells and myeloma cells [73, 82, 85, 86], resulting in neoplastic B-cell death.

11.5 BLyS Signaling Pathway in NHL-Bs

In order to understand mechanisms involving BLyS functions in B-cell survival and proliferation, the BLyS signaling pathway has been studied by several groups. To date, multiple signaling pathways have been reported to be involved in BLyS-induced normal B cell or malignant B-cell survival and proliferation [87, 88]. Among these pathways, the central signaling pathways involving the NF-κB superfamily of critical transcription factors (TF) , appear to be a major one.

11.6 NF-κB Signaling Pathways

NF-κB is a family of transcription factors that plays a critical role in the immune system in many biological processes such as immune and inflammatory responses, developmental processes, cellular growth, and apoptosis [89–91]. NF-κB also regulates the expression of genes outside of the immune system and can influence multiple aspects of normal and disease cell physiology [91–93]. There are five members in the mammalian NF-κB family: p65 (RelA), RelB, c-Rel, p50/p105 (NF-κB1), and p52/p100 (NF-κB2). Most stimuli that activate NF-κB, modulate the activity of the IκB kinase (IKK) complex, which is comprised of three subunits: IKKα, IKKβ, and IKKγ (NEMO). IKKα and IKKβ are both catalytic kinases, while IKKγ is a regulatory component for the IKK complex. It is now clear that there are at least two NF-κB pathways: the classic (canonical) pathway, involved in rapid (e.g., inflammatory) responses, and a slower alternative pathway, with less clear functions. Upon stimulation, the IKK complex phosphorylates the IκB proteins, which then undergo rapid ubiquitination and proteasome-mediated proteolysis and degradation. The activity of classic (canonical) NF-κB is primarily regulated through interaction with inhibitory IκB proteins. There are several IκB proteins, including p105, p100, IκBα, IκBβ, IκBγ, IκBϵ, IkBζ, Bcl-3, and the Drosophila Cactus protein. This process induces the release of the NF-κB family member complexes from their inhibitory cytoplasmic interaction, and the translocation from cytoplasm to nucleus. This NF-κB signaling pathway is called classical (or canonical) pathway, which is mainly comprised of p65, c-rel, and p50 [91]. Another NF-κB pathway can also be activated through a second signaling mechanism called NF-κB2 (or alternative) pathway, in which the NF-κB2 precursor p100 protein is proteolytically partially degraded to form an active p52 protein in a manner critically dependent on the NF-κB inducing kinase (NIK) and IKKα. Both mechanisms of NF-κB activation play important roles in B-cell survival [89–97], but interaction (crosstalk) between the two NF-κB pathways may be important, which is poorly understood. In quiescent cells NF-κB proteins exist as cytoplasmic homo-or heterodimers bound to IκB family proteins that have strong nuclear export signals. This interaction between NF-κB protein and IκB protein covers the nuclear localization sequence (NLS), blocking the ability of NF-κB to bind to DNA by sequestering NF-κB TFs, impeding DNA binding, and resulting in the NF-κB complex being primarily found in the cytoplasm [89–93].

The initial cloning of the NF-κB p50/p105 subunit cDNA revealed homology to the cellular homolog (c-Rel) of the oncoprotein (v-Rel) of the avian reticuloendotheliosis virus, suggesting a possible link between NF-κB and oncogenesis [98, 99]. Subsequently, it was shown that the NF-κB p52/p100 subunit, encoded by the NF-κB2 gene, undergoes structural alterations in certain T-cell lymphomas, chronic lymphocytic leukemias, myelomas, and B-cell lymphomas [100]. Similarly, the gene for the IκB family member Bcl-3 is also found to be translocated and overexpressed in some B-cell leukemias. In addition, c-Rel gene amplification has been detected in several types of B-cell lymphoma [100–103]. Constitutive activation of NF-κB family members is a molecular "signature" that has been reported in several malignant B-cell types [90, 96, 100, 103], but the pathophysiologic mechanism for constitutive maintaining NF-κB activation is not fully established.

Previous studies have also shown that BLyS can activate NF-κB2 through the ubiquitin-mediated proteolysis of p100 [104, 105] and the NF-κB1 pathway through the proteolysis of IκBα in vitro [84, 106, 107]. These studies suggest that activation of NF-κB family members plays a major role in critical BLyS signaling pathways in the mature B-cell lineage, but downstream molecules of the major receptor of BLyS BAFF-R have not been elucidated, although two negative regulators of NF-κB pathway, TNF receptor associate factor 3 (TRAF3) and Act1, have been reported interacting with BAFF-R [108, 109].

11.7 NFAT Signal Pathway

The nuclear factor of activated T cells (NFAT) is another important family of transcription factors (TFs) that have key roles in the transcriptional activation of T-lymphocyte cytokines such as IL-2, IL-4, CD40L, and Fas ligand [110]. In resting T lymphocytes, these transcription factors are maintained in the cytoplasm in the phosphorylated state. In response to increasing intracellular concentrations of calcium, the N-terminus of NFAT protein is dephosphorylated by a calcium-sensitive calmodulin-related protein phosphatase, calcineurin, that induces a molecular conformation change of leading to exposure of the NFAT protein NLS, which leads to NFAT TFs translocation from the cell cytoplasm into the nucleus [110–113]. Recent studies in our laboratory and others have shown that the NFAT family members NFATc1 and NFATc2 are constitutively activated in malignant B cells [30, 114], suggesting that both NFAT and NF-κB activation likely contribute to and in some cases interact in transcriptional mechanisms involved in malignant B-cell survival.

11.8 BLyS Expression Regulation

Several groups of researchers have reported that BLyS is expressed by various cell types in the immune system [64–66, 74, 75], as well as in multiple types of malignant B cells, including multiple myeloma cells [73, 81, 84–86]; however, the nature of BLyS protein regulation and its expression is still incompletely understood. A number of reports indicate that in aggressive NHL-B cells, NF-κB is constitutively activated and mediates neoplastic B-cell growth and cell survival [90, 96, 100–103]. Recent studies from our laboratory and others have shown that activated NFAT expression is also detected in malignant B cells [30, 114], suggesting an important role for NFAT in malignant B-cell growth and survival. Although it has been previously shown that NF-κB, NFAT, and BLyS are important components that maintain B-cell survival and proliferation, the mechanism(s) of their persistent activation and expression, as well as putative interactions between these proteins in malignant B cells, are still unclear. Previous studies have also shown that CD40L (CD154) stimulation induces expression of BLyS protein in normal G0 B cells and increases BLyS expression in Burkitt's lymphoma cells [82], but the underlying mechanism of this observation remains unclear. Another study showed that BLyS expression was increased in EBV LMP1 positive BJAB cells due to constitutive NF-κB activation [115]. In our study, we showed that constitutive BLyS expression was regulated by both NF-κB and NFAT transcription factors (Fig. 11.2). The NF-κB family

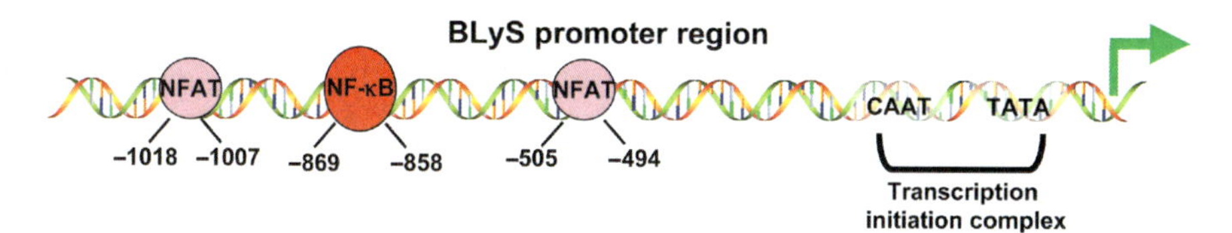

Fig. 11.2 BlyS gene promoter diagram (GenBank accession no. AY129225) showing the putative binding sites for NF-κB (*red*) and NFAT (*pink*). There are one putative NF-κB-binding site and two putative NFAT-binding sites found on the BLyS promoter

members p50, p52, p65, and c-Rel are major components that bind to the BLyS promoter (Fig. 11.3a). NFATc1 and NFATc2 are two components in NFAT-binding complex in the BLyS promoter (Fig. 11.3b). Inhibition of NF-κB or NFAT binding to the BLyS promoter decreases BLyS promoter activity as well as BLyS protein expression. Our study also showed that stimulation with exogenous CD40L (CD154) and anti-IgM-induces BLyS expression in normal human B cells and activates both NF-κB and NFAT binding to the BLyS promoter. This finding supports our contention that NF-κB and NFAT are important regulators of BLyS protein

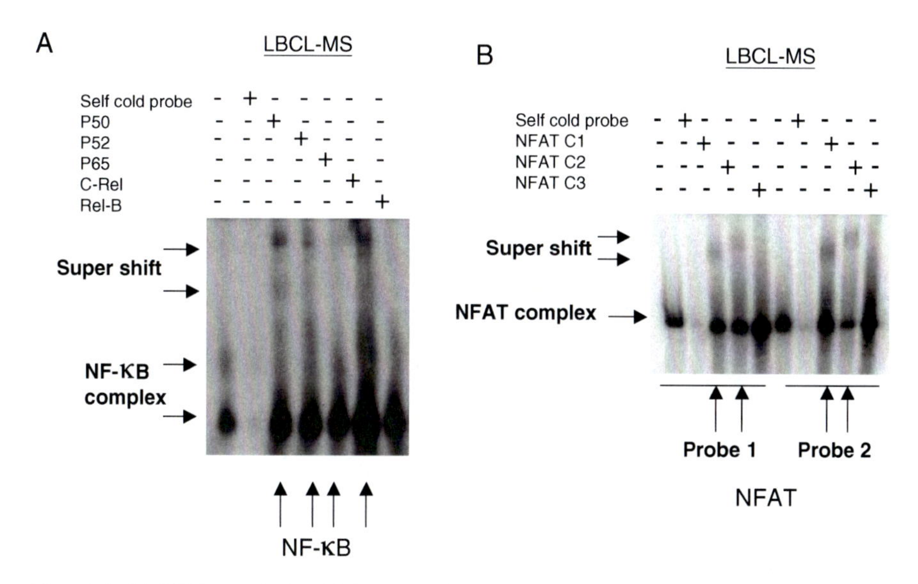

Fig. 11.3 (**a**) EMSA analysis of NF-κB binding to the *BLyS* promoter. Nuclear extracts from NHL-B (MS) cells were incubated with BlyS-NF-κB-binding site oligonucleotides. BLyS-NF-κB cold probe and antibodies to p50, p52, p65, c-rel, or rel-B were added to the binding reaction mixtures. *Arrows* indicate the DNA–protein and super-shifted complex. The NF-κB-binding complex includes p50, p52, c-rel and less p65 (**b**) EMSA analysis of NFAT binding to BLyS promoter. Nuclear extracts from NHL-B (LBCL-MS) cells were incubated with oligonucleotides for the two different BlyS-NFAT-binding sites. BLyS-NFAT cold probes and antibodies to NFATc1, NFATc2, or NFATc3 were added to the binding reaction mixtures. *Arrows* indicate the DNA–protein and super-shifted complexes. The NFAT-binding complex includes NFATc1 and NFATc2

expression and delineates a mechanism by which BLyS expression is induced in normal B cells in vitro after CD40L (CD154) stimulation. Similarly, we found that BLyS expression can be induced in vitro by CD40L and anti-IgM stimulation in grade I follicular B-cell lymphoma patient biopsy cells, suggesting that in this low-grade (indolent) B-cell lymphoma, NF-κB and NFAT activation also can induce BLyS expression. These findings also suggest that this type of low-grade NHL-B also depends on exogenous activation of the CD40/NF-κB pathway (perhaps from epigenetic/lymphoid stroma microenvironment), retaining similarity in this respect to normal B cells, that require exogenous activation rather than endogenous (constitutive) activation of BLyS, as observed in aggressive NHL-B cells [84].

One NF-κB-binding site (–1018~–1007) and two NFAT-binding sites (–869~–858, –505~–494) (Fig. 11.3) in the BLyS promoter have been identified. Mutation of these binding sites reduces NF-kB or NFAT components binding to BLyS promoter, decreases BLyS promoter activity and lowers BLyS protein expression. These findings indicate that constitutive activation of NF-κB and NFAT family members is critical for the transcriptional regulation of the BLyS survival pathway in both normal and malignant B cells. Recognizing this pathway's involvement in the pathophysiology of NHL-B may provide a useful future therapeutic target in aggressive NHL-B. Beside NF-κB and NFAT, a study from our laboratory also found that CD40, one of TNF receptor family members, associates with NF-κB component c-Rel and binds to BLyS promoter to promote BLyS expression [116, 117]. Other transcription factors possibly involved in regulation of BLyS expression still need to be defined.

11.9 Mechanisms of BLyS Signaling in NHL-B-Cell Survival and Proliferation

The full neoplastic phenotype in NHL-B cells cannot be accounted for by the translocation or mutation of a single dominant oncogene and/or the inactivation of a tumor suppressor gene(s), characteristic of these lymphomas, suggesting that undefined cellular/genetic (mutational) abnormalities in addition to the well-defined nonrandom chromosomal translocations are required for the transformed state and involved in the pathogenesis [118, 119]. We have previously shown that abnormal CD40L/CD40 signaling and constitutive NF-κB activation are involved in the pathogenesis of aggressive NHL-B in an interdependent manner [54, 103]. Several research groups have reported that in normal mature murine and human B cells, BLyS signaling stimulates both the NF-κB1 and NF-κB2 pathways, through one or more of the three BLyS receptors: BAFF-R, TACI, and BCMA [65, 70, 77, 79–86, 106, 107, 115]. Studies from our laboratory and others have shown that BLyS and all three BLyS receptors are expressed in NHL-B cells at various levels [73, 84]. These observations suggest that in aggressive NHL-B cells, BLyS induces cell survival through an autonomous signaling pathway(s). In vitro BLyS treatment can stimulate follicular lymphoma cell proliferation with anti IgM co-stimulation

[81]. The downstream pathways involved in NHL-B cells proliferation and survival induced by BLyS have been studied by several groups. One study showed that blocking BLyS binding to follicular lymphoma and LBCL cells by BCMA-Ig treatment decreases this malignant B-cell proliferation. Similar results were observed in NHL-B cells treated with a steroid NF-κB inhibitor, dexamethasone [82]. Further study indicates that constitutive expression of BLyS is important for maintaining endogenous activation of both the NF-κB1 and NF-κB2 signaling pathways in NHL-B cells because inhibition of BLyS protein expression by BLyS siRNA decreases the DNA-binding activity of various NF-κB components such as p50, p52, c-Rel, and the levels of pIκB and p52 protein were also reduced [84]. NF-κB-induced anti-apoptotic and proliferative gene expression is decreased when BLyS bioactivity is blocked in NHL-B cells, such as BCL-2 family proteins BCL-2, BCL-xL, and cell cycle genes c-myc and cyclin D1 [82, 84].

Based on the above evidence, we suggest that in addition to constitutive CD40L expression, in aggressive NHL-B cells, abnormal BLyS signaling and NF-κB activity form a positive feedback loop. NF-κB activation induces BLyS expression, while the upregulated BLyS ligand is in turn involved in maintaining constitutive NF-κB activation, after cognate receptor binding. We hypothesize that such a positive feedback loop then contributes to sustained NHL-B-cell survival and neoplastic cell proliferation (Fig. 11.3). Delineating the specific molecular mechanisms involved in NHL-B proliferation and survival may be crucial in elucidating further mechanism(s) of NHL-B pathogenesis, and could provide further therapeutic targets for aggressive NHL-B.

Besides NF-κB pathway, other pathways involved in BlyS-induced cell proliferation and survival also have been found in murine and human normal B cells or in myeloma cells including the AKT/mTOR [88, 120], Pim2 [88], ERK/MAPK [85, 121, 122], and PAX5 pathways [123], but how these pathways function in BlyS-mediated NHL-B-cell survival and proliferation is not yet clear.

11.10 BLyS As a Therapeutic Target in NHL-B Treatment

Since several studies show that BLyS plays an important role in NHL-B-cell survival and proliferation, BLyS becomes a promising target for future NHL-B therapy. In 2003, a study from Riccobene et al. [124] reported that radiolabeled BLyS showed specific and rapid targeting to lymphoid tissues and B-cell tumors in mice. Unlike monoclonal antibodies, which have long plasma half-lives and considerable liver uptake, BLyS has distinct pharmacokinetic and biodistribution properties that may offer advantages compared with antibody-based radioimmunotherapy. Later in 2006, Halpern et al. [125] reported that Belimumab, a BLyS antagonist, was safe and well tolerated in repeat-dose toxicology studies at 5–50 mg/kg for up to 26 weeks. Monkeys exposed to belimumab had significant decreases in peripheral blood B lymphocytes after 13 weeks of exposure, continuing into the recovery period, despite total lymphocyte counts similar to the controls. There were concomitant decreases in spleen and lymph node B-lymphocyte representation after 13 or 26 weeks of treatment with belimumab. Microscopically, monkeys treated with belimumab for 13 or 26 weeks had decreases in the number and size of lymphoid follicles in the white pulp of the spleen. All findings were generally reversible within a 34-week recovery period. These data confirm the specific pharmacologic activity of belimumab in reducing B lymphocytes in the cynomolgus monkey. The favorable safety profile and lack of treatment-related infections also support continued clinical development of belimumab. In addition, the recombinant Gelonin/BLyS fusion toxin was found to specifically bind to MCL and LBCL cells and induce these malignant cells apoptosis in vitro (IC_{50} = 2–5 pmol/L and 0.001–5 nmol/L for MCL and LBCL, respectively) [31]. Similar effects have been found in BAFF-R-positive CLL cells [126]. Furthermore, the monoclonal antibody of BAFF-R, the major receptor of BLyS, is being studied, which may be used in the treatments of autoimmune disease and B-cell malignancies [127]. All these studies suggest that BLyS and the molecules in its downstream pathway are promising targets for efficient therapy for certain NHL-B, such as MCL and LBCL (Figs. 11.4 and 11.5).

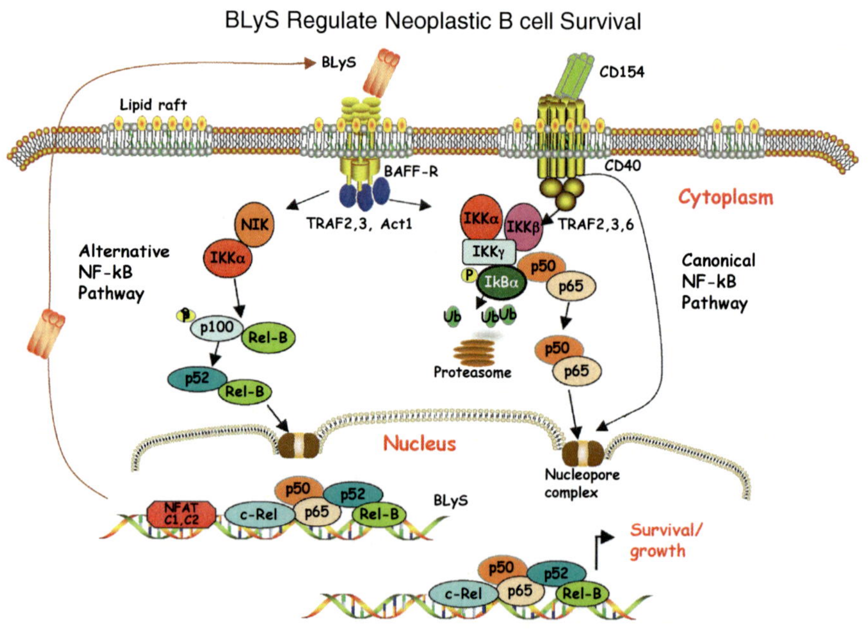

Fig. 11.4 A schematic model of our current conceptual hypothesis of how the BAFF-R signaling is able to generate unrestrained cell growth in aggressive NHL-B cells. In this model, the expression of the BLyS gene is dysregulated by constitutively activated NF-κB and NFAT transcription factors. BLyS protein continually binds to its cognate receptor BAFF-R, BCMA, and TACI on the plasma membrane of lymphoma cell, providing continuous signaling for NF-κB pathway activation. NF-κB activation and dysregulated BLyS expression form a positive feedback loop to promote lymphoma cells proliferation and survival

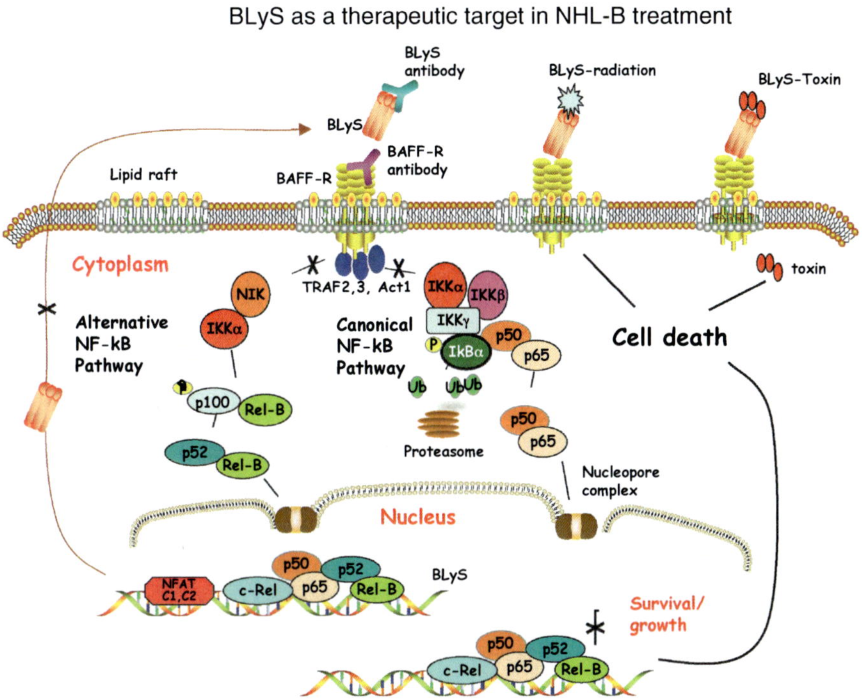

Fig. 11.5 A schematic model of current approaches for using BLyS as a therapeutic target in NHL-B treatment. Either BLyS antibody or BAFF-R antibody to specifically block BLyS signaling in NHL-B cell, radiation-labeled recombinant BlyS, or recombinant BLyS–toxin fusion protein can induce NHL-B-cell death

References

1. Jemal A, Siegel R, Ward E, Hao Y, Xu J, Thun MJ. Cancer statistics. CA Cancer J Clin 2009; 59(4):225–49.
2. Chan JK. The new World Health Organization classification of lymphomas: the past, the present and the future. Hematol Oncol 2001;19:129–50.
3. Armitage JO, Weisenburger DD. New approach to classifying non-Hodgkin's lymphomas: clinical features of the major histologic subtypes. Non-Hodgkin's Lymphoma Classification Project. J Clin Oncol 1998;16:2780–95.
4. Alizadeh AA, Eisen MB, Davis RE, et al. Distinct types of diffuse large B-cell lymphoma identified by gene expression profiling. Nature 2000;403:503–11.
5. Kuppers R. Mechanisms of B-cell lymphoma pathogenesis. Nat Rev Cancer 2005;5:251–62.
6. Shaffer AL, Rosenwald A, Staudt LM. Lymphoid malignancies: the dark side of B-cell differentiation. Nat Rev Immunol 2002;2:920–32.
7. Kramer MH, Hermans J, Wijburg E, et al. Clinical relevance of BCL2, BCL6, and MYC rearrangements in diffuse large B-cell lymphoma. Blood 1998;92:3152–62.
8. Phan RT, Dalla-Favera R. The BCL6 proto-oncogene suppresses p53 expression in germinal-centre B cells. Nature 2004;432:635–9.
9. Ladanyi M, Offit K, Jhanwar SC, Filippa DA, Chaganti RS. MYC rearrangement and translocations involving band 8q24 in diffuse large cell lymphomas. Blood 1991;77:1057–63.

10. Gronbaek K, Nedergaard T, Andersen MK, et al. Concurrent disruption of cell cycle associated genes in mantle cell lymphoma: a genotypic and phenotypic study of cyclin D1, p16, p15, p53 and pRb. Leukemia 1998;12:1266–71.
11. Muschen M, Rajewsky K, Kronke M, Kuppers R. The origin of CD95-gene mutations in B-cell lymphoma. Trends Immunol 2002;23:75–80.
12. Ichikawa A, Kinoshita T, Watanabe T, et al. Mutations of the p53 gene as a prognostic factor in aggressive B-cell lymphoma. N Engl J Med 1997;337:529–34.
13. Kerbauy FR, Colleoni GW, Saad ST, et al. Detection and possible prognostic relevance of p53 gene mutations in diffuse large B-cell lymphoma. An analysis of 51 cases and review of the literature. Leuk Lymphoma 2004;45:2071–8.
14. Bea S, Colomo L, Lopez-Guillermo A, et al. Clinicopathologic significance and prognostic value of chromosomal imbalances in diffuse large B-cell lymphomas. J Clin Oncol 2004;22:3498–506.
15. Houldsworth J, Olshen AB, Cattoretti G, et al. Relationship between REL amplification, REL function, and clinical and biologic features in diffuse large B-cell lymphomas. Blood 2004;103:1862–8.
16. Feuerhake F, Kutok JL, Monti S, et al. NFkappaB activity, function, and target-gene signatures in primary mediastinal large B-cell lymphoma and diffuse large B-cell lymphoma subtypes. Blood 2005;106:1392–9.
17. Rosenwald A, Wright G, Chan WC, et al. The use of molecular profiling to predict survival after chemotherapy for diffuse large-B-cell lymphoma. N Engl J Med 2002;346: 1937–47.
18. Bea S, Zettl A, Wright G, et al. Diffuse large B-cell lymphoma subgroups have distinct genetic profiles that influence tumor biology and improve gene-expression-based survival prediction. Blood 2005;106:3183–90.
19. Elias L, Portlock CS, Rosenberg SA. Combination chemotherapy of diffuse histiocytic lymphoma with cyclophosphamide, adriamycin, vincristine and prednisone (CHOP). Cancer 1978;42:1705–10.
20. Reff ME, Carner K, Chambers KS, et al. Depletion of B cells in vivo by a chimeric mouse human monoclonal antibody to CD20. Blood 1994;83:435–45.
21. Maloney DG, Smith B, Rose A. Rituximab: mechanism of action and resistance. Semin Oncol 2002;29:2–9.
22. Mounier N, Briere J, Gisselbrecht C, et al. Rituximab plus CHOP (R-CHOP) overcomes bcl-2-associated resistance to chemotherapy in elderly patients with diffuse large B-cell lymphoma (DLBCL). Blood 2003;101:4279–84.
23. Philip T, Guglielmi C, Hagenbeek A, et al. Autologous bone marrow transplantation as compared with salvage chemotherapy in relapses of chemotherapy-sensitive non-Hodgkin's lymphoma. N Engl J Med 1995;333:1540–5.
24. Sausville EA, Arbuck SG, Messmann R, et al. Phase I trial of 72-hour continuous infusion UCN-01 in patients with refractory neoplasms. J Clin Oncol 2001;19:2319–33.
25. Smith PG, Wang F, Wilkinson KN, et al. The phosphodiesterase PDE4B limits cAMP-associated PI3K/AKT-dependent apoptosis in diffuse large B-cell lymphoma. Blood 2005;105:308–16.
26. Abramson JS, Shipp MA. Advances in the biology and therapy of diffuse large B-cell lymphoma: moving toward a molecularly targeted approach. Blood 2005;106:1164–74.
27. Schenkein D. Proteasome inhibitors in the treatment of B-cell malignancies. Clin Lymphoma 2002;3:49–55.
28. Lam LT, Davis RE, Pierce J, et al. Small molecule inhibitors of IkappaB kinase are selectively toxic for subgroups of diffuse large B-cell lymphoma defined by gene expression profiling. Clin Cancer Res 2005;11:28–40.
29. Saito B, Shiozawa E, Usui T, et al. Rituximab with chemotherapy improves survival of non-germinal center type untreated diffuse large B-cell lymphoma. Leukemia 2007;21: 2563–6.

30. Pham LV, Tamayo AT, Yoshimura LC, Lin-Lee YC, Ford RJ. Constitutive NF-kappaB and NFAT activation in aggressive B-cell lymphomas synergistically activates the CD154 gene and maintains lymphoma cell survival. Blood 2005;106:3940–7.
31. Lyu MA, Cheung LH, Hittelman WN, Marks JW, Aguiar RC, Rosenblum MG. The rGel/BLyS fusion toxin specifically targets malignant B cells expressing the BLyS receptors BAFF-R, TACI, and BCMA. Mol Cancer Ther 2007;6:460–70.
32. Mohammad RM, Goustin AS, Aboukameel A, et al. Preclinical studies of TW-37, a new nonpeptidic small-molecule inhibitor of Bcl-2, in diffuse large cell lymphoma xenograft model reveal drug action on both Bcl-2 and Mcl-1. Clin Cancer Res 2007;13:2226–35.
33. O'Connor OA, Hamlin P. New drugs for the treatment of advanced-stage diffuse large cell lymphomas. Semin Hematol 2006;43:251–61.
34. Campo E, Raffeld M, Jaffe ES. Mantle-cell lymphoma. Semin Hematol 1999;36:115–27.
35. Swerdlow SH, Williams ME. From centrocytic to mantle cell lymphoma: a clinicopathologic and molecular review of 3 decades. Hum Pathol 2002;33:7–20.
36. Jaffe ES, Harris, NL, Stein, H. World Health Organization Classification of Tumours: pathology and genetics of tumors of haematopoietic and lymphoid tissures. Lyon, France, 2001.
37. Lardelli P, Bookman MA, Sundeen J, Longo DL, Jaffe ES. Lymphocytic lymphoma of intermediate differentiation. Morphologic and immunophenotypic spectrum and clinical correlations. Am J Surg Pathol 1990;14:752–63.
38. Bosch F, Lopez-Guillermo A, Campo E, et al. Mantle cell lymphoma: presenting features, response to therapy, and prognostic factors. Cancer 1998;82:567–75.
39. Argatoff LH, Connors JM, Klasa RJ, Horsman DE, Gascoyne RD. Mantle cell lymphoma: a clinicopathologic study of 80 cases. Blood 1997;89:2067–78.
40. Bernard M, Gressin R, Lefrere F, et al. Blastic variant of mantle cell lymphoma: a rare but highly aggressive subtype. Leukemia 2001;15:1785–91.
41. Bosch F, Jares P, Campo E, et al. PRAD-1/cyclin D1 gene overexpression in chronic lymphoproliferative disorders: a highly specific marker of mantle cell lymphoma. Blood 1994;84:2726–32.
42. de Boer CJ, van Krieken JH, Kluin-Nelemans HC, Kluin PM, Schuuring E. Cyclin D1 messenger RNA overexpression as a marker for mantle cell lymphoma. Oncogene 1995;10:1833–40.
43. Pinyol M, Hernandez L, Cazorla M, et al. Deletions and loss of expression of p16INK4a and p21Waf1 genes are associated with aggressive variants of mantle cell lymphomas. Blood 1997;89:272–80.
44. Pinyol M, Cobo F, Bea S, et al. p16(INK4a) gene inactivation by deletions, mutations, and hypermethylation is associated with transformed and aggressive variants of non-Hodgkin's lymphomas. Blood 1998;91:2977–84.
45. Dreyling MH, Bullinger L, Ott G, et al. Alterations of the cyclin D1/p16-pRB pathway in mantle cell lymphoma. Cancer Res 1997;57:4608–14.
46. Serrano M, Hannon GJ, Beach D. A new regulatory motif in cell-cycle control causing specific inhibition of cyclin D/CDK4. Nature 1993;366:704–7.
47. Rosenwald A, Wright G, Wiestner A, et al. The proliferation gene expression signature is a quantitative integrator of oncogenic events that predicts survival in mantle cell lymphoma. Cancer Cell 2003;3:185–97.
48. Pinyol M, Hernandez L, Martinez A, et al. INK4a/ARF locus alterations in human non-Hodgkin's lymphomas mainly occur in tumors with wild-type p53 gene. Am J Pathol 2000;156:1987–96.
49. Hernandez L, Bea S, Pinyol M, et al. CDK4 and MDM2 gene alterations mainly occur in highly proliferative and aggressive mantle cell lymphomas with wild-type INK4a/ARF locus. Cancer Res 2005;65:2199–206.
50. Fernandez V, Hartmann E, Ott G, Campo E, Rosenwald A. Pathogenesis of mantle-cell lymphoma: all oncogenic roads lead to dysregulation of cell cycle and DNA damage response pathways. J Clin Oncol 2005;23:6364–9.

51. Witzig TE. Current treatment approaches for mantle-cell lymphoma. J Clin Oncol 2005;23:6409–14.
52. Kiss TL, Mollee P, Lazarus HM, Lipton JH. Stem cell transplantation for mantle cell lymphoma: if, when and how? Bone Marrow Transplant 2005;36:655–61.
53. Berinstein NL, Mangel J. Integrating monoclonal antibodies into the management of mantle cell lymphoma. Semin Oncol 2004;31:2–6.
54. Pham LV, Tamayo AT, Yoshimura LC, Lo P, Ford RJ. Inhibition of constitutive NF-kappa B activation in mantle cell lymphoma B cells leads to induction of cell cycle arrest and apoptosis. J Immunol 2003;171:88–95.
55. Zelenetz AD. Mantle cell lymphoma: an update on management. Ann Oncol 2006;17(Suppl 4):iv12–4.
56. Neelapu SS, Kwak LW, Kobrin CB, et al. Vaccine-induced tumor-specific immunity despite severe B-cell depletion in mantle cell lymphoma. Nat Med 2005;11:986–91.
57. Neelapu SS, Lee ST, Qin H, Cha SC, Woo AF, Kwak LW. Therapeutic lymphoma vaccines: importance of T-cell immunity. Expert Rev Vaccines 2006;5:381–94.
58. Martinez N, Camacho FI, Algara P, et al. The molecular signature of mantle cell lymphoma reveals multiple signals favoring cell survival. Cancer Res 2003;63:8226–32.
59. Thieblemont C, Nasser V, Felman P, et al. Small lymphocytic lymphoma, marginal zone B-cell lymphoma, and mantle cell lymphoma exhibit distinct gene-expression profiles allowing molecular diagnosis. Blood 2004;103:2727–37.
60. Bennaceur-Griscelli A, Bosq J, Koscielny S, et al. High level of glutathione-S-transferase pi expression in mantle cell lymphomas. Clin Cancer Res 2004;10:3029–34.
61. Hofmann WK, de Vos S, Tsukasaki K, et al. Altered apoptosis pathways in mantle cell lymphoma detected by oligonucleotide microarray. Blood 2001;98:787–94.
62. Brody J, Advani R. Treatment of mantle cell lymphoma: current approach and future directions. Crit Rev Oncol Hematol 2006;58:257–65.
63. Chaganti RS, Nanjangud G, Schmidt H, Teruya-Feldstein J. Recurring chromosomal abnormalities in non-Hodgkin's lymphoma: biologic and clinical significance. Semin Hematol 2000;37:396–411.
64. Schneider P, MacKay F, Steiner V, et al. BAFF, a novel ligand of the tumor necrosis factor family, stimulates B cell growth. J Exp Med 1999;189:1747–56.
65. Mukhopadhyay A, Ni J, Zhai Y, Yu GL, Aggarwal BB. Identification and characterization of a novel cytokine, THANK, a TNF homologue that activates apoptosis, nuclear factor-kappaB, and c-Jun NH2-terminal kinase. J Biol Chem 1999;274:15978–81.
66. Moore PA, Belvedere O, Orr A, et al. BLyS: member of the tumor necrosis factor family and B lymphocyte stimulator. Science 1999;285:260–3.
67. Batten M, Fletcher C, Ng LG, et al. TNF deficiency fails to protect BAFF transgenic mice against autoimmunity and reveals a predisposition to B cell lymphoma. J Immunol 2004;172:812–22.
68. Mackay F, Woodcock SA, Lawton P, et al. Mice transgenic for BAFF develop lymphocytic disorders along with autoimmune manifestations. J Exp Med 1999;190:1697–710.
69. Khare SD, Sarosi I, Xia XZ, et al. Severe B cell hyperplasia and autoimmune disease in TALL-1 transgenic mice. Proc Natl Acad Sci USA 2000;97:3370–5.
70. Gross JA, Johnston J, Mudri S, et al. TACI and BCMA are receptors for a TNF homologue implicated in B-cell autoimmune disease. Nature 2000;404:995–9.
71. Zintzaras E, Voulgarelis M, Moutsopoulos HM. The risk of lymphoma development in autoimmune diseases: a meta-analysis. Arch Intern Med 2005;165:2337–44.
72. Laabi Y, Strasser A. Immunology. Lymphocyte survival: ignorance is BLys. Science 2000;289:883–4.
73. Novak AJ, Grote DM, Stenson M, et al. Expression of BLyS and its receptors in B-cell non-Hodgkin lymphoma: correlation with disease activity and patient outcome. Blood 2004;104:2247–53.

74. Scapini P, Nardelli B, Nadali G, et al. G-CSF-stimulated neutrophils are a prominent source of functional BLyS. J Exp Med 2003;197:297–302.
75. Nardelli B, Belvedere O, Roschke V, et al. Synthesis and release of B-lymphocyte stimulator from myeloid cells. Blood 2001;97:198–204.
76. Shu HB, Hu WH, Johnson H. TALL-1 is a novel member of the TNF family that is down-regulated by mitogens. J Leukoc Biol 1999;65:680–3.
77. Mackay F, Browning JL. BAFF: a fundamental survival factor for B cells. Nat Rev Immunol 2002;2:465–75.
78. Mackay F, Schneider P, Rennert P, Browning J. BAFF AND APRIL: a tutorial on B cell survival. Annu Rev Immunol 2003;21:231–64.
79. Mackay F, Ambrose C. The TNF family members BAFF and APRIL: the growing complexity. Cytokine Growth Factor Rev 2003;14:311–24.
80. Batten M, Groom J, Cachero TG, et al. BAFF mediates survival of peripheral immature B lymphocytes. J Exp Med 2000;192:1453–66.
81. Briones J, Timmerman JM, Hilbert DM, Levy R. BLyS and BLyS receptor expression in non-Hodgkin's lymphoma. Exp Hematol 2002;30:135–41.
82. He B, Chadburn A, Jou E, Schattner EJ, Knowles DM, Cerutti A. Lymphoma B cells evade apoptosis through the TNF family members BAFF/BLyS and APRIL. J Immunol 2004;172:3268–79.
83. Ogden CA, Pound JD, Batth BK, et al. Enhanced apoptotic cell clearance capacity and B cell survival factor production by IL-10-activated macrophages: implications for Burkitt's lymphoma. J Immunol 2005;174:3015–23.
84. Fu L, Lin-Lee YC, Pham LV, Tamayo A, Yoshimura L, Ford RJ. Constitutive NF-kappaB and NFAT activation leads to stimulation of the BLyS survival pathway in aggressive B-cell lymphomas. Blood 2006;107:4540–8.
85. Moreaux J, Legouffe E, Jourdan E, et al. BAFF and APRIL protect myeloma cells from apoptosis induced by interleukin 6 deprivation and dexamethasone. Blood 2004;103:3148–57.
86. Kern C, Cornuel JF, Billard C, et al. Involvement of BAFF and APRIL in the resistance to apoptosis of B-CLL through an autocrine pathway. Blood 2004;103:679–88.
87. Woodland RT, Schmidt MR, Thompson CB. BLyS and B cell homeostasis. Semin Immunol 2006;18:318–26.
88. Woodland RT, Fox CJ, Schmidt MR, et al. Multiple signaling pathways promote B lymphocyte stimulator dependent B-cell growth and survival. Blood 2008;111:750–60.
89. Ghosh S, May MJ, Kopp EB. NF-kappa B and Rel proteins: evolutionarily conserved mediators of immune responses. Annu Rev Immunol 1998;16:225–60.
90. Garg A, Aggarwal BB. Nuclear transcription factor-kappaB as a target for cancer drug development. Leukemia 2002;16:1053–68.
91. Li Q, Verma IM. NF-kappaB regulation in the immune system. Nat Rev Immunol 2002;2:725–34.
92. Bonizzi G, Karin M. The two NF-kappaB activation pathways and their role in innate and adaptive immunity. Trends Immunol 2004;25:280–8.
93. Ben-Neriah Y, Schmitz ML. Of mice and men. EMBO Rep 2004;5:668–73.
94. Hayden MS, Ghosh S. Signaling to NF-kappaB. Genes Dev 2004;18:2195–224.
95. Ghosh S, Karin M. Missing pieces in the NF-kappaB puzzle. Cell 2002;109(Suppl):S81–96.
96. Richmond A. Nf-kappa B, chemokine gene transcription and tumour growth. Nat Rev Immunol 2002;2:664–74.
97. Pasparakis M, Schmidt-Supprian M, Rajewsky K. IkappaB kinase signaling is essential for maintenance of mature B cells. J Exp Med 2002;196:743–52.
98. Chen IS, Wilhelmsen KC, Temin HM. Structure and expression of c-rel, the cellular homolog to the oncogene of reticuloendotheliosis virus strain T. J Virol 1983;45:104–13.
99. Wilhelmsen KC, Temin HM. Structure and dimorphism of c-rel (turkey), the cellular homolog to the oncogene of reticuloendotheliosis virus strain T. J Virol 1984;49:521–9.

100. Courtois G, Gilmore TD. Mutations in the NF-kappaB signaling pathway: implications for human disease. Oncogene 2006;25:6831–43.
101. Orlowski RZ, Baldwin AS, Jr. NF-kappaB as a therapeutic target in cancer. Trends Mol Med 2002;8:385–9.
102. Basseres DS, Baldwin AS. Nuclear factor-kappaB and inhibitor of kappaB kinase pathways in oncogenic initiation and progression. Oncogene 2006;25:6817–30.
103. Pham LV, Tamayo AT, Yoshimura LC, et al. A CD40 Signalosome anchored in lipid rafts leads to constitutive activation of NF-kappaB and autonomous cell growth in B cell lymphomas. Immunity 2002;16:37–50.
104. Claudio E, Brown K, Park S, Wang H, Siebenlist U. BAFF-induced NEMO-independent processing of NF-kappa B2 in maturing B cells. Nat Immunol 2002;3:958–65.
105. Kayagaki N, Yan M, Seshasayee D, et al. BAFF/BLyS receptor 3 binds the B cell survival factor BAFF ligand through a discrete surface loop and promotes processing of NF-kappaB2. Immunity 2002;17:515–24.
106. Do RK, Hatada E, Lee H, Tourigny MR, Hilbert D, Chen-Kiang S. Attenuation of apoptosis underlies B lymphocyte stimulator enhancement of humoral immune response. J Exp Med 2000;192:953–64.
107. Hatada EN, Do RK, Orlofsky A, et al. NF-kappa B1 p50 is required for BLyS attenuation of apoptosis but dispensable for processing of NF-kappa B2 p100 to p52 in quiescent mature B cells. J Immunol 2003;171:761–8.
108. Ni CZ, Oganesyan G, Welsh K, et al. Key molecular contacts promote recognition of the BAFF receptor by TNF receptor-associated factor 3: implications for intracellular signaling regulation. J Immunol 2004;173:7394–400.
109. Qian Y, Qin J, Cui G, et al. Act1, a negative regulator in CD40- and BAFF-mediated B cell survival. Immunity 2004;21:575–87.
110. Klemm JD, Beals CR, Crabtree GR. Rapid targeting of nuclear proteins to the cytoplasm. Curr Biol 1997;7:638–44.
111. Shibasaki F, Price ER, Milan D, McKeon F. Role of kinases and the phosphatase calcineurin in the nuclear shuttling of transcription factor NF-AT4. Nature 1996;382:370–3.
112. Chow CW, Rincon M, Cavanagh J, Dickens M, Davis RJ. Nuclear accumulation of NFAT4 opposed by the JNK signal transduction pathway. Science 1997;278:1638–41.
113. Beals CR, Clipstone NA, Ho SN, Crabtree GR. Nuclear localization of NF-ATc by a calcineurin-dependent, cyclosporin-sensitive intramolecular interaction. Genes Dev 1997;11:824–34.
114. Marafioti T, Pozzobon M, Hansmann ML, et al. The NFATc1 transcription factor is widely expressed in white cells and translocates from the cytoplasm to the nucleus in a subset of human lymphomas. Br J Haematol 2005;128:333–42.
115. He B, Raab-Traub N, Casali P, Cerutti A. EBV-encoded latent membrane protein 1 cooperates with BAFF/BLyS and APRIL to induce T cell-independent Ig heavy chain class switching. J Immunol 2003;171:5215–24.
116. Lin-Lee YC, Pham LV, Tamayo AT, et al. Nuclear localization in the biology of the CD40 receptor in normal and neoplastic human B lymphocytes. J Biol Chem 2006;281:18878–87.
117. Zhou HJ, Pham LV, Tamayo AT, et al. Nuclear CD40 interacts with c-Rel and enhances proliferation in aggressive B-cell lymphoma. Blood 2007;110:2121–7.
118. Dyer MJ, Heward JM, Zani VJ, Buccheri V, Catovsky D. Unusual deletions within the immunoglobulin heavy-chain locus in acute leukemias. Blood 1993;82:865–71.
119. Willis TG, Dyer MJ. The role of immunoglobulin translocations in the pathogenesis of B-cell malignancies. Blood 2000;96:808–22.
120. Craxton A, Draves KE, Gruppi A, Clark EA. BAFF regulates B cell survival by downregulating the BH3-only family member Bim via the ERK pathway. J Exp Med 2005;202:1363–74.
121. Xu LG, Wu M, Hu J, Zhai Z, Shu HB. Identification of downstream genes up-regulated by the tumor necrosis factor family member TALL-1. J Leukoc Biol 2002;72:410–6.

122. Lesley R, Xu Y, Kalled SL, et al. Reduced competitiveness of autoantigen-engaged B cells due to increased dependence on BAFF. Immunity 2004;20:441–53.
123. Hase H, Kanno Y, Kojima M, et al. BAFF/BLyS can potentiate B-cell selection with the B-cell coreceptor complex. Blood 2004;103:2257–65.
124. Riccobene TA, Miceli RC, Lincoln C, et al. Rapid and specific targeting of 125I-labeled B lymphocyte stimulator to lymphoid tissues and B cell tumors in mice. J Nucl Med 2003;44:422–33.
125. Halpern WG, Lappin P, Zanardi T, et al. Chronic administration of belimumab, a BLyS antagonist, decreases tissue and peripheral blood B-lymphocyte populations in cynomolgus monkeys: pharmacokinetic, pharmacodynamic, and toxicologic effects. Toxicol Sci 2006;91:586–99.
126. Nimmanapalli R, Lyu MA, Du M, Keating MJ, Rosenblum MG, Gandhi V. The growth factor fusion construct containing B-lymphocyte stimulator (BLyS) and the toxin rGel induces apoptosis specifically in BAFF-R-positive CLL cells. Blood 2007;109:2557–64.
127. Lin WY, Gong Q, Seshasayee D, et al. Anti-BR3 antibodies: a new class of B-cell immunotherapy combining cellular depletion and survival blockade. Blood 2007;110: 3959–67.

Chapter 12
Tipping the Scales of Survival: The Role of BLyS in B-Cell Malignancies

Anne J. Novak and Stephen M. Ansell

Abstract Some of the genetic changes that lead to B-cell transformation, such as chromosomal rearrangements, have been discovered, but many of the other contributing mechanistic details underlying transformation events are not yet known. It is clear, however, that dysregulation of the balance between cell proliferation and programmed cell death is a central feature. While many factors have been shown to regulate the normal balance of B-cell growth and apoptosis, one molecule in particular, BLyS, has been recently shown to play a central role in regulating B-cell homeostasis. The similarities between BLyS transgenic mice and B-cell malignancies, e.g., elevated Bcl-2 levels, apoptosis resistance, and prevalence of autoimmune disorders, begged the question of whether or not BLyS was involved in the pathogenesis of B-cell malignancies. In this chapter, we will review and discuss the effects of BLyS on the growth and survival of B cells from patients with non-Hodgkin lymphoma, chronic lymphocytic leukemia, multiple myeloma, and Hodgkin lymphoma. Additionally, we will discuss the clinical relevance and regulatory control of BLyS expression in B-cell malignancies.

Keywords BLyS · BAFF · TNFSF13B · TACI · BCMA · BAFF-R · Lymphoma

12.1 Introduction

12.1.1 B-Cell Growth and Transformation

B-cell malignancies are serious and frequently fatal illnesses that typically arise from B lymphocytes in the bone marrow, spleen, and lymph nodes. Often, malignant cells disseminate from their site of origin and traffic throughout the peripheral lymphoid tissues and can circulate through the blood. Nonlymphoid tissue

A.J. Novak (✉)
Division of Hematology, Mayo Clinic, Rochester, Minnesota 55905, USA

M.P. Cancro (ed.), *BLyS Ligands and Receptors,* Contemporary Immunology,
DOI 10.1007/978-1-60327-013-7_12,

can also be involved and B-cell lymphomas can be found in the salivary glands, lungs, intestinal tract, thyroid, and conjunctiva. Some of the genetic changes that lead to B-cell transformation, such as chromosomal rearrangements, have been discovered, but many of the other contributing mechanistic details underlying transformation events are not yet known. It is clear, however, that dysregulation of the balance between cell proliferation and programmed cell death is a central feature.

Under normal conditions, lymphocytes must strictly regulate growth and apoptosis to provide adequate immunologic defence against infection and yet not overwhelm the organism with inappropriate cell numbers. Tipping the balance of normal B-cell homeostasis can have detrimental effects, such as the inappropriate survival of autoreactive B cells resulting in the development of autoimmune disease [1]. A large number of signaling molecules have been implicated in homeostatic control of B cells. Activation of the B-cell receptor generates second messengers, including calcium influx and activated kinases, leading to the activation of transcription factors NF-AT [2], NF-κB, AP1, and others [3]. Costimulation through TNFR family members such as CD40 is also critical determinant of B-cell function.

In addition to CD40, another TNFSF member that has received a lot of attention from B-cell biologists is B-lymphocyte stimulator, BLyS [4] (TNFSF13B, also called B-cell activating factor (BAFF) [5], TNF homolog that activates apoptosis, nuclear factor NF-κB, and c-Jun NH_2-terminal kinase (THANK) [6], TNF and apoptosis ligand-related leukocyte-expressed ligand 1 (TALL-1) [7], or zTNF4 [8]), which has been shown to be critical for maintenance of normal B-cell development and homeostasis [9–11]. BLyS shares significant homology with another TNF superfamily member, a proliferation-inducing ligand (APRIL) or TNFSF13, which has been found to stimulate tumor cell growth [12] as well as proliferation of primary lymphocytes [13].

12.1.2 BLyS and Its Receptors in B-Cell Biology

Initial studies on the effect of BLyS on B-cell physiology suggested that it costimulated B-cell proliferation and immunoglobulin secretion [4, 5]. Subsequently, a role for BLyS in controlling cell survival was proposed, as B cells isolated from BLyS transgenic mice have elevated Bcl-2 levels and prolonged survival compared to wild-type B cells [8, 14]. Moreover, it was found that BLyS increased the survival of resting or activated B cells and cooperated with CD40L to attenuate B-cell apoptosis [15]. Transgenic overexpression of BLyS in mice results in elevated numbers of mature B cells in the spleen and periphery and development of autoimmune-like manifestations reminiscent of systemic lupus erythematosus [14, 16]. In addition, as BLyS transgenic mice age, they develop a secondary pathology, Sjögren's syndrome [17], which is associated with intense B-cell activity and a predisposition to development of B-cell lymphomas [18, 19]. In parallel to the transgenic studies, BLyS-deficient mice have also revealed a significant role for this molecule in

B-cell development. Peripheral B-cell maturation is arrested and examination of secondary lymphoid organs from BAFF-deficient mice revealed an almost complete loss of follicular and marginal zone B lymphocytes [10, 20]. Taken together, the in vivo studies indicate that BLyS is a critical component of B-cell development and survival.

The underlying mechanism responsible for the effect of BLyS on B cells remains poorly understood, in part because of the complexity introduced by multiple receptors. Study of signal transduction through these receptors is in its relative infancy; however, there is growing literature demonstrating a role for each receptor in activation of NF-κB (reviewed in [21]) and members of the PI-3 kinase and MAP kinase family [22, 23]. Another downstream target of BLyS is Pim2, an oncogene that has been shown to be overexpressed in multiple myeloma, leukemia, and lymphoma [24–26]. In the presence of high BLyS levels, Pim2 expression is elevated in autoreactive B cells [1] and treatment of B cells with BLyS results in upregulation of Pim2 expression [23]. Pim2 can act collaboratively with c-myc to induce lymphomagenesis [27] and is thought to act through an NF-κB-dependent pathway [28]. Overexpression of BLyS in the malignant scenario may result in elevated Pim2 expression enhancing the transformation process.

BLyS has also been shown to influence B-cell function by controlling the cell cycle. BLyS, in concert with p18 and cyclin D2, regulates cell-cycle entry and early G1 progression [29]. Elevated serum BLyS levels may therefore aberrantly upregulate B-cell entry into the cell cycle and result in unchecked cell proliferation. The ability of BLyS to modulate critical aspects of B-cell survival and growth through activation of downstream targets highlights its significance in B-cell biology. Alterations in this system leading to over abundance of BLyS may tilt the subtle balance between normal cell growth and death and create an environment where B cells escape the normal mechanism of apoptosis.

Three receptors (Fig. 12.1), B-cell maturation antigen (BCMA, TNFRSF17) [30], transmembrane activator and CAML interactor (TACI, TNFRSF13B) [31], and BAFF-R (TNFRSF13C) [32] have been identified as receptors for BLyS. BCMA and BAFF-R are predominantly expressed on B lymphocytes, whereas TACI can be found on B cells and activated T cells [33]. TACI and BCMA can also bind to APRIL, whereas BAFF-R is specific for BLyS [32, 34]. The three receptors are structurally unusual compared to other TNFR family members, since they have just one cysteine-rich domain (CRD) (or a partial CRD, in the case of BAFF-R) that contacts a single ligand subunit, rather than multiple CRDs that interact with the boundary between two ligand subunits [35–37]. In addition to the known receptors, several lines of evidence supported the existence of a third APRIL-specific receptor. APRIL modestly stimulated the proliferation of tumor cell lines, but none of these cell types express TACI or BCMA [12, 38]. In addition, BLyS does not bind to APRIL-responsive tumor cells, nor can it induce cell proliferation [38, 39]. These data supported the idea that a novel APRIL-specific receptor could mediate these effects. After a long search for a third TNF-R-like receptor that specifically binds APRIL, two groups recently demonstrated that APRIL binds to heparan sulfate proteoglycans (HSPGs) [39, 40]. However, the biological purpose for APRIL binding to HSPG is not yet clear.

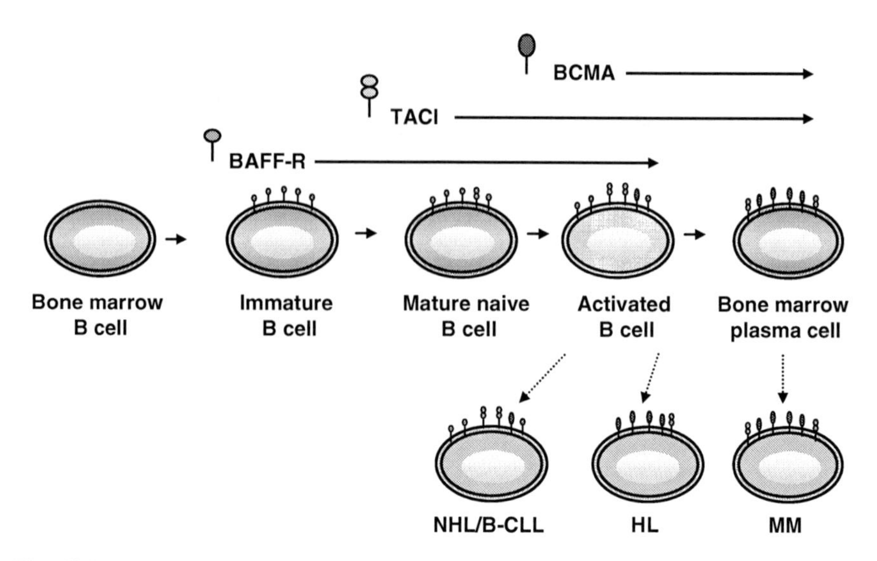

Fig. 12.1 Expression of BLyS receptors on normal and malignant B cells. Throughout human B-cell development, the BLyS receptor profile is ever changing. BAFF-R is the first receptor to be expressed on immature B cells and it is expressed throughout B-cell development until the plasma cell stage when its surface expression is lost. Low levels of TACI expression can be detected on naïve immature B cells and its expression is upregulated upon activation and maintained throughout the plasma cell stage. BCMA expression is first detected on activated B cells and is highly expressed on plasma cells. The BLyS receptor profile on NHL/B-CLL cells is similar to that seen in activated B cells and MM cells look similar to their normal plasma cell counterparts, but have variable TACI expression. HL B cells express BCMA and TACI and lack BAFF-R

A key role for BAFF-R in BLyS-mediated survival has been suggested by studies demonstrating that A/WySnJ mice, which carry a mutation in BAFF-R, have a loss of follicular and marginal zone B cells in secondary lymphoid organs, a phenotype similar to BLyS-deficient mice [20, 32, 34]. Independent of its role as a pro-survival molecule, BAFF-R also mediates isotype switch in B cells [41, 42]. The significance of BCMA in B-cell function remains to be fully elucidated, but there is growing evidence that this receptor is important in normal and malignant plasma cell biology [43–45]. BCMA deficiency leads to impaired survival of long-lived BM plasma cells [46] and BCMA is now believed to be most relevant for later stages of B-cell differentiation or survival, since its expression increases on human plasmablasts and GC B cells [47] and on mouse plasma cells [46]. An important role for TACI in B-cell homeostasis has been implicated as TACI-deficient mice were found to have an accumulation of B cells, splenomegaly, and TACI$^{-/-}$ B cells have increased immunoglobulin production [47–50]. More recently, two groups have recently identified TACI as an important factor in common variable immunodeficiency (CVID) and IgA deficiency (IgAD) [51, 52]. Mutations in TACI were identified in patients with familial CVID and IgAD and found to result in low surface expression levels and defective signaling responses. Of note, CVID patients are at higher risk for development of NHL [53].

12.1.3 Summary

The similarities between BLyS or APRIL transgenic mice and B-cell malignancies, e.g., elevated Bcl-2 levels, apoptosis resistance, and prevalence of autoimmune disorders, begged the question of whether or not BLyS and APRIL were involved in the pathogenesis of disease. Over the past 5 years there have been numerous reports on the role of BLyS in B-cell-related disorders. In addition to BLyS, there is also a large literature on the role of APRIL in B-cell disorders, much of which has been reviewed by Dillon et al. [54]. In this article, we will specifically focus our discussion on the effects of BLyS on the growth and survival of malignant B cells, including non-Hodgkin lymphoma, chronic lymphocytic leukemia, multiple myeloma, and Hodgkin lymphoma. Additionally, we will discuss the clinical relevance and regulatory control of BLyS expression in B-cell malignancies.

12.2 BLyS in B-Cell Malignancies: Growth and Survival

12.2.1 B-Cell Non-Hodgkin's lymphoma (NHL)

B-cell NHL encompasses a broad spectrum of B-cell malignancies that arise from mature B cells that are distinguished by unique immunophenotypic and histopathologic features. NHL is the fifth most common type of cancer in the United States with an estimated 60,000–65,000 new cases being diagnosed annually in the United States [55]. Follicular lymphoma and diffuse large B-cell lymphoma account for the majority of new diagnoses. While some of the genetic changes that lead to NHL B-cell transformation have been discovered, many of the other contributing mechanisms underlying transformation events are not yet known. However, overexpression of Bcl-2 [56, 57] and activation of the NF-κB pathway [58, 59] are commonly found.

Initial studies on the role of BLyS on malignant B cells showed that B cells isolated from NHL patients bound soluble BLyS, and that BLyS enhanced IgM-mediated cell proliferation [60]. In a subsequent study, we characterized the cell surface expression of TACI, BCMA, and BAFF-R in various subtypes of human NHL cells (Fig. 12.1) [61]. The NHL samples examined were diagnosed with small lymphocytic lymphoma (B-CLL/SLL), follicular lymphoma, diffuse large B-cell lymphoma (DLBCL), mantle-cell lymphoma (MCL), or marginal zone lymphoma (MZBL). Similar to memory and naïve B cells [43], NHL B cells express BAFF-R and express low to no BCMA. TACI expression was variable on all histologic subtypes examined. Regardless of the BLyS receptor profile, all samples tested had the capacity to bind soluble BLyS [61]. Data published by He et al. also showed that by PCR, CLL, FL, MCL, DLBC, and Burkitt lymphoma B cells express all three BLyS receptors. A similar pattern was also found in central nervous system lymphomas, where TACI, BCMA, and BAFF-R were all detected by PCR [62]. Unlike some of the other NHL subtypes, Waldenstrom macroglobulinemia (WM) B cells, which have a lymphoplasmacytic phenotype and secrete significant amounts of monoclonal IgM protein, express BCMA on their cell surface [63].

The ability of BLyS to support B-cell survival and homeostasis in B-cell-rich regions raised the possibility that BLyS expression within a tumor microenvironment may attenuate apoptosis and support dysregulated growth of malignant B cells. Using various methods of detection, including PCR and immunohistochemistry, variable levels of BLyS were detected in all NHL histologic subtypes examined, the lowest levels been seen in follicular lymphoma [61, 64]. Because NHL tissue consists of various cells types including B cells, T cells, monocytes, and stroma, it has been difficult to precisely identify the source of BLyS in the malignant microenvironment. Is it expressed by the malignant B cell or another cell type within the tissue? The use of antibodies with different specificities, some recognizing the membrane-bound form of BLyS and others recognizing the soluble form, complicated the initial analysis. There is strong evidence suggesting that lymphoma cell lines can express BLyS and that it can be regulated in part by CD40L and IL-4 [64, 65]. Immunohistochemical and western blot analysis performed on patient NHL B cells also suggest that they can constitutively express the membrane-bound form of BLyS. However, other studies were unable to confirm this data in fresh patient specimens by flow cytometry. An additional complicating factor is the recent finding showing that freshly isolated peripheral B cells display significant levels of prebound BLyS on their cell surface and do not have detectable levels of transmembrane BLyS. Therefore, it is possible that the BLyS detected in the patient NHL B cells may have been prebound to the cells or it may be endogenously expressed. These data highlight the need for a thorough analysis of BLyS expression by NHL B cells, but regardless of the results, all studies to date indicate that BLyS is present in the malignant environment.

The biologic impact of BLyS on NHL B cells has been found to be quite similar to that seen in normal human B cells. BLyS co-stimulates IgM-mediated proliferation of NHL B cells and promotes malignant B-cell survival [60, 61, 64, 65]. In NHL B cells, both autocrine and paracrine expression of BLyS results in degradation and phosphorylation of IκBα and activation of NF-κb, p50–p65, and p50-c-rel [64]. Incubation of NHL B cell with BLyS also resulted in upregulation of anti-apoptotic proteins, Bcl-2 and Bcl-xL, downregulation of the anti-apoptotic protein Bax, and inhibited PARP degradation [64]. Using and siRNA approach to inhibit BLyS expression, Fu et al. were able to show that constitutive expression of BLyS in NHL B-cell lines results in expression of Bcl-2, Bcl-xL, c-myc, and cyclin D1 and activation of the NF-κb complex [65]. Additional studies went on to show that BLyS-mediated induction of NF-κb and NFAT upregulates BLyS expression resulting in a positive feedback loop that contributes NHL B-cell survival [65]. In addition to promoting cell growth and viability, BLyS has also been shown to enhance IgM secretion by WM cells [63].

12.2.2 B-Cell Chronic Lymphocytic Leukemia (CLL)

CLL is the most common leukemia in the Western world and is defined by the accumulation of mature-appearing B lymphocytes with prolonged survival in the periphery and bone marrow [66, 67]. It is predominantly a disease of older individuals; the vast majority of patients are more than 50 years of age at the time of initial diagnosis [68, 69]. B-CLL cells are defined by a unique set of cell surface molecules: $CD19^+$, $CD5^+$, $CD23^+$, and low levels of surface immunoglobulin (Ig). However, additional biological features permit further subcharacterization of B-CLL, including Ig heavy chain variable region somatic mutation status and/or CD38 expression levels, cytogenetic abnormalities, and the expression of ZAP-70 [66, 67]. The clinical course of CLL is heterogeneous; some patients remain stable for a long time without need of therapy, while others progress rapidly to a more advanced disease and die despite aggressive treatment. While numerous advances have been made in our understanding of B-CLL, the molecular pathogenesis of this disease remains to be fully elucidated. However, it has been demonstrated that certain molecular alterations, including overexpression of anti-apoptotic proteins and recurring chromosomal abnormalities, can often be found [66, 67].

Some of the earliest work on the role of BLyS in B-cell malignancies was done using CLL B cells as model system. Our initial work [70] demonstrated that CLL B cells had the ability to bind soluble BLyS and expressed TACI and BAFF-R, while only a small subset expressed BCMA (Fig. 12.1). Similar to normal B cells, B-CLL cells had enhanced cell survival and reduced PARP cleavage and caspase-3 activity in the presence of soluble BLyS. It was also shown that B-CLL cells from a subset of patients aberrantly express BLyS and APRIL while these molecules were undetectable in normal B cells. Quantitative analysis of BLyS expression in CLL B cells revealed that BLyS mRNA was found at low levels compared to HL60 cells and that it was virtually undetectable in the B-cell lymphoma lines. The finding that B-CLL cells express autocrine BLyS was further supported by detection of low cell surface expression of BLyS on CLL B cells from some patients and the ability of a BCMA-Fc decoy receptor to increase apoptosis. The expression of BLyS receptors as well as autocrine expression of BLyS and APRIL by CLL B cell was further confirmed by Kern et al. [71]. However, as previously mentioned, the BLyS protein detected on CLL B cells may have been prebound and it remains to be clearly defined whether CLL B cell truly express significant levels of autocrine BLyS. Nurse-like cells, which are a subset of adherent cells isolated from CLL blood mononuclear cells, were also found to express BLyS at levels similar to $CD14^+$ monocytes and may therefore be an additional source of BLyS in the malignant environment [72].

Stimulation of CLL B cells with BLyS results in attenuated cell death and downregulation of caspase-3 activity and PARP cleavage [70]. Signaling through

BLyS receptors on CLL B cells was also shown to activate members of the NF-κB complex including phosphorylation and degradation of IκBα (classical pathway) and processing of p100–p52 (alternative pathway) [72, 73]. Blocking studies further suggested that signaling through BCMA or TACI resulted in activation of the classical pathway, while BAFF-R activated the alternative pathway. Inhibition of BLyS-BAFF-R binding with anti-BAFF-R prevented processing of p100, but had little effect on cell survival suggesting that the alternative pathway may not be required for the pro-survival effect of BLyS [73].

12.2.3 Multiple Myeloma (MM)

MM is a clonal B-cell neoplasm characterized by accumulation of malignant plasma cells in the bone marrow. MM cells are in close contact with cytokine-secreting stromal cells creating a tumor microenvironment that supports MM cell growth and survival. MM affects 1–5 per 100,000 individuals each year worldwide with a higher incidence in the West [74]. Currently, there is no curative treatment for this disease underscoring the need to achieve a better understanding of the mechanisms underlying myeloma cell growth and survival control.

The study of BLyS binding and receptor characterization has been well characterized in MM and is somewhat unique compared to NHL and CLL. In our initial studies, it was found that both freshly isolated patient and MM cell lines bound soluble BLyS and expressed BCMA and TACI, with variable expression of BAFF-R being seen (Fig. 12.1) [43, 75]. In accordance with this data, developmental regulation of TACI and BCMA was supported by gene profiling studies of B-cell subsets where they were found to be upregulated during the late stages of normal B-cell differentiation and are highly expressed on MM cells [25, 44, 76]. Regulation of TACI, BCMA, and BAFF-R expression on immunoglobulin-secreting cells has also been described in vitro, where it was found that activation of memory B cells with polyclonal stimuli results in upregulation of TACI and BCMA and downregulation of BAFF-R [77]. Moreaux et al. have further characterized the significance of TACI expression in MM. Using a hierarchical clustering of gene expression profile from 65 MM patients, it was determined that the level of TACI gene expression is associated with a specific gene expression profile; $TACI^{hi}$ myeloma cells displayed a mature plasma cell gene signature, while the $TACI^{lo}$ group had a gene signature of plasmablasts [78]. They also found that $TACI^{hi}$ cells had increased expression of c-maf, cyclin-D1, and integrin β7 and that stimulation of $TACI^{hi}$ cells with BLyS results in upregulation of c-maf, cyclin-D1, and β7 expression [78, 79]. Overexpression of TACI in a subset of MM cells has been confirmed using gene expression profiling, array CGH, cIg-FISH [80].

Expression of BLyS in both the normal and the malignant bone marrow microenvironment has been described. Immunohistochemical studies showed that bone marrow sections from MM patients stained brightly for BLyS, while there was low diffuse staining in normal marrows. The level of BLyS varied between patients, with some staining very highly while others had diffused distributions of ligand. These findings correlated with BLyS PCR studies in that there was variable expression of BLyS mRNA in purified CD138^{+} MM patient cells [43]. Autocrine expression of BLyS by MM cell lines has also been described, a number of MM cell lines express BLyS mRNA and membrane-bound BLyS could be detected by both flow cytomety and immunohistochemistry [43, 75]. Autocrine expression of BLyS by MM cells was supported by Fc receptor-blocking studies showing that culture of MM cell lines or patient specimens with TACI-Fc resulted in reduced cell growth and survival [75]. In addition to the MM cells, it has been further shown that bone marrow stromal cells [81] as well as osteoclasts [78, 82] are a rich source BLyS in the bone marrow microenvironment. Additionally, co-culture of MM cells with bone marrow stromal cells or osteoclasts triggers BLyS expression and may therefore provide a positive feed back loop for MM cell growth and survival [81, 83].

While the BLyS receptor profile is unique on MM cell, some of the biologic effect seems to be quite similar to that seen in other B-cell types. BLyS was found to promote cell survival and enhance cytokine-, IGF-I-, and IL-6-mediated cell proliferation [43]. BLyS has also been shown to rescue MM cells from drug-induced apoptosis [75, 81] and promote adhesion of MM cells to bone marrow stromal cells [81]. Signaling through BLyS receptors on MM cells results in phosphorylation of Akt, NF-κB activation, and upregulation of the anti-apoptotic proteins Bcl-2 and Mcl-1 [75, 81].

12.2.4 Hodgkin's Lymphoma (HL)

HL is a lymphoid neoplasm that is characterized by a small number of clonally related germinal center B cells, the Hodgkin and Reed-Sternberg (HRS) cells, surrounded by a large population of inflammatory cell including lymphocytes and histiocytes [84, 85]. Both the HRS and surrounding cellular infiltrate express growth factors and cytokines that are conducive to cell proliferation and survival. On the basis of morphological, immunohistochemical, and clinical criteria, HL can be differentiated into two distinct subtypes, the more common classical HL and the rare lymphocyte-predominant HL [84, 85]. The incidence of HL has a bimodal pattern with the highest incidence seen in young adults as well as in elderly patients with approximately 7,000–8,000 new cases being diagnosed annually in the United States [86].

The rarity of HL and the lack of significant amounts of fresh patient tissue have made the study of this disease challenging. While not perfect, there are a number of HRS cell lines available and they have provided researchers with a model system in which to study the biology of HL. Chie et al. have characterized the role of BLyS and HL and found that malignant HSR cells express variable levels of TACI and BCMA but little to no BAFF-R. They also found that BLyS was expressed by both HRS cells and infiltrating inflammatory cells. Similar to the other malignant

B cells, BLyS was found to promote HRS growth and survival and upregulate Bcl-2, Bcl-xL, and c-Myc. Stimulation of HRS cells with BLyS also results in activation of NF-κB [87].

12.2.5 Summary

The ability of BLyS to promote malignant B-cell growth and proliferation and to activate numerous signaling cascades strongly suggests that this molecule has significant clinical relevance in B-cell malignancies. Dysregulated control of BLyS expression, either by an autocrine or paracrine mechanism, clearly promotes the survival of malignant B cells and may therefore contribute to the development or progression of B-cell tumors (Fig. 12.2).

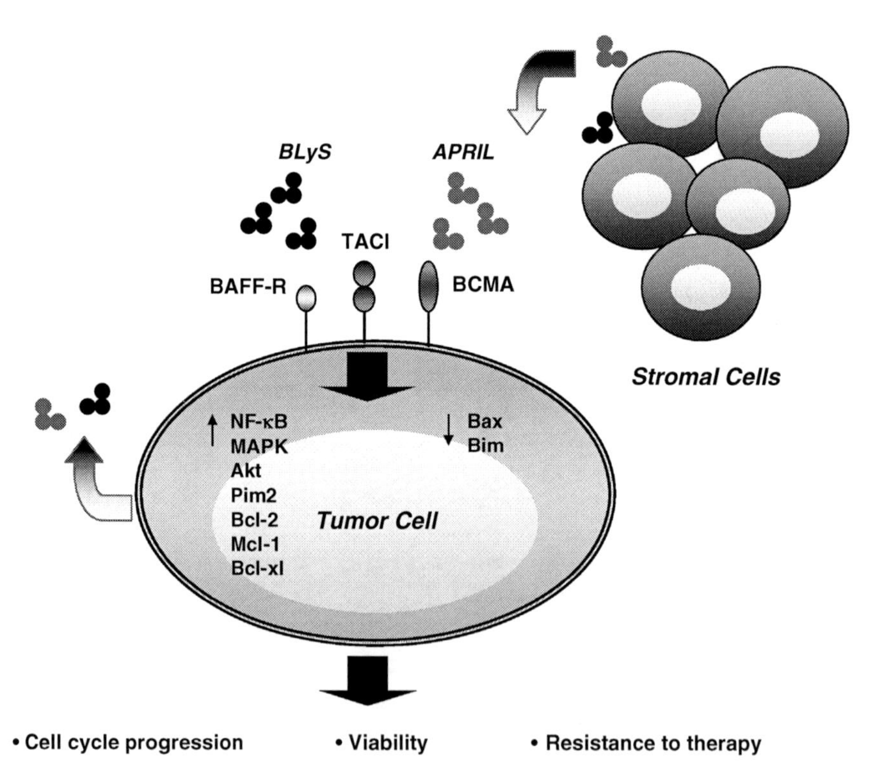

Fig. 12.2 Expression of BLyS in the tumor microenvironment. BLyS can be expressed in an autocrine manner by malignant B cells and by stromal cells within the tumor microenvironment. In addition to BLyS, APRIL has also been shown to have a similar expression pattern, and together, they bind and signal through TACI, BCMA, and BAFF-R. Activation of BLyS receptors results in activation of downstream signals and promotes survival and growth of malignant B cells

12.3 BLyS in B-Cell Malignancies: Regulatory Control of BLyS Expression and Its Clinical Relevance

12.3.1 Regulation of BLyS Expression

Serum BLyS levels have been found to be elevated in B-cell-related disorders, including malignancy [61, 88]. The mechanisms that control the homeostatic level of BLyS are not defined, but they may indirectly effect the number and composition of the of B pool. In support of this, Lesley et al. showed in a murine system, that elevated BLyS levels resulted in rescue of auto-antigen-engaged B cell from elimination, predisposing to autoimmune disease [1]. This data suggests that alterations in BLyS levels can have significant effects on B cells, and in addition to autoimmune disorders, may contribute to the development of malignant disease. In mouse models, transgenic overexpression of BLyS results in elevated numbers of mature B cells and development of autoimmune disorders, but these mice do not develop B-cell tumors [14, 16]. However, when BLyS transgenic mice are crossed with $TNF^{-/-}$ mice, 35% of the mice develop B-cell lymphomas [89], suggesting that BLyS may work in combination with other inflammatory cytokines to promote development of B-cell tumors.

BLyS is expressed by monocytes, macrophages, dendritic cells, neutrophils, and malignant B cells [43, 64, 70, 71, 75, 90–92] and its expression has been shown to be regulated by IFN-γ, IFN-α GM-CSF, and IL-10 [90, 91]. Recently, Fu et al. published work suggesting that both the NFAT and NF-κB-binding sites of the BLyS promoter contribute to BLyS expression [65]. However, a detailed analysis of BLyS promoter activity, both constitutive and inducible, remains to be done. The environmental milieu that drives BLyS expression in disease is not known and thorough study of the environmental factors that control BLyS expression will hopefully shed some light on how BLyS is regulated.

In addition to cytokines, there is increasing evidence that elevated BLyS levels may be due in part to genetic variation with the BLyS gene. In a study of BLyS levels in CLL patients, we found that serum BLyS levels were higher in CLL patients with a family history of lymphoproliferative diseases, suggesting a genetic component [93]. To determine if there was a common underlying genetic event influencing BLyS expression we sequenced the BLyS promoter in patients with B-CLL and normal controls and identified a polymorphic site, –871 C/T (dbSNP ID: rs9514828), that was previously reported to be expressed at increased frequency in SLE patients of Japanese descent and was associated with elevated monocyte BLyS mRNA levels [94]. The –871T polymorphism was shown to be more prevalent in individuals with high BLyS levels and familial incidence of CLL. Additionally, using a reporter assay, it was found that the presence of a T at -871 resulted in increased BLyS expression. While these data suggest that variation in the BLyS gene may be important in regulating BLyS expression and developing CLL, the study size was small. A large-scale analysis of the –871 SNP, as well as other SNPs within the BLyS gene, will hopefully give insight into its significance in disease.

12.3.2 Clinical Significance of BLyS Levels

Numerous groups have measured serum BLyS levels and it has been found to be elevated in patients with NHL [61, 63, 95, 96], CLL [71, 93], MM [75], and HL [97, 98]. The mean serum BLyS level in normal healthy controls was been reported to be between 2.7 and 6.68 ng/ml, and up to 4.2-fold higher in patients [61, 75, 98]. The variability seen in serum BLyS levels is likely due to the use of different assay methods by each group. Serum BLyS levels have also been found to correlate with clinical parameters. In NHL, elevated serum BLyS levels were found to associate with transformation of indolent lymphomas to aggressive lymphomas, the presence of B symptoms, a poor response to therapy, and an inferior overall survival rate [61]. A similar trend has been seen in HL where elevated BLyS levels were found to correlate with resistance to therapy and shorter progression-free survival [97], as well as advanced disease, the presence of B symptoms, and extranodal disease involvement [98].

Many patients with B-cell malignancies are treated with therapies that specifically deplete B cells, such as the rituximab, a monoclonal antibody that binds to CD20. The impact of B-cells depletion on the circulating BLyS levels is not fully understood, but early indications suggest that it may have significant clinical implications. When arthritis patients were treated with rituximab-based therapies, there was a marked increase in serum BLyS levels that correlated with a decrease in the number of circulating $CD19^{+}$ B cells [99]. We have seen a similar trend in NHL, when patients with follicular lymphoma were treated with rituximab, BLyS levels significantly increased after treatment [100]. The mechanism of how this occurs is unclear, but two potential models may be that either the body responds to the reduced number of B cell by synthesizing more BLyS or BLyS levels may increase due to the lack of B cells to metabolize it. Regardless of the cause, the rise in BLyS levels may significantly contribute to the growth and survival of any B cells that remain after therapy and may contribute to relapse.

12.3.3 Clinical Targeting of BLyS

Biotechnology companies are developing therapeutic agents designed to neutralize BLyS, including Lymphostat-B (Belimumab), Atacicept (TACI-Ig), AMG 623, and BR3-Fc, and some of these agents are currently in pre-clinical and early phase trials in B-cell malignancies. Atacicept, a recombinant fusion protein that inhibits BLyS and APRIL, has shown promising preclinical results [83] and has now completed phase 1 trials in NHL and MM. In the phase 1 trial of patients with plasma cell disorders, a total of 16 patients (12 MM and 4 WM) entered the trial. No dose limiting toxicity and no serious adverse events related to the drug were observed. Five MM patients and three WM patients had stable disease after the first treatment cycle; but the rest of the patients progressed. One WM patient had a minor response to treatment (M-protein decreased by greater than 25%).

Most of patients showed a decrease of polyclonal immunoglobulins and B lymphocytes [101]. Atacicept has also shown promising results in a mouse model of myeloma [83].

In the phase 1 trial of atacicept in B-cell lymphomas, 15 patients with relapsed and refractory diffuse large B-cell lymphoma [7], follicular lymphoma [4], small lymphocytic lymphoma/chronic lymphocytic leukemia [2], and mantle-cell lymphoma [2] were treated with escalating doses of atacicept and it was found to be well tolerated at all doses. Five weekly doses produced low to moderate accumulation of free total atacicept and atacicept/BLyS complex in serum. IgA, IgG, and IgM concentrations decreased in a dose-related pattern with a mean decrease of 15–40% from baseline after 4 weeks of atacicept. However, no objective responses were observed [102].

12.4 Concluding Remarks

The significance of the BLyS and TACI/BCMA/BAFF-R ligand/receptor interaction has been well documented and there are numerous citations in the literature studying this system. The evidence for a role of these proteins in malignant disease is clear, and targeted disruption of BLyS binding to its receptors could have significant clinical implications. Therapeutic approaches to inhibiting BLyS are currently in progress and have been reviewed in Dillon et al. [54]. However, it should be noted that clear gaps still exist in our understanding of how each molecule contributes in the malignant scenario both mechanistically and in human populations. We know that BLyS is present at high levels, we know that each of the three receptors can be expressed by malignant B cells, and we know that they promote growth and survival. Beyond that, it is still not clear how and when these ligands and receptors are critically important. One hypothesis is that elevated serum BLyS levels throughout life predispose to development of autoimmune disease and malignancy. Furthermore, we also know very little about the regulatory controls that mediate the expression of these proteins. Altered expression of any one of these proteins may alter the balance between cell survival and death, potentially tipping to scale toward uncontrolled growth. Gaining a better understanding of how this system functions and how it is controlled will hopefully give us insight into their role in B-cell malignancy etiology and prognosis and will provide us with an opportunity for therapeutic intervention, improved clinical management, and perhaps novel therapeutic approaches.

References

1. Lesley R, Xu Y, Kalled SL, et al. Reduced competitiveness of autoantigen-engaged B cells due to increased dependence on BAFF. Immunity 2004;20(4):441–53.
2. Feske S, Giltnane J, Dolmetsch R, Staudt LM, Rao A. Gene regulation mediated by calcium signals in T lymphocytes. Nat Immunol 2001;2(4):316–24.

3. Gugasyan R, Grumont R, Grossmann M, et al. Rel/NF-kappaB transcription factors: key mediators of B-cell activation. Immunol Rev 2000;176:134–40.
4. Moore PA, Belvedere O, Orr A, et al. BLyS: member of the tumor necrosis factor family and B lymphocyte stimulator. Science 1999;285(5425):260–3.
5. Schneider P, MacKay F, Steiner V, et al. BAFF, a novel ligand of the tumor necrosis factor family, stimulates B cell growth. J Exp Med 1999;189(11):1747–56.
6. Mukhopadhyay A, Ni J, Zhai Y, Yu GL, Aggarwal BB. Identification and characterization of a novel cytokine, THANK, a TNF homologue that activates apoptosis, nuclear factor-kappaB, and c-Jun NH2-terminal kinase. J Biol Chem 1999;274(23):15978–81.
7. Shu HB, Hu WH, Johnson H. TALL-1 is a novel member of the TNF family that is down-regulated by mitogens. J Leukoc Biol 1999;65(5):680–3.
8. Gross JA, Johnston J, Mudri S, et al. TACI and BCMA are receptors for a TNF homologue implicated in B-cell autoimmune disease. Nature 2000;404(6781):995–9.
9. Schneider P, Takatsuka H, Wilson A, et al. Maturation of marginal zone and follicular B cells requires B cell activating factor of the tumor necrosis factor family and is independent of B cell maturation antigen. J Exp Med 2001;194(11):1691–7.
10. Schiemann B, Gommerman JL, Vora K, et al. An essential role for BAFF in the normal development of B cells through a BCMA-independent pathway. Science 2001;293(5537):2111–4.
11. Thompson JS, Schneider P, Kalled SL, et al. BAFF binds to the tumor necrosis factor receptor-like molecule B cell maturation antigen and is important for maintaining the peripheral B cell population. J Exp Med 2000;192(1):129–35.
12. Hahne M, Kataoka T, Schroter M, et al. APRIL, a new ligand of the tumor necrosis factor family, stimulates tumor cell growth. J Exp Med 1998;188(6):1185–90.
13. Yu G, Boone T, Delaney J, et al. APRIL and TALL-I and receptors BCMA and TACI: system for regulating humoral immunity. Nat Immunol 2000;1(3):252–6.
14. Mackay F, Woodcock SA, Lawton P, et al. Mice transgenic for BAFF develop lymphocytic disorders along with autoimmune manifestations. J Exp Med 1999;190(11):1697–710.
15. Do RK, Hatada E, Lee H, Tourigny MR, Hilbert D, Chen-Kiang S. Attenuation of apoptosis underlies B lymphocyte stimulator enhancement of humoral immune response. J Exp Med 2000;192(7):953–64.
16. Khare SD, Sarosi I, Xia XZ, et al. Severe B cell hyperplasia and autoimmune disease in TALL-1 transgenic mice. Proc Natl Acad Sci USA 2000;97(7):3370–5.
17. Groom J, Kalled SL, Cutler AH, et al. Association of BAFF/BLyS overexpression and altered B cell differentiation with Sjogren's syndrome. J Clin Invest 2002;109(1):59–68.
18. Royer B, Cazals-Hatem D, Sibilia J, et al. Lymphomas in patients with Sjogren's syndrome are marginal zone B-cell neoplasms, arise in diverse extranodal and nodal sites, and are not associated with viruses. Blood 1997;90(2):766–75.
19. Zulman J, Jaffe R, Talal N. Evidence that the malignant lymphoma of Sjogren's syndrome is a monoclonal B-cell neoplasm. N Engl J Med 1978;299(22):1215–20.
20. Gross JA, Dillon SR, Mudri S, et al. TACI-Ig neutralizes molecules critical for B cell development and autoimmune disease. impaired B cell maturation in mice lacking BLyS. Immunity 2001;15(2):289–302.
21. Mackay F, Schneider P, Rennert P, Browning J. BAFF and APRIL: a tutorial on B cell survival. Annu Rev Immunol 2003;21:231–64.
22. Hatzoglou A, Roussel J, Bourgeade MF, et al. TNF receptor family member BCMA (B cell maturation) associates with TNF receptor-associated factor (TRAF) 1, TRAF2, and TRAF3 and activates NF-kappa B, elk-1, c-Jun N-terminal kinase, and p38 mitogen-activated protein kinase. J Immunol 2000;165(3):1322–30.
23. Woodland RT, Fox CJ, Schmidt MR, et al. Multiple signaling pathways promote B lymphocyte stimulator (BLyS)-dependent B cell growth and survival. Blood 2007;111(2): 750–60.
24. Yoshida S, Kaneita Y, Aoki Y, Seto M, Mori S, Moriyama M. Identification of heterologous translocation partner genes fused to the BCL6 gene in diffuse large B-cell lymphomas: 5′-RACE and LA – PCR analyses of biopsy samples. Oncogene 1999;18(56):7994–9.

25. Claudio JO, Masih-Khan E, Tang H, et al. A molecular compendium of genes expressed in multiple myeloma. Blood 2002;100(6):2175–86.
26. Amson R, Sigaux F, Przedborski S, Flandrin G, Givol D, Telerman A. The human protooncogene product p33pim is expressed during fetal hematopoiesis and in diverse leukemias. Proc Natl Acad Sci USA 1989;86(22):8857–61.
27. Allen JD, Verhoeven E, Domen J, van der Valk M, Berns A. Pim-2 transgene induces lymphoid tumors, exhibiting potent synergy with c-myc. Oncogene 1997;15(10):1133–41.
28. Hammerman PS, Fox CJ, Cinalli RM, et al. Lymphocyte transformation by Pim-2 is dependent on nuclear factor-kappaB activation. Cancer Res 2004;64(22):8341–8.
29. Huang X, Di Liberto M, Cunningham AF, et al. Homeostatic cell-cycle control by BLyS: induction of cell-cycle entry but not G1/S transition in opposition to p18INK4c and p27Kip1. Proc Natl Acad Sci USA 2004;101(51):17789–94.
30. Madry C, Laabi Y, Callebaut I, et al. The characterization of murine BCMA gene defines it as a new member of the tumor necrosis factor receptor superfamily. Int Immunol 1998;10(11):1693–702.
31. von Bulow GU, Russell H, Copeland NG, Gilbert DJ, Jenkins NA, Bram RJ. Molecular cloning and functional characterization of murine transmembrane activator and CAML interactor (TACI) with chromosomal localization in human and mouse. Mamm Genome 2000;11(8):628–32.
32. Thompson JS, Bixler SA, Qian F, et al. BAFF-R, a newly identified TNF receptor that specifically interacts with BAFF. Science 2001;293(5537):2108–11.
33. von Bulow GU, Bram RJ. NF-AT activation induced by a CAML-interacting member of the tumor necrosis factor receptor superfamily. Science 1997;278(5335):138–41.
34. Yan M, Brady JR, Chan B, et al. Identification of a novel receptor for B lymphocyte stimulator that is mutated in a mouse strain with severe B cell deficiency. Curr Biol 2001;11(19):1547–52.
35. Bodmer JL, Schneider P, Tschopp J. The molecular architecture of the TNF superfamily. Trends Biochem Sci 2002;27(1):19–26.
36. Hymowitz SG, Patel DR, Wallweber HJ, et al. Structures of APRIL-receptor complexes: like BCMA, TACI employs only a single cysteine-rich domain for high affinity ligand binding. J Biol Chem 2005;280(8):7218–27.
37. Liu Y, Hong X, Kappler J, et al. Ligand-receptor binding revealed by the TNF family member TALL-1. Nature 2003;423(6935):49–56.
38. Rennert P, Schneider P, Cachero TG, et al. A soluble form of B cell maturation antigen, a receptor for the tumor necrosis factor family member APRIL, inhibits tumor cell growth. J Exp Med 2000;192(11):1677–84.
39. Hendriks J, Planelles L, de Jong-Odding J, et al. Heparan sulfate proteoglycan binding promotes APRIL-induced tumor cell proliferation. Cell Death Differ 2005;12:637–48.
40. Ingold K, Zumsteg A, Tardivel A, et al. Identification of proteoglycans as the APRIL-specific binding partners. J Exp Med 2005;201(9):1375–83.
41. Castigli E, Wilson SA, Scott S, et al. TACI and BAFF-R mediate isotype switching in B cells. J Exp Med 2005;201(1):35–9.
42. Litinskiy MB, Nardelli B, Hilbert DM, et al. DCs induce CD40-independent immunoglobulin class switching through BLyS and APRIL. Nat Immunol 2002;3(9):822–9.
43. Novak AJ, Darce JR, Arendt BK, et al. Expression of BCMA, TACI, and BAFF-R in multiple myeloma: a mechanism for growth and survival. Blood 2004;103(2):689–94.
44. Tarte K, Zhan F, De Vos J, Klein B, Shaughnessy J, Jr. Gene expression profiling of plasma cells and plasmablasts: toward a better understanding of the late stages of B-cell differentiation. Blood 2003;102(2):592–600.
45. Avery DT, Kalled SL, Ellyard JI, et al. BAFF selectively enhances the survival of plasmablasts generated from human memory B cells. J Clin Invest 2003;112(2):286–97.
46. O'Connor BP, Raman VS, Erickson LD, et al. BCMA is essential for the survival of long-lived bone marrow plasma cells. J Exp Med 2004;199(1):91–8.

47. Ng LG, Sutherland AP, Newton R, et al. B cell-activating factor belonging to the TNF family (BAFF)-R is the principal BAFF receptor facilitating BAFF costimulation of circulating T and B cells. J Immunol 2004;173(2):807–17.
48. von Bulow GU, van Deursen JM, Bram RJ. Regulation of the T-independent humoral response by TACI. Immunity 2001;14(5):573–82.
49. Seshasayee D, Valdez P, Yan M, Dixit VM, Tumas D, Grewal IS. Loss of TACI causes fatal lymphoproliferation and autoimmunity, establishing TACI as an inhibitory BLyS receptor. Immunity 2003;18(2):279–88.
50. Yan M, Wang H, Chan B, et al. Activation and accumulation of B cells in TACI-deficient mice. Nat Immunol 2001;2(7):638–43.
51. Castigli E, Wilson SA, Garibyan L, et al. TACI is mutant in common variable immunodeficiency and IgA deficiency. Nat Genet 2005;37(8):829–34.
52. Salzer U, Chapel HM, Webster AD, et al. Mutations in TNFRSF13B encoding TACI are associated with common variable immunodeficiency in humans. Nat Genet 2005;37(8): 820–8.
53. Kinlen LJ, Webster AD, Bird AG, et al. Prospective study of cancer in patients with hypogammaglobulinaemia. Lancet 1985;1(8423):263–6.
54. Dillon SR, Gross JA, Ansell SM, Novak AJ. An APRIL to remember: novel TNF ligands as therapeutic targets. Nat Rev Drug Discov 2006;5(3):235–46.
55. Jemal A, Siegel R, Ward E, Murray T, Xu J, Thun MJ. Cancer statistics. CA Cancer J Clin 2007;57(1):43–66.
56. Graninger WB, Seto M, Boutain B, Goldman P, Korsmeyer SJ. Expression of Bcl-2 and Bcl-2-Ig fusion transcripts in normal and neoplastic cells. J Clin Invest 1987;80(5): 1512–5.
57. Korsmeyer SJ, McDonnell TJ, Nunez G, Hockenbery D, Young R. Bcl-2: B cell life, death and neoplasia. Curr Top Microbiol Immunol 1990;166:203–7.
58. Sanchez-Beato M, Sanchez-Aguilera A, Piris MA. Cell cycle deregulation in B-cell lymphomas. Blood 2003;101(4):1220–35.
59. Staudt LM. Gene expression profiling of lymphoid malignancies. Annu Rev Med 2002;53:303–18.
60. Briones J, Timmerman JM, Hilbert DM, Levy R. BLyS and BLyS receptor expression in non-Hodgkin's lymphoma. Exp Hematol 2002;30(2):135–41.
61. Novak AJ, Grote DM, Stenson M, et al. Expression of BLyS and its receptors in B-cell non-Hodgkin lymphoma: correlation with disease activity and patient outcome. Blood 2004;104(8):2247–53.
62. Krumbholz M, Theil D, Derfuss T, et al. BAFF is produced by astrocytes and up-regulated in multiple sclerosis lesions and primary central nervous system lymphoma. J Exp Med 2005;201(2):195–200.
63. Elsawa SF, Novak AJ, Grote DM, et al. B-lymphocyte stimulator (BLyS) stimulates immunoglobulin production and malignant B-cell growth in Waldenstrom's macroglobulinemia. Blood 2005;107(7):2882–8.
64. He B, Chadburn A, Jou E, Schattner EJ, Knowles DM, Cerutti A. Lymphoma B cells evade apoptosis through the TNF family members BAFF/BLyS and APRIL. J Immunol 2004;172(5):3268–79.
65. Fu L, Lin-Lee YC, Pham LV, Tamayo A, Youshimura L, Ford RJ. Constitutive NF-{kappa}B and NFAT activation leads to stimulation of the BLyS survival pathway in aggressive B cell lymphomas. Blood 2006;107(11):4540–8.
66. Caligaris-Cappio F. Biology of chronic lymphocytic leukemia. Rev Clin Exp Hematol 2000;4(1):5–21.
67. Kipps TJ. Chronic lymphocytic leukemia. Curr Opin Hematol 2000;7(4):223–34.
68. Palma M, Kokhaei P, Lundin J, Choudhury A, Mellstedt H, Osterborg A. The biology and treatment of chronic lymphocytic leukemia. Ann Oncol 2006;17(Suppl 10): x144–54.

69. Foon KA, Rai KR, Gale RP. Chronic lymphocytic leukemia: new insights into biology and therapy. Ann Intern Med 1990;113(7):525–39.
70. Novak AJ, Bram RJ, Kay NE, Jelinek DF. Aberrant expression of B-lymphocyte stimulator by B chronic lymphocytic leukemia cells: a mechanism for survival. Blood 2002;100(8):2973–9.
71. Kern C, Cornuel JF, Billard C, et al. Involvement of BAFF and APRIL in the resistance to apoptosis of B-CLL through an autocrine pathway. Blood 2004;103(2):679–88.
72. Nishio M, Endo T, Tsukada N, et al. Nurselike cells express BAFF and APRIL, which can promote survival of chronic lymphocytic leukemia cells via a paracrine pathway distinct from that of SDF-1alpha. Blood 2005;106(3):1012–20.
73. Endo T, Nishio M, Enzler T, et al. BAFF and APRIL support chronic lymphocytic leukemia B-cell survival through activation of the canonical NF-kappaB pathway. Blood 2007;109(2):703–10.
74. Parkin DM, Bray F, Ferlay J, Pisani P. Global cancer statistics, 2002. CA Cancer J Clin 2005;55(2):74–108.
75. Moreaux J, Legouffe E, Jourdan E, et al. BAFF and APRIL protect myeloma cells from apoptosis induced by interleukin 6 deprivation and dexamethasone. Blood 2004;103(8):3148–57.
76. Tarte K, De Vos J, Thykjaer T, et al. Generation of polyclonal plasmablasts from peripheral blood B cells: a normal counterpart of malignant plasmablasts. Blood 2002;100(4): 1113–22.
77. Darce JR, Arendt BK, Wu X, Jelinek DF. Regulated Expression of BAFF-binding receptors during human B cell differentiation. J Immunol 2007;179(11):7276–86.
78. Moreaux J, Cremer FW, Reme T, et al. The level of TACI gene expression in myeloma cells is associated with a signature of microenvironment dependence versus a plasmablastic signature. Blood 2005;106(3):1021–30.
79. Moreaux J, Hose D, Jourdan M, et al. TACI expression is associated with a mature bone marrow plasma cell signature and C-MAF overexpression in human myeloma cell lines. Haematologica 2007;92(6):803–11.
80. Keats JJ, Fonseca R, Chesi M, et al. Promiscuous mutations activate the noncanonical NF-kappaB pathway in multiple myeloma. Cancer Cell 2007;12(2):131–44.
81. Tai YT, Li XF, Breitkreutz I, et al. Role of B-cell-activating factor in adhesion and growth of human multiple myeloma cells in the bone marrow microenvironment. Cancer Res 2006;66(13):6675–82.
82. Abe M, Kido S, Hiasa M, et al. BAFF and APRIL as osteoclast-derived survival factors for myeloma cells: a rationale for TACI-Fc treatment in patients with multiple myeloma. Leukemia 2006;20(7):1313–5.
83. Yaccoby S, Pennisi A, Li X, et al. Atacicept (TACI-Ig) inhibits growth of TACI(high) primary myeloma cells in SCID-hu mice and in coculture with osteoclasts. Leukemia 2007;22:406–13.
84. Braeuninger A, Kuppers R, Strickler JG, Wacker H-H, Rajewsky K, Hansmann M-L. Hodgkin and Reed-Sternberg cells in lymphocyte predominant Hodgkin disease represent clonal populations of germinal center-derived tumor B-cells. Proc Natl Acad Sci 1997;94(17):9337–42.
85. Knowles DM. Neoplastic Hematopathology. New York: Lippincott Williams & Wilkins, 2001.
86. Glaser SL, Jarrett RF. The epidemiology of Hodgkin's disease. Baillieres Clin Haematol 1996;9(3):401–16.
87. Chiu A, Xu W, He B, et al. Hodgkin lymphoma cells express TACI and BCMA receptors and generate survival and proliferation signals in response to BAFF and APRIL. Blood 2007;109(2):729–39.
88. Stohl W. B lymphocyte stimulator protein levels in systemic lupus erythematosus and other diseases. Curr Rheumatol Rep 2002;4(4):345–50.

89. Batten M, Fletcher C, Ng LG, et al. TNF deficiency fails to protect BAFF transgenic mice against autoimmunity and reveals a predisposition to B cell lymphoma. J Immunol 2004;172(2):812–22.
90. Scapini P, Nardelli B, Nadali G, et al. G-CSF-stimulated neutrophils are a prominent source of functional BLyS. J Exp Med 2003;197(3):297–302.
91. Nardelli B, Belvedere O, Roschke V, et al. Synthesis and release of B-lymphocyte stimulator from myeloid cells. Blood 2001;97(1):198–204.
92. Craxton A, Magaletti D, Ryan EJ, Clark EA. Macrophage- and dendritic cell-dependent regulation of human B-cell proliferation requires the TNF family ligand BAFF. Blood 2003;101(11):4464–71.
93. Novak AJ, Grote DM, Ziesmer SC, et al. Elevated serum B-lymphocyte stimulator levels in patients with familial lymphoproliferative disorders. J Clin Oncol 2006;24(6):983–7.
94. Kawasaki A, Tsuchiya N, Fukazawa T, Hashimoto H, Tokunaga K. Analysis on the association of human BLYS (BAFF, TNFSF13B) polymorphisms with systemic lupus erythematosus and rheumatoid arthritis. Genes Immun 2002;3(7):424–9.
95. Briones J, Timmerman JM, Panicalli DL, Levy R. Antitumor immunity after vaccination with B lymphoma cells overexpressing a triad of costimulatory molecules. J Natl Cancer Inst 2003;95(7):548–55.
96. Oki Y, Georgakis GV, Migone TS, Kwak LW, Younes A. Elevated serum BLyS levels in patients with non-Hodgkin lymphoma. Leuk Lymphoma 2007;48(9):1869–71.
97. Oki Y, Georgakis GV, Migone TS, Kwak LW, Younes A. Prognostic significance of serum B-lymphocyte stimulator in Hodgkin's lymphoma. Haematologica 2007;92(2):269–70.
98. Tecchio C, Nadali G, Scapini P, et al. High serum levels of B-lymphocyte stimulator are associated with clinical-pathological features and outcome in classical Hodgkin lymphoma. Br J Haematol 2007;137(6):553–9.
99. Cambridge G, Stohl W, Leandro MJ, Migone TS, Hilbert DM, Edwards JC. Circulating levels of B lymphocyte stimulator in patients with rheumatoid arthritis following rituximab treatment: relationships with B cell depletion, circulating antibodies, and clinical relapse. Arthritis Rheum 2006;54(3):723–32.
100. Ansell SM, Novak AJ, Ziesmer S, et al. Serum cytokine levels predict time to progression after rituximab as initial therapy in patients with follicular grade 1 non-hodgkin lymphoma. ASH Annual Meeting Abstracts 2007;118(11):999A.
101. Rossi J-F, Moreaux J, Rose M, et al. A phase I/II study of atacicept (TACI-Ig) to neutralize APRIL and BLyS in patients with refractory or relapsed multiple myeloma (MM) or active previously treated waldenstrom's macroglobulinemia (WM). ASH Annual Meeting Abstracts 2006;108(11):3578.
102. Ansell S, Witzig TE, Novak A, et al. Phase 1 clinical study of atacicept in patients with relapsed and refractory B-cell lymphoma. ASH Annual Meeting Abstracts 2006;108(11):2722.

Index

M.P. Cancro (ed.), *BLyS Ligands and Receptors*, Contemporary Immunology,
DOI 10.1007/978-1-60327-013-7,

C